THE LETTERS OF DOMINIQUE CHAIX, BOTANIST-CURÉ

ARCHIVES INTERNATIONALES D'HISTOIRE DES IDÉES

INTERNATIONAL ARCHIVES OF THE HISTORY OF IDEAS

151

THE LETTERS OF DOMINIQUE CHAIX, BOTANIST–CURÉ

by

ROGER L. WILLIAMS

THE LETTERS OF DOMINIQUE CHAIX, BOTANIST-CURÉ

by

ROGER L. WILLIAMS

Distinguished Professor of History Emeritus,
University of Wyoming,
Laramie, Wyoming, U.S.A.

SPRINGER SCIENCE+BUSINESS MEDIA, B.V.

A C.I.P. Catalogue record for this book is available from the Library of Congress.

ISBN 978-94-010-6310-4   ISBN 978-94-011-5490-1 (eBook)
DOI 10.1007/978-94-011-5490-1

Printed on acid-free paper

CONTENTS

MAP 1

The Prøvince of Dauphiné, showing the boundaries of the present
departments of Isère (Grenoble), Hautes-Alpes (Gap), and Drôme (Valence).

MAP 2

Localities often cited by Dominique Chaix

PREFACE

Between 1772 and 1799, Dominique Chaix wrote 170 letters to Dominique Villars. None of the letters that Villars wrote in response have survived, and there is evidence to indicate that Chaix simply did not retain incoming letters once they had served their informative purpose. Villars, blessed with more ample facilities, kept incoming letters; and those from Chaix are now preserved in the Bibliothèque Municipale de Grenoble in three volumes under the number R10073. A transcription of them for public use was made earlier in this century under the supervision of Georges de Manteyer [Georges-Barthélemy-Marie Pinet de Manteyer] when he was archivist of the Department of Hautes-Alpes. I am greatly indebted to Mme Marie-Françoise Bois-Delatte, Conservateur des Fonds Dauphinoise at the Bibliothèque Municipale d'Etude et d'Information in Grenoble, not merely for making these letters available to me, but for her eagerness to see someone take an interest in Dauphinois botanists.

I met a similar friendly interest at the Archives du Département des Hautes-Alpes in Gap. I thank M. Pierre-Yves Playoust, Directeur des Services, and members of his archival staff for the cordial assistance in my search for materials documenting the career of the abbé Chaix. I am also obligated to the Hunt Institute for Botanical Documentation at Carnegie Mellon University in Pittsburgh for providing me photocopies of Villars' letters in their Allioni collection. They relate to the composition of a flora for Dauphiné and are revealing of Villars' character.

The Chaix letters reflect an intellectual collaboration of roughly thirty years, during which period Villars' **Histoire des plantes du Dauphiné** was written and published in three volumes, the first flora for the province. Chaix contributed more than the catalogue of plants for the region around Gap that appeared in the first volume. The letters reveal a running discussion of species in Dauphiné, a concern to correct Villars' shaky Latin, and, ultimately, dismay over the numerous typographical errors that appeared in the published text.

It does not appear that the letters were ever consulted by botanists during the 19th century who worked on the flora of Dauphiné: neither by Auguste Mutel nor by J.-B. Verlot. In more recent years, the letters have been examined on rare occasions for specific or limited purposes, whether botanical or clerical. But citations to them, or conclusions drawn from them, I have not found to be invariably reliable. This has occurred not so much from misreading a particular passage as from being misled by the implications of a sentence or paragraph that did not say precisely what the writer had meant. The correct meaning sometimes only becomes clear within the context of several letters that may not be contiguous. Moreover, Chaix was never as concerned for literary style in hasty correspondence as he could be on occasion in more formal expression. In truth, literary style for Chaix was Latin style, whether that of botanical authors, theologians, or the writers of antiquity.

Many of the letters contain long botanical passages that Chaix endeavored to shorten through frequent use of abbreviations: authors' names, book titles, not to speak of plant names. The reader will find these fleshed out, in brackets, for intelligibility; as even the experienced botanist may not today immediately recognize references more commonplace in the 18th century. The same technique has been used to identify a plant name for the reader that has since become a synonym and may not be familiar. In the absence of herbarium specimens in most instances, it must be understood that such identifications are probabilities.

The translation of botanical words from French to English is relatively uncomplicated as the botanical terminology in both languages is substantially Latinate. Chaix's prose is another matter. He was wordy and long-winded, in part a desire to make himself understood and to observe the conventions of elaborate salutations and complimentary endings, most of which I have omitted as tedious repetitions. But Chaix also lived in great solitude: celibate and in a remote region amongst parishioners to whom he could not talk about the wonders of vegetation. That frustration was relieved through long epistolary conversations with Villars, the great and closest friend of his life. The problem has been to translate them meaning for meaning, rather than word for word, while retaining the essential character of the original.

A word about spelling and pronunciation needs to be added. I have used *Villars* throughout to avoid confusion, as that is how he is known to botanists. But he was born *Villar* and unquestionably used that spelling until a mistake was made on a document in Paris that he chose not to contest, simply adapting to *Villars*. One of the unfortunate results is that his **Prospectus de l'histoire des plantes de Dauphiné** was correctly published by *Villar* in 1779; whereas the name had changed to *Villars* by 1786 when the first volume of the flora was published. As for Chaix, the spelling offers no problems, but the pronunciation does. In Paris it may sound like *Shay*, but in the Midi it becomes *Shex*, unfortunate though that sound may be, and where the *x* is still sounded.

The Chaix letters, if provincial in scope, were part of a phenomenal upsurge in French interest in flowers and botany between 1750 and 1815 that they themselves recognized as botanophilia. The passion was really one facet of the heightened interest in natural history common to Western Europe in the 18th century; in the course of which opinion or fashion veered from the classical or baroque, the mechanical or artificial, to the *natural*, the sentimental, the rustic, and the individual. Early in the 20th century, that movement was thoroughly catalogued by Daniel Mornet in two seminal studies, **Le Sentiment de la nature en France de J.-J. Rousseau à Bernardin de Saint Pierre** (1907), and **Les Sciences de la nature en France au XVIIIe siècle** (1911).

Citing Mornet's examination of 500 private libraries, D. G. Charlton, in **New Images of the Natural in France 1750-1800** (1984), was convinced that both the comte de Buffon's **Histoire naturelle** and the abbé Pluche's **Le Spectacle de la nature** were more widely purchased by the public in that era than the most successful literary works, including Rousseau's **La Nouvelle Héloïse**. The *public* in those days was essentially the literate. Evidence of the new fashion has also been found in the popularity of private collections of natural artifacts, known as natural history cabinets, that enthralled the literate and crossed all social lines. The Chaix letters reveal a network of provincial physicians and clergy in particular who were eager plant collectors and exchanged specimens for personal herbaria.

This new intellectual enthusiasm contributed forceably to the foundation of provincial libraries, academies, learned societies, and newspapers in 18th-century France, usually the collective efforts of laymen. In prior centuries, such innovations would more likely have been sponsored by bishops or religious orders. Daniel Roche, in his **Le Siècle des lumières en province** (1978), concluded that between the end of the reign of Louis XIV and 1789 a "republic of letters" became increasingly vibrant as reflected in the foundation and growth of provincial academies. He classified Grenoble as a late-comer into that cultural republic. Its municipal library dated from 1772, the year the young Villars reached the city. The library directors would declare themselves to be a literary society in 1787; and by letters patent from the king in 1789, they were formally recognized as the Académie Delphinale. By then, Dominique Villars was prominent among the founding fathers.

This new intellectual excitement in the service of natural history may be found in the instructions to bold explorers from government, whether royal, republican, or imperial, for the collection of exotic plants. Such instructions were usually inspired by the savants at the Jardin du roi, later the Jardin des plantes, and usually communicated by André Thouin, the chief gardener. Thouin had also inaugurated a remarkable correspondence with botanists within France, which sponsored a widespread exchange of seeds and roots between Paris and the remote corners of the kingdom.

The consequent influx and distribution of plants previously unknown to Europe provoked a crisis in plant nomenclature and classification, which the reader will find constantly reflected in the Chaix letters. The immense work of Carl Linnaeus produced a rational order: first, by reforming botanical nomenclature for universal use; second, by renaming all known species according to his new (binomial) nomenclature; third, by producing a relatively simple and rational classificational system meant to accommodate all known species.

Linnaeus set the stage for the emergence of botany as an independent science in the 18th century, independent, that is, from traditional medicine and herbalism, thus contributing to the prestige of the sciences in general and to making botany a fashionable pastime. Because the classificational system of Linnaeus depended solely upon an observation of the sexual organs, the pistil and the stamens, it was recognized by botanists from the outset as being *artificial.*

The next step, begun by Linnaeus himself, was to seek a more *natural* classification that encompassed *all* parts of a plant. That quest for the *natural* helped to define the intellectual and aesthetic climate of the 18th century: both opinion and fashion within "the republic of letters," the canons of flower-painting, the new principles behind gardening and landscape architecture, and even a new sensitivity to the ravages of deforestation that threatened the vitality of the soils and the productivity of French agriculture, and about which we get occasional insights from the Chaix letters.

The primary focus of the letters was upon botanical issues. Only in 1789 did political matters begin to intrude and provoke religious tensions, also raising doubts about the future of scholarship when royal patronage vanished. The letters give insight into the development of botany into a science in the 18th century, notably outside of Paris. They illustrate in particular in what ways the spirit of the Enlightenment had penetrated a rural, isolated region: the passion to extend public knowledge and to contribute to public well-being. In a word, to be *useful* to society. The letters contribute evidence about botany, not only as a fashionable intellectual pursuit in the later 18th century, but as a realm where Christians and skeptics could meet in the interest of science, if not always in great congeniality. They give us, finally, the life of a faithful rural parish priest, both before and after 1789, who never for a moment believed he had compromised his faith by taking the oath to uphold the Civil Constitution of the Clergy. Nor did he retract it even after the Revolution had destroyed his vocation, which dissolved his morale. Given his origin, his insulation, and his meager resources, his achievement remains admirable.

THE SETTING

Dominique Chaix was born on 8 June 1730 at a grange or farm call Berthaud in an alpine region still remote today. The property may be found northwest of Gap, just above Rabou, at La Crotte along the Torrente de La Crotte, one of the several tumbling tributaries of the Petit Buëch. The Chaix family, originally from the commune of Chaudun, not only farmed the Berthaud property, but also a higher farm called La Grangette under a great crag known as the Pic de Bure. Sheep-raising accounted for more than three-quarters of their agricultural income.[1]

The continguent Bois des Donnes, now a part of the Forêt Domaniale des Sauvas, provides a hint of the historical ownership of that region. A Carthusian convent had been founded in the quartier de Berthaud in 1188, an area roughly enclosing Rabou, La Roche-des-Arnauds, Manteyer, and the Dévoluy: rugged country lost for five months of the year in snow and ice. Immediately to the west was the Carthusian monastery of Durbon. Both houses were under a single administration. On occasion the nuns sought authorization to move to a milder location, but in vain. In 1446 (some say 1448), a great fire destroyed the convent and all its records. The nuns took refuge at Durbon, a retreat that became a convenience for more than a century. The bishop of Gap, during a pastoral visit in 1599, was surprised to find both sexes under the same roof and thought very poorly of it.

His order for a separation was honored, the last nun being transferred to another house in 1601. The monks of Durbon acted quickly to unite the possessions of Berthaud to their own, and that is how matters stood when Chaix was born in 1730. The monastic house enjoyed *tout droit de haute seigneurie, haute, moyenne, et basse justice.* A financial statement surviving from 1760 reveals that Durbon had drawn a revenue of 577 livres from La Grange de La Crotte, and 802 livres from La Grange de La Grangette. Nothing is said about what was left for the Chaix family.

But it is clear from Chaix's later letters that he had immense respect and affection for the monks at Durbon, visiting them periodically. Not only did they have a library of more than a thousand volumes, in which the natural sciences were well represented, but several of the monks took a serious interest in botany. It is a curious fact that the several Carthusian houses in Dauphiné were known to welcome any botanist as a guest when on a collecting trip. The monastic library at Durbon was scattered because of the Revolution, and Chaix would have the grief of seeing the monastery evacuated in July of 1790, the municipality of St.-Julien-en-Beauchêne taking statements from members of the monastic community as to preferences about their future status.[2]

~

[1] Villars, "Notice historique sur Dominique Chaix, botaniste." **Bull. Soc. Etudes Hautes-Alpes** 3, no. 1 (1884): 297-298.
[2] Charles Charronnet, **Monastères de Durbon et de Berthaud (Diocèse de Gap). Documents historiques** (Grenoble: Alphonse Merle, 1863), pp. 47-49, 64, 69-70; Paul Guillaume, **Clergé ancien et modern du diocèse de Gap. Abbés, prieurs, curés, vicaires, chapelains . . . de toutes les paroisses du diocèse actuel** (Gap: Jean & Peyrot, 1909), p. 184.

Chaix began his study of Latin in 1743 at the age of thirteen, studying with a local priest, the only practical instructional route for a peasant boy in the eighteenth century, and which led not infrequently to a clerical career. At sixteen, in 1746, he was sent to Grenoble to attend the local *collège* that was controlled by Jesuits. We may infer from several later letters that he was sheltered while in Grenoble by the Claude Bruno family, whose children he tutored in exchange, and who were uncommonly kind to him. He remembered them gratefully ever after, although apparently never again saw any member of the Bruno family.

After ordination in 1755 at the age of twenty-five, Chaix received appointment as a vicar in Gap, attached to the cathedral chapter, where he came under the eye of the bishop, Pierre-Annet de Pérouse, a learned man with a large personal library.[3] One of Chaix's duties was to say mass in a local convent dedicated to the instruction of young girls. The superior, Marguerite de Colvin, cultivated medicinal plants in her garden. Chaix always remembered her as an admirable person, and her interest in plants caught his attention. It was she who inspired his first interest in botany.

Chaix was twenty-eight when he was named curé des Baux on 4 July 1758, a small parish of perhaps 150 inhabitants about 16 kilometres from Gap. Their spiritual requirements left him much free time. During the winters of his first few years, he was actively evangelical and would join several friends, François Gaude of Oze and Barthélémy Faure of Châteauneuf-d'Oze in particular, on missions into more populous parishes where instruction was needed. Such missions could last from forty to fifty days, bringing in as much as 120 to 150 livres for three missionaries. It appears that it was during just such a mission that Dominique Villars first met and came to know Chaix in 1765.

During the summers, meanwhile, Chaix worked his garden of about 800 square meters of surface, too small to contain a great number of plants, but proportionate to his physical and financial resources. No more than one-quarter of it was his kitchen garden. The remaining three-quarters was devoted to unusual plants or those new or unknown to botanists, generally forty to sixty new plants each year, whose development was recorded in his notes, and which were then confined to his collection of dried plants to make room for new plantings. His experience and reflection, when added to his study of Linnaeus, made him in time a gardener as skillful in raising plants and in caring for seed as he was an observer of their vegetations, their variations, their diseases, and of the changes worked by culture.

Shortly after Villars met Chaix, the latter was appointed to be a curé in Gap by Bishop François de Narbonne-Lara on 17 September 1767, amounting to a substantial increase in duties and income. To occupy the new post meant taking further training in theology at the university in Avignon. Chaix initially resisted the suggestion of promotion but soon submitted to his bishops's direction. The young Villars feared that he would lose the guidance of his new friend. But Chaix, who botanized on the route to and from Avignon, returned not only with a degree and new plants, but with a determination to sacrifice the new post and its revenues to remain in Les Baux with its leisure time to pursue botany.

∽

[3] For the bishops of Gap in the 18th century, see Timothy Tackett, **Priest & Parish in Eighteenth-Century France** (Princeton: Princeton University Press, 1977), pp. 27-31.

Apparently Narbonne-Lara did not press him further.[4]

Dominique Villars, born 14 November 1745, was fifteen years younger than Dominique Chaix. His birthplace is usually given as Le Noyer, a village high on the west flank of the valley of the Drac in Champsaur. To be precise, one should say that he was born on a farm in the hamlet of Le Villard, apart from Le Noyer. There remains clear archival evidence that his family spelled its name *Villar*. He, alone, would adopt *Villars* some years later, the spelling by which he is known to botanists. His father owned a small property appraised at 15,000 livres and grazed sheep on it for his principal income. He also served Le Noyer as village clerk, evidently a man of some education and merit.

To read the introductory pages of Villars' flora would lead to the conclusion that he had been a lonely child, born to solitude and given to reflection, developing an interest in plants by the age of twelve quite on his own. It needs to be added that, by the age of seven, he had been taught to read by his father, and that, thereafter, his instruction had been confided to Jean-Louis-Honoré Arnaud, curé du Noyer, which accounts for his introduction to writing, arithmetic, and Latin. This instruction ended at the age of twelve when he began to assist his father in the compilation of the village records.

Villars would say that the first person he had known with an interest in plants was Antoine Gentillon-Médaille, about 1757, a herbalist who collected plants around St.-Bonnet, an intelligent but unsophisticated and superstitious man who had studied an illuminated Matthiolus, on the basis of which he composed a variety of remedies for various ailments.[5] It has since been asserted that Villars, in fact, had been introduced to botany by the abbé Arnaud when botanizing on the mountain above Le Noyer.[6]

Unequivocal is the fact that Villars' father died on 10 April 1760, his grandfather on 26 March 1761, leaving the boy at age fifteen the head of a family of eight children with only the mother to guide him. Having little sympathy for his interest in natural science, she placed him as a clerk in the office of a *notaire* in Gap where he might learn something of practical matters that would enable him to manage the small patrimony just inherited.

During the several years he spent as a clerk in Gap, he came to the attention of several kindly people who encouraged his study of natural science. One was Marguerite de Colvin, the same superior of the Couvent de la Charité who had inspired Chaix. The other, Jean-Balthazar Laugier, a physician originally from nearby Tallard, who had trained at Montpellier and was establishing his practice at Corps, lent him medical books. Villars would say that he acquired a passion for medicine as the way to be most useful to humanity from Laugier and, therein, made the connection between botany and medicine.[7]

~

[4] Villars, "Notice historique sur Dominique Chaix," pp . 298-300, 315-316; Guillaume, **Clergé ancien et moderne du diocèse de Gap**, p. 34, 98; Adolphe Rochas, **Biographie du Dauphiné** (Paris: Charavay, 1856-1860), 1: 194; Félix Allemand, **Dictionnaire biographique des Hautes-Alpes** (Gap: Alpine, 1911), pp. 141-143. Allemand is not reliable in every detail.
[5] Pietro Andrea Mattioli (1500-1577), a physician of Siena, who published commentaries on Dioscorides that were translated into many languages.
[6] B. Joyeux and A. Dejarnac, "Le Médecin Dominique Villars (1745-1814)." **Bull. Soc. Etudes Hautes-Alpes**, Année 1969: 122, 126-127.
[7] Villars, **Hist. Pl. Dauph.** 1:viii-x (1786).

Advised by the abbé Arnaud that nothing was likely to deflect Villars from his medical and scientific interests, his mother responded by arranging his marriage to force domestic responsibility upon him. It is said that he acquiesced out of respect for his mother whose religious faith he shared.[8] On 8 June 1763, at the age of eighteen, he was married to Jeanne Disdier, also eighteen. She brought a dowry of 2000 livres, a small fortune by the rural standards of Le Noyer. He then assumed his late father's position as *secrétaire-greffier* of the community of Le Noyer.[9]

It seems probable that Villars would have been sentenced to his father's life had it not been for the chance meeting with the abbé Chaix in 1765. Villars had learned about Chaix from Madame de Colvin as someone who could teach him, but their paths had never before crossed. The details of their collaboration thereafter are obscure until 1769 when they made their first field trip together in the region near Gap: On the mountains around Les Baux, Rabou, and Chaudun; on the Col Bayard; and eastward up the Drac through Ancelle as far as Orcières. They were intent upon collecting not only specimens to be dried, but seeds and live plants for Chaix's garden for later observation.

Subsequently, but still in the summer of 1769, two botanists from Grenoble, Dr. Pierre Clappier and Pierre Liottard, were completing their fieldwork in the Champsaur and around Gap: an odd pair in that Clappier had taken his medical degree at Montpellier, where he had studied botany under Antoine Gouan, Pierre Cusson, and Philibert Commerson; whereas Liottard was barely literate, if knowledgeable about plants. Villars joined them for several days in the field. He found that Clappier was primarily interested in precise plant identification, and that he had acquired (as had Liottard) a detailed knowledge of Linnaean terminology. So familiar was he with the works of Linnaeus that he experienced difficulties in the field only when he came upon unknown species or those peculiar to Dauphiné.

When the fine season returned in 1770, Chaix and Villars made a circular field trip, beginning in the lower Valgaudemar. They passed through La Chapelle and Rif-du-Sap to Le Clot-Jocelme; then turned south, crossing into the valley of the Drac Blanc. They reached the Champsaur through both Champoléon and Orcières, then returning to their respective homes. Both men continued to botanize independently, building personal collections from their native region. Villars, that year, also visited Clappier and Liottard in Grenoble to examine their herbaria and to see a private garden they cultivated.[10]

As Villars was never specific about the matter, one can only infer that it was Dr. Clappier, about whom very little is known, who encouraged Villars to take some training in surgery in Grenoble. Clappier had become a local notable, especially since his recent friendship with Jean-Jacques Rousseau, and could well have been the person who brought

∾

[8] Dr. Victor Bally, **Notice historique sur la vie et les travaux du docteur Villars, naturaliste, correspondant de l'Institut** (Grenoble: Imprimerie de Maisonville, 1858), pp. 2-8.
[9] Georges de Manteyer, "Les Origines de Dominique Villars, le botaniste (1555-1814)." **Bull. Soc. Etudes Hautes-Alpes**, ser. 4, 40 (1921): 129-137.
[10] Villars, Hist. Pl. Dauph. 1: x-xiv (1786). The garden was evidently owned by Dr. Joseph-François Prunelle de Lière, a physician who did not practice medicine. See Rochas, **Biographie du Dauphiné** 2: 308n.

Villars to the attention of the royal intendant for Dauphiné, Christophe Pajot de Marcheval, who had been in Grenoble since 1761.

It is a pity that neither Clappier nor Liottard left any recollections of their association with Rousseau whose visit to Grenoble in 1768 was a brief encounter with local botany. Rousseau scholars are well aware that his interest in the subject dawned about 1763 or 1764 during his later years; and that the eight "elementary" letters on the subject that he wrote to Mme Etienne Delessert of Lyon between 1771 and 1773 became, in published form, an enormous stimulation to popularizing botany. They also proved to be the inspiration for her son's taste, and Benjamin Delessert's immense herbarium became the basis for the great collections in Geneva.[11]

For all that, Rousseau was no naturalist, nor did he aspire to be one. As he wrote to Malesherbes, he did not want to become an expert, but simply to learn about botany as a pleasurable and salutary way of life. It was a distraction, an amusement, albeit sometimes a fatiguing one, having no practical value. Using the Linnaean method, he became proficient at identifying plants. But he sought no order in nature, nor did he dwell upon the significance of phenomena he observed in nature, nor had he the slightest concern for the possible medicinal use of plants. In the end, nature became the sanctuary for his sentimental mind in flight from society.[12]

The episode in Grenoble really began in Lyon where Rousseau arrived on 18 June 1768. Although Rousseau had made numerous sojourns in Lyon in his younger days, only on this occasion did he establish friendly relations with Claret de La Tourrette and the abbé Rozier, naturalists and pillars of the science section of the Académie de Lyon.[13] On 7 July 1768, the three of them left Lyon for botanizing in the Grande-Chartreuse, from which Rousseau descended to Grenoble, arriving somewhat exhausted on the 13th. He remained in Grenoble for a month except for a brief trip to Chambéry.

Although Rousseau rented quarters in the rue des Vieux-Jésuites, he had an unofficial host and companion in Grenoble, Gaspard Bovier, who was delighted to receive him and watch over him. Bovier was an attorney affiliated with the Parlement de Grenoble, who is said to have had a fondness for literature. His gracious intent reflected the warm public reception Rousseau met in Grenoble. But Rousseau's irascibility and his suspicious mind at that stage of his life could defeat the best intentions of others. The only thing about which he wanted to talk was botany, that is to say, the aesthetic aspects of plants; whereas poor Bovier valued plants only for their medicinal properties. As a consequence, Rousseau much preferred the companionship of the simple gardener, Pierre Liottard, and would

∼

[11] John Briquet, "Jean-Jacques Rousseau, botaniste." **Bull. Inst. Natl. genevois** 41 (1914): 134-136.
[12] See Sir Gavin de Beer, "Jean-Jacques Rousseau botanist." **Ann. Sci.** 10, no. 3 (1954): 189-208; Jean Starobinski, "Rousseau and Buffon." **Jean-Jacques Rousseau, Transparency and Obstruction** (Chicago: University of Chicago Press, 1988), pp. 235-238.
[13] Marc-Antoine-Louis Claret de Fleurieu de La Tourrette (1729-1793), judge at the fiscal court of Lyon until 1771, permanent secretary of the Académie in Lyon from 1767 to 1793. The Abbé Jean-François Rozier (1734-1793), director of the Ecole vétérinaire de Lyon and in charge of its botanical garden. See Dr. Antoine Magnin, **Les Botanistes Lyonnais I. Claret de La Tourrett** (Lyon: H. Georg, 1885), pp. 6-10, 214-216.

reward Bovier for his attentions with an unflattering portrayal in the seventh promenade of his **Rêveries du promeneur solitaire**, calling Bovier his bodyguard.[14]

Rousseau gives us a Gaspard Bovier, so timidly deferential, so stupidly respectful, that he dared not warn Rousseau that a fruit he was tasting was poisonous. Rousseau knew, to the contrary, that the berry was edible; and there were no dire consequences. Bovier did not discover the calumny until twenty years later and, in outrage, set down the true version of the incident in a manuscript, "Journal du séjour de Jean-Jacques Rousseau à Grenoble sous le nom de Renou." It is apparent from his description that Rousseau had invited him to taste the bright red berries of barberry, *Berberis vulgaris* L., which Bovier mistook for the yellowish-orange berries of sea buckthorn, *Hippophae rhamnoides* L. The latter are also edible, but popular knowledge in Dauphiné held them to be frightfully poisonous. Barberry, as Rousseau explained, was sometimes used in ragouts when one did not have a lemon. When Bovier declined to taste one, he may have offended Rousseau's botanical expertise; and the incident was recast to make a fool of Bovier.[15]

Before Roussseau left Grenoble in some haste for Bourgoin on 13 August 1768, he urged Pierre Liottard to write to him. And, indeed, one can find, in the published correspondence of Rousseau, letters not only to Liottard, but to Dr. Clappier and to Claret de La Tourrette, dating from late 1768 into early 1770: the botanical aftermath of the acquaintances made in Lyon and Grenoble. One of them indicates that the suggestion for the **Dictionnaire de botanique** was made to Rousseau by Claret de La Tourrette.[16]

It was Villars' good fortune, therefore, that he impressed Dr. Clappier and Liottard favorably at a moment when botany enjoyed some notoriety in Grenoble. But whereas Villars, with the encouragement of Clappier, only expected to stay in Grenoble for about six months, time enough to learn a little surgery and about bleeding, so that he could devote his life to the sick and take greater advantage of the knowledge he had acquired of the

∽

[14] Rousseau, **Rêveries du promeneur solitaire** (Paris: A. Desrez, 1837), 1: 441. What little we know of Pierre Liottard depends upon a memorial written by a notable legal expert in Grenoble who took an interest in botanic gardens: Jacques Berriat-Saint-Prix, "Notice historique sur Pierre Liottard, botaniste, lue à l"Académie de Grenoble les 6 et 17 août 1799." **Magasin encyclopédique, ou Journal des sciences, des lettres et des arts**, no. 8 (1 Fructidor Year VI): 504-510; and upon Villars, **Notice sur la vie et les talens de Pierre Liottard, excellent botaniste, mort à Grenoble, le 29 Germinal l'an 4me**. Grenoble: Emile Baratier, 1887. Liottard was born in Saint-Egrève, a village near Grenoble, in 1729, into a poor working family. He received the most rudimentary instruction and was destined to be a weaver. Toward the age of twenty, he sought to escape that fate by enlisting in the army; but he was wounded in 1756 during the French assault on Port Mahon and forced to return to his vocation. Thereafter, he found employment with an uncle, Claude Liottard, an herb merchant. Their trips to gather plants awoke his taste for botany, and he learned its principles from Dr. Clappier. By the time Rousseau visited Grenoble, Liottard had developed a reputation as a botanical fanatic.

[15] Auguste Ducoin, **Trois mois de la vie de Jean-Jacques Rousseau. Juillet-Septembre 1768** (Paris: Dentu, 1852), pp. 8, 31, 36-38, 60-63; Ernest Jovy, "Un Document inédit sur le séjour de Jean-Jacques Rousseau à Grenoble en 1768." **Soc. sci. arts Vitry-le-François** 19 (1899): 117-120; Rochas, **Biographie du Dauphiné** 1: 172-174.

[16] Rousseau to Claret de La Tourrette, Monquin, 26 January 1770. **Correspondance complète** (Geneva: Voltaire Foundation, 1965-1984) 37: 211-212. Also see Gustave Vallier, **Lettres inédites de Jean-Jacques Rousseau** (Grenoble: Prudhomme, 1863), pp. 5-12.

medicinal virtues of plants, his great promise had been reported to the royal intendant. And the latter, Christophe Pajot de Marcheval, summoned Villars for an interview concerning his aspirations. The intendant, who had won a reputation as a man of simple manners and of excessive indulgence bordering on weakness,[17] responded by granting Villars a pension of 500 livres a year from provincial funds.

Such an income would permit Villars to take the full three-year apprenticeship at the Ecole de Chirurgie de Grenoble in the local hospital, just established that year by the initiative of the intendant. The entire income from his property could be devoted to the support of his wife and two infant sons in Le Noyer. He did not think to move them to Grenoble, as he meant to practice in his home country once the apprenticeship (1771-1774) should be completed. The hospital order was the Brothers of St. Jean de Dieu de la Charité, and the chief surgeon the Reverend Father Dominique Durand.[18]

Between 1765 and 1771, Villars had been able to visit Chaix for botanical instruction during the months of good weather, for the walk between Le Noyer and Les Baux could have been accomplished within a day. The move to Grenoble meant a far greater separation; and, for the next twenty-eight years of Chaix's life, the two saw each other infrequently. Long letters became the substitute for visits, Chaix always regarding his letters as conversation.

Chaix was originally the teacher. The intendant's patronage enabled Villars to give full time to learning, not surgery alone, but the natural sciences, Latin, and Greek. If his facilites soon far surpassed those possessed by Chaix, and if his experience soon made him the senior figure, the two remained loyal and selfless collaborators until the end. More than botany bonded them. Both were of peasant origin. If surpassing that state because of education and profession, neither man ever escaped the humility of his birth. Humility becomes a curé, to be sure; but the extravagant deference of Villars to his social betters would go quite beyond the requirements of courtesy, as illustrated in his letters to Carlo Allioni of Turin. In Chaix he had a comfortable friend, an equal no matter their differences.

[17] Ducoin, **Trois mois de la vie de Jean-Jacques Rousseau**, p. 66.
[18] Bally, **Notice historique sur la vie et les travaux du docteur Villars**, pp. 9-11; Joyeux and Dejarnac, "Le Médecin Dominique Villars," p. 129 ; Villars, **Hist. Pl. Dauph.** 1: xiv-xv (1786); Albin Gras, "Historiques des institutions médicales de la ville de Grenoble." **Bulletin de la Société de statistique, des sciences naturelles et des arts industriels du Département de l'Isère**, sér. 1, 3 (1844): 264-265.

Letters one to ten are addressed to Monsieur Dominique Villars, student of surgery at the Hospital of the Fathers of Charity in Grenoble. [1772-1774]

1

Les Baux, 26 July 1772

I have shared in every possible way the grief that the premature death of your younger child has caused you.[19] Everything that concerns you touches me deeply; but in accidents that are not within our power, the Christian must recognize the hand of the sovereign arbiter. We must say with St. Francis of Borgia, a great servant of God, when learning of the death of his daughter: God had lent her to me, He taketh her back, blessed be his holy name.

The plants that you had the kindness to send me when returning my box . . . have taken root very well, and I am delighted with the very fortunate success. Namely, *Circaea alpina* [L.], *Circaea lutetiana* [L.], *Artemisia pontica* [L.], *Geranium inquinans* [L.], *Iris florentina* [auct.], *Chrysocoma linosyris* [L.], *Cactus opuntia* [L.], *Salsola?* or *Soda?* I shall examine it. *Stachys*, I do not see why you call it *alpina* [L.]. Have you made a mistake? Not having been given its flower, I cannot say anything. *Senecio sarracenicus* [L.]: This will unlikely be anything else as our forests are full of it. Your plant charms me by the elegance of its leaves. . . . Your would-be campanula, which has seemed to you to have the habit of [*Specularia*] *speculum-veneris* [L.], I am determining for you, without having seen the flower, to be *Lysimachia nemorum* [L.]. It does not have the milky character of the campanulas. A delightful syngenesia, which I had seen last year toward Grenoble, will reveal itself next year as a beautiful garlic onion. I shall talk about it when it appears. The garlic with a two-leaf spathe I believe to be *Allium oleraceum* [L.] that grows spontaneously in my garden. . . . Of all your plants, only one of them perished, one that you do not seem to have labeled. It had two stalks, quite woody, leaves sparse and smooth, resembling those of flax, but quite firm. See if you can recall it and try to replace it in time. As for *Delphinium elatum* [L.], it has produced a few seeds for me. I ask again for *Dichromus fraxinella*[20] that will be easy for you to procure by sowing the seed immediately after its collection.

[Dr.] Clappier[21] remained here for two weeks. He had intended making a longer sojourn in order to visit a number of cantons in our Alps, and I had promised to accompany him for another twelve days. But after certain remarks, which appeared to me to go beyond propriety, I broke openly with him. I have known too well his way of thinking to have preserved a high opinion of it. He understands medecine, I believe that. He is a botanist, I acknowledge that, but too headstrong in his opinions and too difficult. The truth must be sought, but will one only find it by doubting everything? I knew that he has been acquainted with Renou,[22] but I did not know that he is as . . . as [Renou]; and that he regards the practices of the Church as fanatical without room for religious toleration. I count on your prudence not to confide this on paper to any other person.

~

[19] Dominique, born 5 November 1770, died 22 June 1772. Georges de Manteyer, **Les Origines de Dominique Villars le botaniste, 1555-1814** (Gap: L. Jean & Peyrot, 1922), pp. 133-134.
[20] See *Paspalum* L. in Gramineae.
[21] Pierre Clappier of Grenoble.
[22] Jean-Jacques Rousseau.

A long time ago Monsieur Liottard[23] made me aware of the disagreement that arose between you two; but I said nothing about it to you, and I would still say nothing in the hope that everything would calm down. You are too religious to have to make all the arguments that define virtue on this subject without sacrificing what you owe to yourself. I shall not conceal from you the fact that this piece of news was very mortifying to me.

I shall continue to dry a number of plants for you. For your part, make a collection of seeds for me that you know I lack. Dr. Clappier assured me that *Leucoium vernum* [L.] is abundant in the meadows and beyond the Drac toward Fontaine on the way to Sassenage; and that *Uvularia amplexifolia* [L.] [=*Streptosus amplexifolius* (L.) DC.] is found in the Grande-Chartreuse. I have only two small bulbs of the first one, and I very much want the second one. I hope that in time you will obtain both of them for me. The liberty that I take with you must be taken by you as testimony to the friendship that I have sworn to you forever. Throughout your serious endeavors, take care of your health. I pray the Lord to protect you in your goals and to lead them to a happy ending. Let me have news from you. I embrace you with all the fondness and sincerity for which you know me.

2

Les Baux, 22 August 1772

I do not know what occurrence prevented your letter of July 30th from reaching me until last Sunday. I had inquired at the stage office during that week, and it had not yet arrived. I was anticipating it, and you have truly gratified my expectation. I find myself compensated for the anxiety that the delay caused me.

Clappier made himself too well known to me for me to have retained the least regard for him. The silence with which he responded to the shipments of plants I had made to him, his sordid zeal for withholding his own (having brought me in exchange only two straws), his unbridled license to speak in barracks-room terms, his neglect of propriety and tact, have rendered him insupportable for me. He has a mind, memory, and talent; but I doubt if he is reliable. I desire the knowledge that he may have much less than I do the excellent authors that he possesses. If ever you should frequent him, it should be to profit from his books rather than from his learning. Your letter has enboldened me to say all of this to you, but I am still counting on your prudence. If he speaks favorably of me to you (even though I am only deficient when he is compelled to treat me in his own fashion), understand that it is not his real opinion.

Botanical Observations

1. You sent me *Erigeron canadensis* L., which this incomparable author cites as having become naturalized in Europe, and which the author of Pilat[24] also cites as spontaneous in the Lyonnais, but which you had designated as a *Senecio* or perhaps a *Solidago*, it being in flower. I have examined it closely. . . .

---

[23] Pierre Liottard, gardener in Grenoble. Among the three letters from Chaix to Liottard preserved by the Bibliothèque Municipale de Grenoble, N. 2046, two date from 1770. But they simply concern exchanges of plants and seeds with no reference to any wrangles.

[24] Marc-Antoine-Louis Claret de Fleurieu de La Tourrette, **Voyage au Mont-Pilat dans la province du Lyonnois**, contenant des observations sur l'histoire naturelle de cette montagne, et des lieux circonvoisins (Avignon et Lyon: Regnault, 1770). [SY= *Conyza canadensis* (L.) Cronq.]

2. Please try to replace for me the *Chrysocoma linosyris* [L.] that did perish. I take two specimens you also sent me to be an aster, and which are doing well. Their leaves have the general appearance of the leaves of *Aster novi-belgii* [L.], which I have grown for a long time, but they are clearly something else.

3. Despite the obstinacy of the *pseudophilosophe*, our *Scheuchzeria* will always be *Scheuchzeria* for your stated reasons. There had already been questions about it between him and me.[25] We shall find *Anthericum calyculatum* [L.] when it shall please the Lord. I did find *Anthericum ramosum* [L.] on [Col] Bayard, which I have united with *A. liliago* [L.] and *A. liliastrum* [L.].[26]

4. Monsieur [Jean-François] Rozier was right to say that Linnaeus put the pimpernel at the foot of the oak, because our common pimpernel, *Poterium sanguisorba* [L.] [SY= *Sanguisorba minor* Scop. ssp. *minor* (1772)] immediately precedes *Quercus* in *Monoecia Polyandria*, the tall one of our meadows being *Sanguisorba officinalis* L.

5. The *Dianthus* with corolla nearly plumose, verging on white, common here, is *Dianthus superbus* L. I had it without knowing it at a time I asked Liottard for it.

6. You characterize *Allium magicum* L. very well, perhaps without having read Dodonaeus.[27] See how the latter described it as *Moly theophrasti* Dod. . . . I am speaking of the tall wild garlic that you sent me.[28]

7. Our *pseudophilosophe* was quite nonplussed when he saw *Anemone narcissiflora* [L.] in the meadows around Orcières, which he had always wanted me to make into some *Anemone trianthon* in Barrelier, separating it scrupulously from a specimen he had given me in Grenoble.[29]

8. Then he recovered when I called *Ranunculus glacialis* [L.] the [buttercup] of Valgaudemar. Said *pseudophilosophe* fixed me with a forced smile so as to make me understand that I am given to extreme stupidity. Yet, knowing only *Ranunculus rutaefolius* [L.], he boldly held that it was chimerical to endeavor to separate *R. glacialis* from *R. rutaefolius*. Not having seen the buttercup in question, what effrontery for him to accuse Linnaeus and Bauhin of useless repetition! And why does he find it bad that one follows these great authors? If he means what he says, he has no grounds.[30]

9. The monkshood from the woods of Chaudun is *Aconitum variegatum* L., which [La Tourrette] brought from Mont Pilat to the garden of the Ecole vétérinaire de Lyon. If you expect to be still in Grenoble next year, I will send it to you for planting if it is not already there.

I shall do my best to fulfill your list [of plants]. I am sorry not to have had it sooner as several of these plants have already disappeared. The umbel from Rabou is *Tordylium maximum* [L.]. The geranium from said Rabou is *Geranium ciconium* [L.]. No longer counting on books from the Doctor [Clappier], I am trying to make comments based upon those I still have: Tournefort [**Institutiones rei herbariae** (1700)], Clusius [**Rariorum**

~

[25] While Chaix does not say so, *Scheuchzeria palustris* L. had earlier been regarded as a *Juncus*.
[26] SY= *Paradisea liliastrum [L.] Bertol. (1840)*.
[27] Rembert Dodoens, 1517-1585, of The Netherlands.
[28] It appears that Villars would publish this as *Allium schoenophrasum* L.
[29] Jacques Barrelier, **Icones Plantarum per Galliam, Hispaniam et Italiam observatae** (1714).
[30] Both species are still held to be distinct.

aliquot stirpium, per Pannoniam, Austriam, & vicinis quasdam provincias observatum historia (1583)], Dodonaeus, **Pemptades** [1616], and Linnaeus, **Classes plantarum** [1738].

Will you not come back to this country for any vacation? Then I would have both the pleasure of seeing you and of giving you the botanical items that I will have. . . .

If you are in communication with Monsieur [Jean-Emannuel] Gilibert, could you not get from him, in exchange for some alpine plants, some exotic seeds, such as *Epimendium* or *Polemonium*? I will soon have a dozen pots. Try to get me some early plants, as from the Lily Family, which live in the Chartreuse.

10. (Item from Botanical Observations): Have you observed well the *Galeopsis galeobdolon* [L.][31] that you earlier obtained for me? It has not shown me its fructification, but its trailing habit, nearly like the grapevine or glechoma, makes it appear to me different from a specimen of the true *G. galeobdalon* that flowered for me, and which I obtained from Monsieur Liottard. If you made a mistake, I should be enriched by one more plant, a *Stachys hirta* L. or *Stachys arvensis* L. . . . The *Stachys* you took in the clay soil between Vizelle and Jarrie is *Stachys cretica* L. . . .[32]

3

Les Baux, 26 September 1772

Your worth and your knowledge have provided you entry into the company of the distinguished in Grenoble. Knowing your talent and discretion in advance, I had quite anticipated that this would happen. Your reputation is demonstrated to me every day, and you have just put unequivocal proof of it before my eyes. The plants from Palestine, the Cape of Good Hope, from Peru, Canada, and western America are not found along the hedgerows of Grenoble. I have been pleased beyond all expectation by the rare collection that you have had the kindness to send me. I can respond only very unequally by a counter-shipment of plants lacking the royal colors in which yours are attired. If only amateur kings from nature, they will be equally valuable to you in fulfilling your goal if you do not have them already. You will recognize most of them by their habit; consequently I have only numbered those that I believe require your particular observation.[33]

I repeat, my dear friend, what I said to you in my last letter: I am sorry not to have had your list of requests earlier. I would have sowed certain plants that you ask me for, and I would have dried certain others that have long since disappeared. If you can cope for the present, we shall supply them next year.

I do not find your *Hypericum* in Tournefort, unless it be *Hypericum alp. humilius, magno flore punctato*.[34] What do you think about the difference I make between *Hypericum*

---

[31] SY= *Lamiastrum galeobolon* (L.) Ehrend. & Polatschek (1966) is an erect perennial.

[32] Villars evidently believed it to be *Stachys germanica* L.

[33] Chaix appended a long list of plant names, most of them not annotated. He also endeavored to work with the exotics that Villars had sent him, asking for correction on any errors made in his notes, which need not detain us here as our focus is the flora of Dauphiné.

[34] *Hypericum androsaemifolium* Vill., **Prosp. Pl. Dauph.** 44 (1779). Also see Vill., **Hist. Pl. Dauph.** 3: 502 (1788). Verlot (1872) believed this to be a more oval-leafed form of *H. richerii* Vill. found at higher elevations.

montanum serraturis glandulosis and *Hypericum pulchrum calcibus serrato-glandulosis*? In addition, the former has *folio ovato, lato*, and the latter *foliis connatis*, but not *latis*.

I have procurred for myself a number of pots in order to protect delicate plants over the winter. If you find it necessary to spend some money in order to obtain plant containers, I would send it to you immediately very gladly. One has to take some trouble in sending rare plants when it is for a destination far from the city. Such care is needed in particular for onion-like flowers. But you must remember that, when you will have left Grenoble, you will be able to see again (if the Lord gives us life) those items at my house that previously delighted you. I always act in matters regarding you with the liberty that your kindness allows me.

1. I was already settled on *Picris hieracioides* [L.], which has some striking characters noted by Linnaeus, as well as on *Anthericum calyculatum* [L.], having seen a drawing of it in Clusius, **Pannoniam**, and having reviewed my dried specimens.

2. As the *pseudophilosophe's* alleged *Andryala sinuata* [L.] can only have a place in *Hieracium*, I believe [it to be] *H. lyratum* L. I believe I have already seen *Andryala sinuata* [L.] in a syngenesia that you labeled in that genus when sending it to me.

3. Being sure of my *Galeopsis galeobdolon* [L.], I find yours of that name to have a very different habit and cannot accept it without seeing the fructification.

4. The handsome stachys I have from you is *Stachys spinosa, ramulis spina terminatis* L. [*Stachys spinosa cretica* Casper Bauhin], so-named from the prickles on the calyx. This confirms again what you had earlier told me that we have many crested plants.

5. How you have pleased me by having made known to me our common wolf's hair under the name *Nardus stricta* [L.].

6. You must have found *Anthoxonthum odoratum* [L.]. I ask you for it as I still do not know it and doubt that it lives in this country.

Every day I pray the Lord for your health and success.

4

Les Baux, 8 November 1772

Your trip in Chartreuse has delighted me, and it is, and will be, very useful to me. You have mortified your plagiarist; but, in not accepting your views, he will not correct himself.

On the 31st of last month, I had the incredible joy of receiving, as my All Saints' Day gift, your letter enclosed with your shipment dated the 15th of the same month. Before responding to you, I must urge you kindly to choose Monsieur Pellissier as messenger for what you will have the goodness to send me in the future. I have no reason to be happy with Reynaud, the alternative messenger, who brought this last package. He always charges double carrying costs, that is, 12 sols for said package. He is negligent about getting them to me. On one occasion he withheld a package of plants, meant for putting into the ground, from Twelfth Night to the end of February. Perhaps he treats me this way in the knowledge that I prefer to use the said Pellissier.

But let us turn to your botany. In your letter you provide some clarifications that convince me. You also propose things about which I have some doubt and about which I am going to give you my opinion. I shall propose [some opinions] myself and ask you for yours.

1. This *Ornithogalum*, very common in the spring, is not *umbellatum* [L.], which I

have never seen, but *O. narbonense* [L.] *filamentis lanceolatis membranceis:* that is certain.[35]

2. *Hieracium lyratum* [L.] is quite probable. The stems are multiflora. In favorable locations it is not so small.[36]

3. This *Astragalus* with vesicular, glabrous siliques, its stems decumbent, but without the nude scape that *A. vesicarius* must have, is *A. alpinus* L. . . .

4. I have found this diapensia on our [Pic de] Bure. [*Androsace diapensia* Vill. SY= *Androsace helvetica* (L.) All.]

5. I sense how much my supposed *Hypochaeris radiata* differs from the true *H. radiata* L. Consequently, I remain quite in doubt about its status. . . . Meanwhile, let us call it the *Hypochaeris* of La Grangette, it having been seen only there.[37]

I come now to your current shipment:

1. Does *Anthoxanthum odoratum* [L.] grow in our meadows in Champsaur? It was unknown to me. I would ask you for it, either seeds or alive. It is a very pretty grass.

2. I found *Bupleurum longifolium* L. in Chaudun and have had it in the garden for a long time, . . the leaves being as long as those of arnica. Your specimen is *B. ranunculoides* L. Its umbels resemble the flowers of ranunculus, a fine species I had not before seen. Note: the fine *Bupleurum* from the mountains of Valgaudemar cannot be *B. stellatum* L. Everything suggests otherwise. I think it better to say that Linnaeus did not include [your plant], and that it is *Bupleu. alp. latifol. minus*, Tournefort, **Insti. rei herb.**, 310; [and] *Perfoliata alpina, latifolia, minor*, Casper Bauhin, **Pinax**, 277. Note: I also do not see in Linnaeus the *Bupleurum* of our mountains with graminaceous leaves and nude stalks that I believe to be *Bupleu. mont. flosculis exiguis* T. above [and] *Perfoliata mont. flosulis parvis* C.B. above.[38]

3. I have three stalks of *Ranunculus alpestris* [L.] in my garden sent by Liottard that he apparently had found in the Chartreuse.

4. This *Potentilla* is not . . . *nivalis* [Lapeyr.], which is digitate, not ternate.[39] Haller, having noted three potentillas with white flowers (*alba* [L.] and *caulescens majus* and *minor* L.) says that he had never seen a fourth one Johann Bauhin called *pentophyllum foliis inferne et superne incanis e Baldo monte pentophyllum alp. petrosum*, Lobelius, which is undoubtedly the one found on the slopes among crags that I have already received from Liottard.

5. The *Rubus* is . . . *fruticosus* [L.], a variety. We have only *R. caesius* [L.] here, brambles in fact.

6. *Hyoseris scabra* [L.] is the one that grows here in soil along roads. You know that we

---

[35] Villars decided he had a new species, calling it *Ornithogalum lacteum* Vill.; but it did not stand and is a synonym for *O. narbonense* L. See Vill., **Hist. Pl. Dauph.** 2:272 (1787).

[36] It appears that Villars, after repeated cultivation of this plant, decided it was different and new: *Hieracium pappoleucon* Vill., **Prosp. Pl. Dauph.** 36 (1779). **Hist. Pl. Dauph.** 3:134 (1788). [SY= *Hieracium intybaceum* All. (1773).

[37] This became *Hypochaeris uniflora* Vill., **Prosp. Pl. Dauph.** 37 (1779), which Villars said had been found by Chaix in La Cluse- en-Dévoluy.

[38] Villars published it as *Bupleurum gramineum* Vill., **Prosp. Pl. Dauph.** 23 (1779). *Hist. Pl. Dauph.* 2: 575 (1787). [= *Bupleurum ranunculoides* L. ssp. *gramineum* (Vill.) Hayek (1927).]

[39] It was probably *Potentilla nitida* L., which is white-flowered and ternate, and which Villars later located in the Grande Chartreuse.

found *Hyoseris minima* [L.] between St. Jacques and St. Maurice; and you earlier sent me *Hyoseris foetida* [L.], which I should like to have alive.[40]

7. This delicate umbel with crosswise leaves is surely *Seseli carvifolia* Vaill. in Linnaeus. Compare the annotation. I had already dried it. If you can find it again, I should like to have it alive.[41]

8. The other umbel, which has flowered in my garden, does not have general involucres and cannot be a *Peucedanum*. It is *Seseli montanum* L. that has affinity with the preceding species.

9. Our small *Silene* from La Chapelle that you have sent, and which grows elsewhere, is *Silene rupestris* [L.]. And the other *Silene* with the unifloral stems, which Clappier has inappropriately called *rupestris*, is certainly *Silene saxifraga* L. from among rocks in the mountains.

10. In my shipment to you, you have *Trifolium alpinum* [L.], whose flowers resemble those of astragalus, and which is abundant around Orcières, and which Clappier did not recognize during your trip. I have also found it in flower. You know that you found it among our first narcissus-anemones.

My paper is going to run out. Kindly keep me in your pleasant mind, for me—and then for the plants. The more I have news from you, the more pleasure you will give me. I shall do everything possible to indicate my gratitude and inviolable affection.

5

Les Baux, 30 April 1773

For your botanical treatise, you grasp all factors that have a connection with this science, proving to me again the richness of your intelligence and the quickness of your imagination. Since you ask for my opinion, here it is:[42]

1. It appears to me that you dwell too much on the agreeable aspects. In your place, I would cut it back or reveal [the pleasant] only in the context of several of your depictions. The botanist who seeks no more than an acquaintance with plants, who pauses at a tapestry, or in a garden, or along a brook, only to become acquainted with the inhabitants, does not really become a botanist but a mere spectator. . . . But neither must he have universal knowledge. It is enough, when he sees a plant, to give its name, its genus, the species, and to know all its parts.

2. You should remove from your diction the conditional particle *si* that you use too often, and that you use moreover in an absolute sense. You should not always lead your listener by the hand. You will surprise him pleasantly with the recital of something unexpected. Vary your style as much as it is possible for you.

3. In demonstrating the usefulness of botany, bring it up in order to explain the

~

[40] SY= *Arnoseris minima* (L.) Schweigger & Koerte and *Aposeris foetida* (L.) Less. respectively.
[41] Taking this cue from Chaix, Villars examined *Seseli carvifolia* L. in **Species Plantarum**, concluding that Linnaeus had confused its description in Bauhin with that of *Seseli saxifragum* L. Thus, he would republish it as *Seseli cardifolium* Vill., **Prosp. Pl. Dauph.** 24 (1779). **Hist. Pl. Dauph.** 2: 586 (1787). [= *Seseli elatum* L. ssp. *cardifolium* (Vill.) P. Fourn. (1937)]
[42] This refers to the botany lessons that Villars began to give in Grenoble in 1773 and would repeat annually thereafter, becoming a justification for his pension.

mechanical art that identifies those plants that provide food and drink, but also indicate that this is not the real business of botany. If we are to understand the utility of botany, it should be sufficient to know that it occupies the serious attention of the most celebrated academics; and that men, most notable through birth and their erudition, make a study of it all their lives. You may cite them.

4. Once you have cut back considerably on the agreeable aspects, try to extend a bit more on the need for botany, which is the principal point to be made in the case of your surgical students with medications to concoct: the dangers from ignorance about medicinal herbs, the properties of various herbs, that they are readily available, and the purpose of the Creator in spreading them around us. Why should this wise provider have placed remedies for Europeans at the extremity of the East Indies? I wish you the happiest successes in this undertaking. If I can be of any more help, let me know. But you know my mental capacities. I may be able to formulate thoughts; but I am clumsy in enunciating them and slow to recall names and such from memory, because I converse so little. That is my lot.

Some specimens at the beginning of your notebook:

1. *Ranunculus arvensis, chaemeli folia, flore minore luteo* Tournefort, I. r. h., . . a plant omitted by Linnaeus.[43]

2. *Cenchrus capitatus* [L.] appears quite polygamous, but I have not seen anything else from the genus. I have planted some clumps of it in the garden to examine it better.[44]

3. A *Carduus* that I have not seen in Linnaeus I have also planted in the garden. . . . You mention *Carduus defloratus* [L.] to me, a picture of which you will see in Daléchamps 1: 492. Our mountain *Carduus* merits close examination. It will not be found in Linnaeus. I am referring to the dried specimen I will show you.

This is your box from Grenoble with the *Helleborus viridis* [L.] for the Rev. Father Apothecary. I shall address a larger box to you with some native plants that you will kindly send back to me at your convenience with Monsieur Pellissier. . . .

I depend upon you giving me your mind. I was recently mortified to be deprived of the pleasure of your company, but your affairs left you no choice. Among friends, one behaves without ceremony. Extend my greetings to your dear family, and take care of your health. I embrace you with all my heart.

6

Les Baux, 24 May 1773

Your friendly letter of May 8th from Le Noyer reached me right away. I thank you, and I am quite touched by the gracious things you say to me. Accept this small shipment of plants that I have not prepared adequately because of the continual rain the past few days, also preventing me from going out to botanize. I will be delighted if something among them will merit a place among those in the rich garden of the Hôpital de la Charité. I have not labeled them in detail as you will recognize them easily. . . . I also send a basket full of *Helleborus viridis* [L.] for the Rev. Father Apothecary. I hope he takes pleasure in them. You may offer him my respects as well as my small services regarding our alpine plants.

∽

[43] Perhaps he meant that Linnaeus omitted Tournefort's plant as a synonym for *Ranunculus arvensis* L.
[44] SY= *Echinaria capitata* (L.) Desf. (1799).

If you can send me any plants in returning my box, try to send it by the first or the next regular mail, and I shall take care to have your shipment picked up at the messenger's office. The great heat, that suffocates everything on the roads, is soon coming. Your two *Uvularia* [*Streptosus*] were dead thanks to the long delay you know about, rotting in fact. Please replace them at your convenience with *Lathyrus nissolia* [L.].

In my garden, I had you look at a stalk of ranunculus that is truly *Ranunculus polyanthemos* L. (*R. polyanthemos simplex* Lob.). . . . If the double ranunculus that you talked about to me, *Ranunculus polyanthemos duplex* Lob., is placed along side said ranunculus in said place, please send it to me. With pleasure will I put it with my *Ranunculus acris flore pleno* [Bauh.] and *Ranunculus aconitifolius flore pleno* [Clus.] that the botanist calls monsters. These are monsters I find to be very beautiful.

Tell me something about your botany course. Let me know when you will be coming back to this country. Those moments that I will spend botanizing with you will be quite valuable for me. I write in haste as my accommodation is about to leave for Gap. Against my inclination, I am forced to finish a discussion I wish could go on forever. Yours from the heart.

7

Les Baux, 25 June 1773

I received your letter of June 17th yesterday. It gave me so great a pleasure that I cannot put off responding to it, even though you have pleased me with the prospect of seeing you this summer.

How delighted I am that my plants pleased you . . . and pleased beyond expectation by the gracious reception you have witnessed in my regard. I shall try always to respond in kind.

As for Liottard, my procedure is one of final attempt, as all your efforts for him have been unfruitful, chilling me, too, in his regard. I shall break with him only when he breaks with me. If he takes offense over a commission that does not entirely square with his ideas, I do not know what to do about it. The matter must be discussed between Monsieur Dheralde[45] and him. I want nothing to do with it.

I share completely in your delight over the *Bulbocodium vernum* L. I will give you my thoughts in the same spirit you gave me yours. Those specimens, which I sent you out of my *Colchicum montanum* L., are precisely those whose collection you made with me during our little outing in the woods that you know. I do not have any more of them, and it is not possible to find any more for examination until a new springtime. I have quite believed them to be the same as those in a small valley on our mountain of Loubet [above Rabou] where, some years ago, I met them for the first time but did not examine them. But, as for those on said Loubet, I have reason to make *Colchicum montanum* L. of them, being unquestionably 6-androus, 1-gynous. . . . In order to determine *Bulbocodium* L. see if your specimens are only 6-androus, 1-gynous but not 3-gynous, whether its leaves are lanceolate, and if the six stamens are inserted at the base of the petals. I have compared my specimens several times with said *Bulbocodium* without getting them to accord. A word

---

45 Pierre-Joseph Dheralde or D'heralde was the leading physician in Gap.

from you on this matter will reassure me and teach me about the treasures in my neighborhood.[46]

The secund wild acanthus of Daléchamps is *Carduus tuberosus* L.[47] *Carduus defloratus* [L.] is the other cirsium from Dodonaeus, Daléchamps 1: 492.[48]

Carduus acanthoides L. is found in the fields around Gap. I recently sent you a half-formed specimen of it from Le Noyer. Since then, I have looked at it closely, especially at the laciniation under the corolla, cut nearly to the base. I have dried some of it for you. It is *Onopyxus* in Dodonaens, Daléchamps 2: 350, an annual.

As for our perennial mountain thistle (the one we have burdened with the name *defloratus*), I know nothing positive about it, except that in our country it disclaims the name *defloratus* and seems to be distinct. I shall examine it when possible, but where is it found in Linnaeus?

I am resuming my letter that I interrupted to visit a property in my parish. This walk provided me the pleasure of finding two stalks of *Lathyrus nissolia* [L.], one near the other in a remote spot: one in bloom, but the other still growing, which I carefully dug up and transplanted to my garden for seed. It is a rare plant for me, never before seen despite all my searching. I saw it by chance, as I was looking at *Adonis aestivalis* [L.]. The specimen of *Lathyrus* that I am adding here, very common in the fields, I take to be *Lathyrus hirsutes* [L.]. Contrary to its disposition by Monsieur Linnaeus, however, I have never see it other than uniflora, and its tendrils are 3-leaved. Am I mistaken? Give me your opinion.[49]

Take care to name the *Carthamus carduncellus* that is illustrated in Daléchamps 2: 348.[50] Without doubt, *Carthamus caeruleus* L. is the species with which nature has favored the sterile slopes of Les Baux. It is remarkable for its black anthers. [SY= *Carduncellus caeruleus* (L.) Presl. (1826)] You must have specimens of it, and I shall dry more of them. Dr. Clappier found it to be very handsome but did not resolve the species to my satisfaction.

Extend my compliments to Monsieur Jullien.[51] All plant material coming from him will give me great pleasure, but anything procured for me for cultivation will be even more welcome and of greater long-term value.

You seem convinced that your indisposition creates difficulty for me. Try to recover and take care of yourself, in particular when you are taking field trips for botanizing.

Give me the satisfaction of a response, by way of the post, as soon as you are able.

~

[46] Linnaeus had placed *Bulbocodium* in Hexandria Monogynia, but *Colchicum* in Hexandria Trigynia. In fact, both are trilocular: but *Bulbocadium* has 1 style, undivided at base, 3-fid near apex, and flowers in the spring. *Colchicum* has 3 styles, free from base, and flowers in autumn. *Colchicum montanum* L. is a SY of *Bulbocodium vernum* L.

[47] Villars would give it as *Onopordum acanthium* L. in Hist. Pl. Dauph 3: 26 (1788).

[48] Villars would rebaptise it as *Carduus cirsioides* Vill., an illegitimate name. Prosp. Pl. Dauph. 30 (1779).

[49] Despite Chaix's observation, Villars would repeat Linnaeus's description. The species is described today as 1-3 (4)-flowered, tendrils 3-leaved.

[50] This had already been named *Carthamus carduncellus* L. SY= *Carduncellus monspelliensium* All.

[51] Curé de Saint-Georges-de-Commiers (on the Drac near Vizille), who botanized with Villars in 1773, sometimes also accompanied by Liottard that summer. Villars, in Hist. Pl. Daup. l:xv (1786), specifically noted the Grande-Chartreuse, Sassenage, Villard-de-Lans, Le Moucherotte, Revel, Uriage-les-Bains, Allevard, and the Carthusian monastery of Prémol, all sites in the vicinity of Grenoble.

Nothing pleases me so much as your letters. I embrace you with the most sincere friendship.

Nota. Mirabilis lusus naturae: onnes quot novi Astragali, dum nocte dormiunt, foliola sursum juncta. Habent: Astragalus vero galegi formes (sub crepusculo hoc vespere observavi), dum dormit, foliola sua deorsum dimittet. Spectator Naturae dornum nunquam redit vacuus. Vale et salve.

8

Les Baux, 10 October 1773

Following your letter of September 20th from Le Noyer, I had the butter picked up for my stock on St. Michael's Day from the indicated person in St. Bonnet. I am very pleased and under obligation to you for it. Your letter introduces me to *Sisymbrium irio* [L.] that, without precise examination, I had been calling *Cardamine parviflora* [L.], presumably the plant you saw near the Porte Colombe in Gap. . . .

I am taking advantage of the next trip by our messenger, Pellissier, to attend to the opinions expressed in your letter of September 25th from Grenoble. Give my respects to the Rev. Father Apothecary when presenting the *Helleborus viridis* [L.] to him. I hope it gives him pleasure.

I believe that I have assembled in this notebook all your dried plants from our botanizing on [Pic de] Bure, or that you kept aside for yourself at my home. In that regard, I say that you collected *Phaca alpina* [L.] and *Astragalus montanus* [L.] (that is certain). The other *Astragalus* with flowers *subalbidis quasi luteis, carina apice subcaerulea*, . . is either *Astragalus vesicarius* [L.] or not in my *Species Plantarum* L. In addition to *Aretia alpina* [L.] [=*Androsace alpina* (L.) Lam.], the other one with affinity to it, *floribus subsessilibus*, which I called *Diapensia helvetica* [L.]—but you did not see the character of that genus. Here is the note that Dr. Clappier made when sending a specimen of it to me from Grenoble on 5 April 1770: "*Diapensia helvetica* L. . . . is a plant that Linnaeus determined only from its habit, and which he very badly judged in consequence, for it belongs in the genus *Androsace*." I shall henceforth recognize our beautiful renunculus, *Ranunculus alpinus fumeriae folio* Tourn. I. r. h. Our discovery is no less striking, as I understand from your letter.[52] Please report to me your observations of our beautiful artemisia when they have been made.[53]

Friend Liottard, in a shipment sent to me, makes no mention of you, but I see your tracks in it. His plants are very fine. Please give him proof of my gratitude for them. You may tell him that I shall endeavor to have sent to him things I think could give him pleasure, and that I would like to be as rich in rare objects as he is in order to be reciprocal.

∼

[52] This reference to their discovery is confusing. The buttercup in question was *Ranunculus rutaefolius* L., and neither Chaix nor Villars claimed it as his own. Each one identified it as collected on the mountains of the Dévoluy. Villars would later remark [**Hist. Pl. Dauph.** 3: 740 (1788)] that the species probably ought to be segregated from *Ranunculus*, which was perhaps their "discovery" in 1773; and it ultimately found a new home in *Callianthemum* C. A. Meyer.

[53] This would become *Artemisia insipida* Vill., Prosp. Pl. Dauph. 32 (1779), found by Chaix in woods belonging to Monsieur Mondet [**Hist. Pl. Dauph.** 3: 249-250 (1788)].

I am reading his copy of [Michel] Adanson [**Familles des plantes** (1763)] and am struck by the liberties he takes with botanical authors. It would be appropriate to read a criticism.

Your plants have overwhelmed me with joy. I have recognized some of them even without labels; about others where I am doubtful, you must reassure me. . . .[54] Seeds, cuttings, or live plants, all are welcome here, at your convenience of course. You know what I have and what I lack.

Tell Dr. Clappier not to be troubled about his paper, that I shall have it sent to him at the earliest. And let me know whether he has received his books, as he is not a man to advise me about it. When you have the opportunity to obtain for me, through purchase, some book that you judge would be useful to me, buy it. You know my sentiments. If you have something to be sent to me in my box, but cannot have it ready for Monsieur Pellissier's return trip, wait for his next trip. I receive things in good shape through him.

Having interrupted my letter at this point for a time, your letter of September 30th intervened. Monsieur Pellissier, who had taken care to get your box to me last Sunday, apparently had overlooked said letter in which you cleared up those questions I have written about above. . . .

As for the trees and shrubs that you should be able to send to me, see if I should be able to preserve them over the winter by using six additional pots that can be brought to me without delay from Sisteron, and do whatever you think to be best. I found the packets of seeds in the aforementioned volume by Adanson, for which my thanks.

Monsieur Liottard says in his letter to me (no doubt under your influence): "The shrub that you have in a pot, which you do not recognize, is *Croton tinctorium* L." [=*Securinega tinctoria* (L.) Rothm.]. I wish it were, but it has no resemblance [to that species]. . . . I am enclosing some leaves from said shrub that you will kindly show to friend Liottard. How I would commend it if, by some happy metamorphosis during its adolescence, it became *Dictamnus fraxinella* [Pers.]! [=*Dictamnus albus* L.].

My niece sends you her respects. You know with what sincerity I am, Monsieur and very dear friend, your very humble and very obedient servant. I also extend compliments to you on behalf of Dom Courrier of Durbon.[55]

9

Les Baux, 7 November 1773

Your letter of October 29th brought me all the more pleasure as I had been hoping for it. I am benefiting from all the observations you have made for me, and I am savoring in advance the impression of pleasant things you have promised me.

During the week of St. Luke, I paid a visit to Father [Jean] Gaillard, former curé de Rabou, now prior-curé de Reynier in Upper Provence, about 7 to 8 leagues from here. I became acquainted with a few plants that I take pleasure in noting for you: 1) *Seseli*

<hr>

[54] He provides a tentative list of identifications that indicate the plants were exotics.

[55] Durbon was one of several Carthusian monasteries in Dauphiné where botanists on field trips were especially welcome. Many of Chaix's collections over the years were made on the rugged route from La Roche-des-Arnauds to Montmaur, La Cluse, and over the ridges to Durbon. The great forest of Durbon, owned and conserved by the monks, was spared the ravages of deforestation common elsewhere in the Alps by the 18th century.

glaucum L. [=*Seseli montanum* L.], which you may have sent to me dried last year without a label. 2) *Centaurea isnardi* L. [=*Centaurea aspera* L. ssp. *aspera*]. 3) *Ruta graveolens* L., perhaps a variety of *Ruta sylvestris major* C.B. [=*Ruta montana* (L.) L.]. 4) *Momordica elaterium* L. [=*Ecballium elaterium* (L.) A. Richard]. 5) *Carduus marianus* L. [=*Silybum marianum* (L.) Gaertner]. 6) *Lepidium graminifolium* L. 7) *Lepidium ruderale* L. 8) *Sisymbrium irio* L. 9) *Lamium maculatum* L. 10) *Aster acris* L. [=*Aster sedifolius* L.]. 11) *Inula germanica* L. 12) *Dracocephalum pardalinaches* L. [=*Dracocephalum austriacum* L.].

During this past season, the mountain of said Reynier, if quite beautiful, did not show me the treasures that it possesses, having been nibbled by the shepherds' flocks from Provence.[56] According to Monsieur Gaillard, *Scilla bifolia* [L.] grows there. I dug up two bulbs of it, but regrettably lost them. Monsieur Gaillard also gave me a bulb that he claimed to be an *Ophrys insectifera* [L.]. I have given the present messenger a shipment for Monsieur Liottard in which I put specimens of all these plants. If you are interested in any of them, go to his place where you will see them. I will keep any of them for you at your pleasure.

By *Doronicum*, I mean the one that is abundant on our [Pic de] Bure and on the high and cold mountains of our Alps, which cannot be *Doronicum pardalianches* L., not having basal cordate leaves. It could be *Doronicum plantagineum* L. as the leaves are decurrent the length of the petiole. It is true that they are not acute, the basal leaves at least, but rather obtuse. It is surely a different species from the former. *Doronicum minus officinarium* Daléchamps, illustrated in 2: 100, also does not match the description any more than the figures of *Aconitum pardalianches* in the same volume do. Give me your opinion about the above.[57]

I have read Adanson. His natural families will never produce a botanist, as he presumes that the reader already knows plants. Not all of his reform is to my liking. I am very glad to read him, but I do not want any more of him. In some places I do not understand him; in other places he offends me; elsewhere his judicious criticisms please me. But I shall ask Monsieur Liottard to permit me to read the second volume when I send back this one to him.

Do not forget the collection you promised me. I leave to your judgment the acquisition of botanical books for me and rely on your usual care for live plants.

I am waiting for your young brother whom you wish to entrust to me for the rudiments of Latinity.[58] I sent word to your mother that she can bring him here on Sunday, the 21st of the current month. I have arranged this delay in order to give myself time to acquire the necessary provisions, some wine in particular that can only be delivered during the week preceding said Sunday, either because of some fairs, or because the grape harvest was quite delayed this year. I shall give him my best care possible.

≈

[56] The annual transhumance of sheep from Provence to Dauphiné had long contributed to the deforestation of the mountains.
[57] Villars evidently ruled out *Doronicum plantagineum* L., declaring it to be a synonym of *D. paralianches* L. (which it is not). It is possible that the species in question here he later published as *Arnica stiriaca* Vill., **Prosp. Pl. Dauph.** 32 (1779). **Hist. Pl. Dauph.** 3: 210 (1788). [=*Doronicum stiriacum* (Vill.) Dalla Torre = *Doronicum clusii* (All.) Tausch.]
[58] Jean-François Villar, born 27 September 1759.

The incident concerning Monsieur Jullien's books is peculiar. In the first place, I am extremely sorry about this loss for him, but also for those of us who would have had the opportunity to read them.

All considered, I do not believe that we will be able to go beyond the **Systema naturae** L. to which the **Mantissa** L. is now added. It will someday be necessary. But we must proceed in a manner appropriate to us and, to do that, to take the surest route. If Monsieur Rozier [of Lyon] answers you, will we not be able to follow his way to that effect? I have never read a flora, but it would be desirable to have one of them. But I leave everything to your prudent discretion. . . .[59] Continue to share the ideas you are proposing with me. I cannot receive pleasanter news than that which comes from you. It will be a great assuagement for me amidst the piles of snow that will surround me during the coming months in my solitude.

10
Les Baux, 26 February 1774
Your valued letter of January 28th reached me only on the 17th of this month, having been held in the post because we rarely go to Gap in the bad season. Your brother, Pierre, who came here on the 7th, the day of the carnival fair, delivered a letter to me written later than the above along with the books you sent for your young student. I gave him my letter of response for you; but, fearing that it has gone astray, I am repeating its contents here.

The **Selectae e profanis scriptoribus historiae** is an indispensable book, an anthology for the more advanced forms in secondary schools. The **Selectae e Veteri Testamento historiae** was a collection expressly made for beginners in secondary schools.[60] Postponing the former book until your brother's second year, I shall only put the latter title into his hands for him to explicate until Easter; and I shall have him translate it, from Easter to the end of the year, to provide his own version. . . . Also, after Easter, until the holidays, I shall explicate the letters of Cicero for him for prose and the fables of Phaedrus for poetry, alternately every day, the former in the afternoons, the latter in the mornings. For his lessons in Latinity this year, in sum, he will study no other books except for the **Rudiment** [the primer of the Latin language]. . . . For the present, these are the books I ask you for. If you do not run across them, I will buy them for him in Gap. The cost for them is moderate.

If you have the opportunity, you may buy in advance for him the **Abrégé des particules** in three parts; and a Virgil, Horace, Quintus Curius, Sallust, and Ovid, all small in-12 and without any translations; as well as several of Cicero's works, either **Orations**, **Philippics**, **De Officiis**, **Laelius** or **De Amacitia**, or **Prosody**, a small brochure.

You give me real pleasure in telling me that you are thinking about undertaking a work on the plants of Dauphiné. That is the point I have wanted to see you reach ever since your first elementary lessons in botany. Since you do not presume to be able to improve upon

~

[59] This is the first apparent reference to a proposal to author a flora for the province.
[60] Jean Heuzet, the French humanist (ca. 1660-1728) was the author of these and other books meant for classroom teaching. The **Selectae e profanis scriptoribus historiae** (1727) was widely used as an anthology of classical wisdom and was divided into five books: De Deo (on divinity); **De prudentia** (on manners); **De justitia** (on justice); **De fortitudine** (on courage); and De temperantia (on temperance).

the Linnaean descriptions for those plants that excellent botanist has observed, you could simply cite the Linnaean description where there is nothing more to be said, along with one or two of the best synonyms. You could make your own observations on those that require them. After having given your definition, you could indicate which ones are little known or unknown or rare. You would indicate their native locale and provide a drawing. A critical observation on the contraindication of the plants' healing virtues would be valued.

The title, **Flore grenobloise**, does not please me. It is too restricted. Grenoble is only one place in Dauphiné. The title, **Essai sur les plantes du Dauphiné**, strikes me as too unpretentious, and you will reject it yourself after you have sought the productions of our province for a few years. Your work must be in Latin for less difficulty in your expressions and for greater precision. Thus, it seems to me that your work could support as a title, or something similar, **Botanicum Delphinense ex Methodo et nomenclatura Linnaeana, in quo rariores stirpes hactenus forte ignotae aut saltem a nemine vix satis conceptae descriptae, iconebres expressae reperiuntus; addita fictarum virium plantarum critica observatione. Opera et labore Dominici Villari Vapincensis chirurgi.** I sense, in fact, my shortcomings in botany, but I would be able to be of some use to you in Latin diction. Try to arrange the means to carry this through. God willing, I am confident of success.

As for my watch, I entrust it to you. Get the highest price for it that you can. I am not concerned about getting the money. When you have got the money out of it, be good enough to buy twelve hands of brown paper that you may give to Dr. Clappier for me. I believe that should replace what he left at my house. I am very far from wanting to be in debt to him on this matter. You may use additional money from said sale for my copy of [Louis Gérard,] **Flora Gallo-Provincialis**, or for other books that you will be good enough to get for me. Also, you could use some of it to have the manuscript notes, which I enclose herein, properly bound for me in sheepskin. In your letter, you promise me a compilation taken from [Johannes] Scheuchzer. You will obliged me doubly if you would take the trouble of inserting that material yourself at the end of my manuscripts so that I could have it all in the same book. Caution the bookbinder not to cut very much from the top and the bottom in order not to spoil the handwriting, to retain the arrangement of the notes, to put in some blank leaves at the end for additional notes, and to engrave on the back of the book as a title: **Excerpta botanica**; and, lower down on said back: **Domin. Chaix**, which you should write out as a model for him. I have additional books to be bound.

Please give the abbé Jullien his copy of **Regnum vegetabile** and convey to him my proper gratitude. At the front of his book I am enclosing a letter for him in order to fulfill that part of my obligation.

When you next send me a shipment of plants, do so in care of Monsieur Pellissier. Wrap the roots and particularly the ends of cuttings in moistened moss so that we do not lose the fruit of our labors. The *Canna indica* [L.], that I wanted so much, and that you sent to me, is dead. Apparently any cold is fatal. Please get it replaced for me, either from your stock or that of friend Liottard. Let me also remind you that Liottard will yield *Uvularia amplexifolia* [L.] and *Helliborus niger* [L.] to you for me. . . .

Please, indeed, give my compliments to dear friend Liottard. Tell him not to forget me in his botanical work, whether in plants or in seeds. Everything that I might be able to do for him will be done wholeheartedly.

Your brother adds a letter of his own to mine here, [as] I have urged him to indicate to you his desire to respond to your kindnesses and to those of his dear mother.

My mother sends you her very humble respects. Accept my most sincere regards.[61]

Letters eleven to twenty-nine were addressed to Monsieur Villars, Surgeon, residing at his home in Le Noyer, but botanizing frequently as authorized, as he prepared for the publication of the prospectus of his flora. He would soon discover that the continuing patronage of the intendant, the renewal of his pension, had been opposed by the religious order where he had interned in Grenoble, and that success breeds jealousies. During the summer of 1774, he made a trip into Lower Dauphiné, Provence, and Languedoc with Dr. Clappier. At Montélimar, they examined the herbaria of Dr. Pierre Garidel and the Chicoyneau family, then possessed by Dr. Jean-Joseph Menuret de Chambaud. They examined the rich collections (both herbaria and library) of Jean-François Séguier at Nîmes. At Montpellier they saw the collections of Antoine Gouan and Pierre Cusson. They were able to botanize around Avignon, Aix, Marseille, Toulon, and Hyères. The long trip accounts for the substantial hiatus in Chaix's correspondence that summer. [1774-1779]

11

Les Baux, 1 October 1774

I have only just received your letter of September 18th this evening along with the small package. The negligence of the messengers has been too much! . . . I have not written lately, expecting every day either the honor of your visit or that of your letters according to your promise. I fear that you may already have sent your package for Montpellier. Therefore, to make haste as much as possible for me, I am sending your brother to you with my messages.

Please have the kindness to read the letters I have addressed to Messieurs Gouan and Cusson, to attach the labels I have designated [to the right plants], and then seal said letters. I have included therein some seeds for them and some specimens of our *Heracleum cussonianum.*[62]

I have taken all the notes I wanted from your **Mantissa 2 [altera]**[63] and observations by Linnaeus, which I am sending back to you. I am very much obliged to you for it, and I am looking forward to the happy moment of your visit here in order to discuss some observations:

~

[62] Antoine Gouan (1733-1821) and Pierre Cusson (1727-1783) were contemporaries at Montpellier. Gouan was internationally known through his publications; but Cusson, though virtually unpublished, had gained a reputation as an authority on umbels, finding in their fruit new bases for segregating the genera within the family, especially upon what he called the *peri-embryum*, the albumen that surrounds the embryo. See Charles-Frédéric Martins, **Le Jardin des plantes de Montpellier** (Montpellier: Chez Boehm, 1854), p. 38. Villars would later say [**Hist. Pl. Dauph.** 2: 640-641 (1787)] that the *Heracleum* in question here had been found on Mont Aurouze in 1769 by Dr. Clappier, and later by both Chaix and Villars. It had been determined to be a *Heracleum* by Cusson, after which Chaix and Villars meant to name it *Heracleum cussonianum*. Even though Cusson told Villars in December of 1774 that the name should be *H. bipinnatum*, and later, in March of 1775, recommended the name *H. delphinense* on the grounds that all the specimens seen had come from Dauphiné, Villars would publish it as *H. pumilum* Vill., **Prosp. Pl. Dauph.** 26 (1779). It is now held to be a synonym of *H. minimum* Lam., **Fl. Fr.** 3: 413 (1778), but Lamarck's accepted priority could be challenged.

[63] Gouan contributed some of the plants that Linnaeus published in **Mantissa altera** (1771).

1) I am cultivating *Erysimum hieracifolium* L. here, from the Champsaur, with siliques appressed and not divergent. . . .

2) I have *Erysimum cheiranthoides* L. here from seeds (from Monsieur Gouan), with spreading siliques, annual, with leaves not quite entire, but subdentate.

3) The tetradynamous plant that you gave me as *Sisymbrium tenuifolium* [L.] is without any doubt *Sinapsis laevigata* [L.]. An autopsy separates it from the sisymbriums and clearly puts it in the sinapes.

4) We have *Carthamus carduncellus* [L.] here and not *Carthamus caeruleus* [L.].[64]

5) *Hieracium pumilum* **Mantissa** 2 is our alleged *Crepis pygmaea* [L.] without doubt.

6) The description of *Buchnera canadensis* **Mantissa** l: 88, fits [*Verbena aubletia* Jacq.] very well. Is it by error that *Buchnera* is in Angiospermia? as it is gymnospermous.[65]

7. You have made an incorrect note in **Mantissa** 1: 79, as the *Ranunculus platanifolius* [L.] given there is the same as our *Ranunculus sylvanticus albus* whose leaves are palmate and not compound, with linear bracts.[66]

8. In addition, you have made a second erroneous note in regard to *Ranunculus aconitifolius* [L.], the double variety of which I have cultivated but have never seen in the wild, with leaves divided up to the petiole and, as a consequence, compound and not palmate, with lanceolate bracts.[67] Linnaeus says [of the latter species] that *Ranunculus platanifolius* is very similar to it, with the exception that *R. platanifolius* (which is our common species) is twice higher.

Your brother will spend several days away in the region for a bit of vacation. . . . As he has not yet seen all the explications of authors that one sees in the fourth form, I should be able to have him read all of them between now and St. Martin's Day. And I hope that, with this help and the authors I shall give to him, he will be ready to enter the third form in the Collège [de Gap] by St. Martin's Day or St. Catherine's Day. This is all I can do for you and him here beyond that time, as I will then be obliged to give my attention to a small nephew who will come to my house. Besides, your brother finds himself alone here and grows weary of this abode, and he will benefit more in a school where the lessons are more frequent.

You will understand that the handsome *Geranium argenteum* [L.], deprived of its element for fourteen days, did not preserve its life. If you still have another live shoot of it, the dear disciple will bring it to me along with whatever else you may be willing to send me. But try to accompany him yourself so that we can enjoy a few days of your presence. In anticipation of that happy event, I embrace you heart and soul.

~

[64] Chaix was correct. The species = *Carduncelles monspelliensium* All.

[65] *Buchnera canadensis* L. [= *Verbena aubletia* Jacq. = *Verbena canadensis* (L.) Brit. = *Glandularia canadensis* (L.) Nutt.] is a North American species both cultivated and naturalized in Europe. The references to Angiospermia and Gymnospermia can be confusing. Linnaeus used the terms to distinguish between plants that had seeds enclosed in capsules and those with seeds exposed in nutlets. In **Species Plantarum**, he had put *Verbena* in Diandria Monogynia, but gave it as Didynamia Angiospermia in **Mantissa** I. As the species has nutlets, Chaix correctly argued here that *Verbena* should be in Didynamia Gymnospermia (putting it adjacent to the mints).

[66] Villars gave Chaix's description in **Hist. Pl. Dauph.** 3: 734 (1788).

[67] The leaves are currently described as palmately 3- to 5-lobed, the middle one free to base.

12

Les Baux, 10 February 1775

Through intimation in your letter, I have sensed with regret the need you have had for your **Pinax**.[68] I shall have it sent back to you when one can get from here to Gap. In the meantime, I shall use it to help put Linnaean nomenclature in my Daléchamps.[69] Whether from a want of opportunity or from negligence on my part, I have denied you the use you could have made of it. I hope for your indulgence. You are aware that impassioned enthusiasts often frustrate their best intentions.

I have read with pleasure your paper on plants to be examined. I have noticed and corrected some mistakes in the Latinity. I have made few observations, not feeling prepared to do more, which you may read at the end of a few species. The amiable Séguier's encounter with my fine *Cucubalus* gives me real pleasure. He merits much more of your attention, and mine, then does the obstinate pseudophilosophe who accused me of failure to see the coronula of the petals, a character of the *Silenes*. Two years ago, for amusement, I planted two stalks of it in my garden, and it has continually appeared to be without appendages, fitting the wording of Clusius used by Séguier. I am calling it with some reason *Cucubalus silenoides*, because, otherwise, it has great affinity with the silenes, in particular its clavate calyx, the character that made me suspect it to be *Cucubalus italicus* [L.]. But it is surely not the latter, which is a perennial. Note the sharp point at the extremity of the leaves and their light pubescence. . . .[70]

The books that you tell me about on behalf of that bookstore in Grenoble must not be allowed to get away. I shall approve with gratitude the use of my bit of money, and I am disposed to add something more to it if it is necessary. Think about Séguier's [**Plantae veronenses** (1754)] or [Bauhin's] **Pinax** for me, etc.

When you get your orders from the Monsignor Intendant,[71] if they include a trip to Paris for you, write to me before your departure. If, on the other hand, you will be returning from Grenoble, try to bring back some plants to put in the ground, whether live plants or seeds, from either your friends or the provost's gardener. I had two young shoots of *Lobelia cardinalis* [L.], which makes me believe that someone in Grenoble has it. Without your help, I shall never again see this beautiful botanical genus with my own eyes.

Persius and Juvenal are the most difficult authors we have in Latin. They are

~

[68] Casper Bauhin, **Pinax theatri botanica** (1623).

[69] Jacques Daléchamps, **Historia generalis plantarum**, 2 vols. (1586-1587). This is often referred to as **Historia plantarum lugdunensis**, as it covered the Lyonnais and was published in Lyon; and in Bauhin's **Pinax** simply as **Lugd**. A French translation was published in Lyon in 1615, the edition that Chaix owned.

[70] *Cucubalus* and *Silene* are very close; but the former produces a berry, the latter a capsule, a distinction not mentioned by Chaix. As he did not find appendages, he believed his species should be *Cucubalus silenoides* Chaix in Vill., **Hist. Pl. Dauph.** 3: 614 (1788) [=*Silene italiae* Pers.].

[71] In both 1775 and 1776, Villars' pension was renewed by the intendant, Pajol de Marcheval, who had originally encouraged the idea of a flora for the province. The intendant had had a notable record for promoting agricultural knowledge. In 1760, when he was organizing the Société d'agriculture de Limoges, he solicited help from the Academy in Paris to combat the "Angoumois caterpillar" that ravaged the grain in his region, obtaining the services of Duhamel and Tillet. See André J. Bourde, **Agronomie et agronomes en France au XVIIIe siècle** (Paris: Ecole Practique des Hautes Etudes, 1967), 1: 265-266.

explicated only in Rhetoric, often in vain even in that class. That was my experience, as a result of which I nearly know them only by name. Your brother will be able to use them in the future.

13

Les Baux, 4 March 1775

Your brother is quite prepared to recognize by himself whether the teaching he will be given furthers the prospects for his advancement. This [past] winter I have explicated for him the second part of **Selectae e Veteri Testamento historiae** for prose and the first two books of the **Aeneid** for poetry. He has completed his versions of the **Selectae e profanis scriptoribus historiae** and his themes on the adventures of Telemachus. If I had continued to teach him, he would have persevered with his themes and translations. I would have explicated for him the third and fourth books of the **Aeneid** for poetry, two or three books from Quintus-Curtius, with some of Cicero's works, either **De Senectute**, or **De Amicitia**, or **De Officiis**, without neglecting him for the composition of verse. And I would have believed that this was sufficient for him to complete the third form. He is now supplied with most of these books, and you will please indicate to me the ones in particular they want him to read that he does not have. I am now sending him my **Telemachus**, . . and everything I have is at your disposition and at his service. I have offered [my books] to him with all my heart. He has left me his **Selectae e Veteri Testamento historiae** for my nephew's use.

Until the end of my life, botany will be my principal pleasure: unproductive for science, if you will, but satisfying. I cannot wish for anything more. To illustrate my bent to you, I have just written to Father Dom Fourmault [at Durbon] and Monsieur Cusson [at Montpellier], begging them to get me some seeds; and I call upon you to exercise all your influence in my favor in regard to [any seeds] addressed to the Monsignor Intendant by said Fourmault. I have a great obligation to you for the care you are taking in having the books I want sent to me. I leave the choice of them to you along with all the bother.

I await a word of response from you about the remarks I put in for you at the end of your manuscript. But I will get it at your leisure, as I learn that you have few moments of respite.

~

While the following eight-month hiatus in the correspondence may suggest missing letters, this was a period of intensive botanizing, arranged and financed by the intendant, to provide Villars the benefit of the company of distinguished naturalists and giving Villars his first link with Paris. Jean-Etienne Guettard (1715-1786) had published in botany, even advancing his own plant classification system;[72] but he was now interested in mineralogy, what we would call geology today, and wanted to explore Dauphiné. Barthélemy Faujas de Saint-Fond (1741-1819), a native of Dauphiné (Montélimar), was a young attorney of independent means much attracted to a study of the mountains. This trip in 1775 was his first work in the field, and he

~

[72] Jean-Etienne Guettard, **Observations sur les plantes** (Paris: Durand, 1747).

formed a high opinion of Villars as a botanist.[73]

The fieldwork, in fact, was accomplished in a series of outings from Grenoble: Beginning 10 July 1775 to Allevard, St.-Hugon, Pré Nouveau, Le Grand Charnier, Pointe de l'Aup-du-Pont, Pointe du Gléyzin, to the Montagne des Sept Laux, country offering what Villars called superb horrors.[74] They set out again on 1 August 1775 for the convent of the Grande-Chartreuse, also visiting the Petit-Som, the Grand-Som, the Charmant-Som, and St.-Pierre- d'Entremont. Villars expressed his gratitude for the ready assistance provided by the monks of the Grande-Chartreuse, who not only gave directions but offered guides and supplies. He would meet the same welcome from the pious solitaries in all the regional Carthusian houses: Prémol, St.-Hugon, Durbon, Bouvante, and La Salette.[75]

They then turned eastward into the higher mountains, proceeding through Vizille up the valley of the Romanche, to visit Allemond, Le Bourg d'Oisans, Venosc, La Bérarde, Mont-de-Lans, Huez, Besse, Auris, Clavans, and La Grave, reaching Briançon by crossing the Col du Lautaret. Villars found the region around Briançon particularly rich in plants: from Névache to Granon and Montgenèvre, southward to Le Bourget and the valley of Cervières. They found the Queyras and its tributary valleys exceptionally interesting, proceeding downward from Mont-Dauphin to Embrun and the woods of Boscodon.

When they reached the neighborhood of Gap, which must have been early September, they paid a call on Dominique Chaix. "From Gap," Guettard wrote, "one can go to Les Baux in three hours. You follow the main road from Veynes, climbing for an hour, during which time you cross two ravines on wood bridges and the ravine of Charance on a stone bridge. To the right is the country house of the Bishop of Gap and a chapel. After crossing the stream, you leave the main road, which goes to the left, and take the path on the right that climbs gently for another hour until reaching a chapel on the right [Chapelle de Sauveterre] and an oratory on the left. At the oratory you turn to the right and go to the hamlet of Courrier [Corréo]. After passing through it, you descend to the [Petit] Buëch, crossing it on a wood bridge, from where you climb up for a quarter of an hour to the village of Les Baux."[76] Guettard was deeply impressed by his meeting with Chaix. About sixty years of age at the time, Guettard was very much a man of the world: a grandson of François Descurain and a physician who bore the title Médecin-Botaniste to the duc d'Orléans, a man much traveled and of a lively character. The contrast with Chaix's isolation and solitude, with his limited resources, provoked not merely admiration but veneration.[77]

In returning to Grenoble, the party passed over Mont-Bayard into Champsaur, visiting Ancelle, Orcières, and Champoléon; then north into the Valgaudemar and down to Corps and La Salette. By then it was mid-September, and the snow was starting to cover the highest mountains. Consequently, for their third trip of the year, beginning 19 September 1775, they sought lower elevations. Their route passed through Les Echelles, Le Pont-de-Beauvoisin,

~

[73] Barthélemy Faujas, **Histoire naturelle de la province du Dauphiné** (Grenoble: Chez la Veuve Giroud, 1781), p. 418. See J.-B.-M.-Georges Bory de Saint Vincent, "Eloge de Faujas de St. Fond." **Ann. gen. sci. phys.** (Brussels) 2 (1819): 22-24; Cap, **Le Muséum d'histoire naturelle** (Paris: L. Curmer, 1854), pp. 68-69.

[74] Villars, **Hist. Pl. Dauph.** l: xix (1786).

[75] Ibid. l: xviii-xxvii (1786).

[76] Jean-Etienne Guettard, **Mémoires sur la minéralogie du Dauphiné** (Paris: Impr. de Clousier, 1779), p. 43.

[77] Villars, "Notice historique sur Dominique Chaix, botaniste," p. 301.

Crémieu, La Côte-St.-André, St. Marcellin, Tain; and then south in the Rhône valley to
Valence, Montélimar, and terminating in Orange.[78]

14
Les Baux, 20 December 1775

I have compared your plants noted from your Latin [edition] of Daléchamps with my
French version, and I have transcribed the numerical citations. I have even added there,
when I believed it necessary to do so, the generic or specific names for their equivalents in
the **Pinax** of Casper Bauhin. I sense that some of my notes contain errors, reflecting errors
that said Bauhin perhaps fell into; but others will appear to you to be accurate. In going
through my Daléchamps, I have noticed many other quite good illustrations in it that
merit being cited in your work. I would be quite delighted if, once you go through my said
Daléchamps with your usual care, you meet something in my notes that can be of use to
you. You can accomplish this review easily here at my house if you will honor me with a
stay of a few days during your next visit. . . .

If, during your field trips, you have not run across *Paederota* [L.], it will not be found
in Dauphiné.[79] The plant I took for it (and perhaps you, too, did earlier) is *Veronica*
teucrium L. . . . Should our beautiful grass at Le Noyer be *Agrostis arundinacea* L.? *Arundo*
agrostis Scop.?

My niece . . . just gave birth a few days ago to a girl for whom I am the godfather. She
offers you her humble respects as does the little Magdelline, my current governess.

If you go to Grenoble, please do not forget our seeds and the plants for me that will
be at your disposition. I have much regretted the loss of the *Lobelia cardinalis* [L.], which
perished during its adolescence. Monsieur Liottard had obtained its seed for me, and you
will know whether he is still an accommodating man.

I enclose here the seeds you told me you could present to Monsieur Séguier. I wish
you and all your family very happy holidays. I am anticipating with these greetings the
beginning of the new year. Honor me with your response, which already gives me
perceptible pleasure in advance.

15
Les Baux, 10 February 1776

On account of the bad weather that made the trip from Gap impracticable, your letter
of January 22nd, enclosed within the great Ray,[80] did not reach me until last Saturday. I
owe you a thousand thanks, both for the said Ray, which, being voluminous, must have
cost you considerable money; and for the sundials that have given both my niece,
Margueritte, and me real pleasure. I extend her thanks to you. They appear to me to be
very accurate, having checked them against my clock.

I read in your letter that I am neglectful in our epistolary relations. From that, I

⸰⸱

[78] Villars, **Hist. Pl. Dauph.** 1: xxviii-xxxi (1786). Guettard's dating of the trips in 1775 may be found in his
Mémoires sur la minéralogie du Dauphiné, pp. 1, 12, 59.
[79] It is only found in the Eastern Alps.
[80] John Ray, **Historiae plantarum** , 2 vols. (1686-1688).

conjecture that either you did not receive my last letter, or yours in response went astray. I had the honor to write you at Christmastime, and the letter was carried by your brother on his way to spend the holidays at your home. I added your package of notes on Daléchamps to my letter. Your above-mentioned letter of January 22nd says not a word about that matter. Having seen your brother in Gap after his return from Le Noyer, who told me he had no response to give me from you, I could have reason to conclude that my letter and package of cards had not been received by you. But I rather presume that one of your letters was lost.

As for the *Avena*, I have *sativa* [L.], *fatua* [L.], *sterilis* [L.], all received from you; *flavescens* [L.], triflora [= *Trisetum flavescens* (L.) Beauv.], collected at Loubet; one other collected in my meadow that is close to the latter, but five-flowered, perhaps a variety of it? I have our oats that I believe to be *Avena sesquitertia* L. [= *Avenula pubescens* (Hudson) Dumort. ssp. *laevigata* (Schur.) J. Holub.], but I find my dried specimen to be only biflora. It was inappropriate for Linnaeus to call his avena *sesquitertia*, which indicates *third and a half*, since, according to his description, his *third* flower is only half-formed. Thus, he should have named it *sesquialtera, second and a half* or *demi-third*. Tell me what you have observed about our oats. I still have the *Avena pubescens* of the Alps as triflora, the calyx verging on violet with silvery margins. But among the latter can be found some with entirely white spikelets, more slender, which might be only a variety of it. . . . I have a specimen from Liottard that I believe to be *Avena pratensis* [L.] [=*Avenula pratensis* (L.) Dumort.], as well as another one from Liottard with a long panicle that I do not know.

If you could obtain for a time the [**Agrostographia sive Graminum**] by Johann Scheuchzer, it could clear up this very thorny family for us. Scapoli will not satisfy you on the *Avena*; he allies them all to his *pilosa*. He appears to me to have worked better on the *Carex*. Nevertheless, I rank Scopoli higher than any other floral author whom I have read, because he has examined objects very closely, and because such examination has enabled him to remove certain plants from one genus and place them in another, and to bring to light certain species omitted by Linnaeus or reduced to simple varieties. I am convinced that you will concur with this author on a number of points. I am sending it back to you, as you have requested, along with Morison.[81] This summer, when you will be taking trips, be good enough to let me have the book again.

All my aims will be fulfilled if I can contribute in some way to the production of a work that will make you known. I shall work to deliver the Catalogue of Plants to you that I have collected with the observations I will have made about them.

When you write to Monsieur Guettard, offer him my very humble respects, and tell him that I beg him to remember that he promised me to procure some seeds for me from the Jardin du roi. You, too, have given me to expect to receive some of those given to you by Father [Dom] Fourmault.

I am undertaking the twenty low masses for the six livres you promise me on behalf of your brother-in-law. I shall be very much obliged when you are able to procure [the money] for me. You are aware that my tiny parish cannot provide [the money] for me.

It is hardly necessary to express the joy that you will give me when you will be able to

***

[81] J. A. Scopoli, **Flora carniolica** (1760); and Robert Morison, **Plantarum historiae universalis** (1680).

take time out from your occupations to intrude yourself into my insignificant solitude. My respects to your dear family.

16
Les Baux, 6 May 1776

Last Wednesday, a day of the fair in Gap, I expected to get news of you from your brother, hoping to see you with him or to receive some letter from you. Having had neither of these satisfactions, I had to settle for addressing you a word of greeting. Last Saturday, your very brief letter of April 22nd was brought to me, informing me of a prior letter that I have not yet received and whose loss I regret extremely. . . . You do repeat something of the contents of that first letter to which I have the honor to respond.

I think as you do on the subject of my four *Seseli*. The one I brought back from beyond the Durance, and which I at first believed to be *S. glaucum* L., will surely prove to be that. In cultivation, it already puts out very glaucous leaves. The taller one with greener leaves, from seeds provided by Monsieur Gouan, will be *S. montanum* [L.]. (Its label was within the shipment).[82] The tall [plant] received alive from Monsieur Liottard, which perished after its fructification, will therefore be *S. annuum* [L.]. I see its petioles as emarginate at the end, that is to say, at the place where the leaf begins to pinnate. Our annual or biennial differs from the latter. I presume it is little known by the authors. Based upon the description of it that I gave to Monsieur Cusson, he responded: "*Yours could well be another seseli.*"

As for willows, I say that *Salix purpurea* L. is unknown to me (perhaps you have it near you). I hold, until provided convincing proof to the contrary, that our small white water willow, very common, is *Salix helix* L. It is probably found in Switzerland, in Dauphiné, along the Rhône, in the Gapençais, and was known to Daléchamps. The illustrious Haller, [**Historia Helvetiae**], no. 1640, speaks of our helix, *Salix monandra* Arduin, without making any mention of the *S. purpurea* L. Secondly, I do not see in our willow either the descriptions or the notes that Linneaus gives for his *S. purpurea*. But, thirdly, I do see his treatment of *S. helix* in our willow. . . .[83]

I am up to Decandria L. in my catalogue of plants of Gap. You will find notes I have included here.

I think you ought not to treat *Galium glaucum* L. I have called the plant under review *hispidum*, the leaves rigid, ash grey, mucronate. I suspect it is only a variety of *Galium album* [Miller].[84] I believe I have found *Galium boreale* L. in wet meadows around Gap. It remains to be learned whether the fruit is hispid; without that, I shall send it back to the

***

[82] The two species are now generally lumped as *Seseli montanum* L., described as a glabrous perennial. But Emile Burnat, **Flore des Alpes Maritimes** 4: 155-158 (1906), suggested recognizing the two forms as varieties of *S. montanum* L.

[83] Villars believed that *Salix helix* L. was a synonym of *Salix monandra* Ard., Hist. Pl. Dauph. 3: 767 (1788). The species in question here may well have been *S. viminalis* L. that Villars included in his flora, but without description, as identical to Haller's no. 1641.

[84] This seems to be the inspiration behind *Galium rigidum* Vill., Prosp. Pl. Dauph. 19 (1779). Hist. Pl. Dauph. 2: 319 (1787), where the new species is described as "little different from *G. album*." SY= *G. lucidum* All. (1773).

rubioides. But its leaves are not scabrous below. I have called the small species with rough leaves that is found in barren, watery places in meadows *Galium uliginosum* [L.].

Seseli pinnatum Cuss.[85] is what I call *Pimpinella seseli.* I think it is the *Bunium* in Daléchamps. It has the odor of *carvi* [caraway], its leaves have something of the appearance of parsley, but do not resemble the lower leaves of coriander. . . .

Let me know if Haller mentions our *Statice* of the Alps. I do not recall finding it in his Nomenclator [ex **Historia plantarum indigenarum Helvetiae** (1769)]. With the exception of Casper Bauhin, I doubt if the other authors have known it well, as they all place it along the seacoast.[86]

I have just found *Orchis fucum referens, colore rubiginoso* Casper Bauhin, very well described by Johann Bauhin. (I have *Orchis facum referens major, foliis superioribus candidis et purpurascentibus* Casper Bauhin in my garden, obtained from Monsieur Gaillard.)[87] I also found *Aristolochia pistolochia* [L.] in the same place. *A. rotunda* [L.] and *A. clematitis* [L.] grow near Gap. Moreover, *Crepis albida*, or so I presume, was found in the same place.[88] I have found *Convallaria polygonatum* L. [= *Polygonatum odoratum* (Miller) Druce] with two flowers on the same peduncle, but the multiflora still escapes me. Perhaps you will catch it yourself.

When you next do me the honor of writing me, address your letters to Monsieur your brother, who, knowing the people in my parish, will have the kindness to get them to me. I beg you to procure me this satisfaction the soonest it is possible for you. I wish for your perfect health.

~

The fieldwork with Guettard and Faujas resumed on 28 May 1776, the fourth trip. Villars neglected to tell Chaix that he would again be absent from Le Noyer for many weeks. The party moved southwestward from Grenoble to Claix, Vif, Monestier-de-Clermont, and passed through the Trièves region to La Croix-Haute, St.-Julien-en-Beauchêne, and Aspremont. They then headed for Le Buis by going through Serres, L'Epine, and Bellecombe. From Le Buis, they explored Mont Ventoux, even though it was beyond Dauphiné, to observe in nature species already reported by Barrelier and Jussieu and to collect material for comparison.

Returning to Le Buis, they moved roughly north to Nyons, Vinsobres, Valréas, Dieulefit, and Crest; then turned eastward to Die, Lesches-en-Diois, Baurières, St.-Pierre-d'Argençon, Laragne-Montéglin, and Sisteron. Turning northward again, they went through La Saulce, Gap, and Veynes to reach the Carthusian monastery of Durbon. They then explored the mountains of the Dévoluy and the Champsaur. As the lateness of the season the previous year

~

[85] This is apparently a species discussed in a letter to Chaix but never published. Villars would publish it as *Seseli cardifolium* Vill., **Prosp. Pl. Dauph.** 2: 586 (1787). Located in Les Baux, Champsaur, and Valgaudemar, it could well be the unknown annual discussed above that was described by Chaix to Cusson. [= *Seseli annuum* L. ssp. *carvifolium* (Vill.) P. Fourn. (1937)].

[86] Villars recognized it as *Statice armeria* L. var. b. [= *Armeria alpina* Willd. = *Armeria pocutica* Pawl. ssp. *alpina* (Willd.) P. Silva (1971). From Mont Aurouze].

[87] Villars quite correctly described these two orchids as variants of *Ophrys insectifera* L. [**Hist. Pl. Dauph.** 2: 49 (1787)] found near Grenoble, Gap, and La Roche-des-Arnauds.

[88] *Crepis albida* Vill., **Prosp. Pl. Dauph.** 37 (1779). **Hist. Pl. Dauph.** 3: 139 (1788): uncommon new species found near Gap, Les Baux, Briançon, Serres, Die, and La Bâtie.

had prevented a visit to the Valbonnais, they now deviated from their intended route to examine valleys of Valjoffry, Le Désert, Le Périet, and Chantelouve before returning to Grenoble.

After a brief respite, they set out on their fifth and last trip on 29 July 1776. Moving through the mountains around Sassenage west of Grenoble, they visited Lans-en-Vercors, Villard-de-Lans, Corrençon-en-Vercors, turning southward to reach Die once again. They then turned northwestward, passing through the region of Quint and into the forest of the Chartreuse de Bouvante, a country they found to be covered with the poisonous Aconitum napellus L. From there they crossed the Royans country to Pont-en-Royans, and then proceeded up the left bank of the Isère to Grenoble. That ended the two seasons of field trips under government orders with Guettard and Faujas. Pierre Liottard was present on at least some of these trips, enabling Villars to form an opinion of his abilities.[89]

Guettard, meanwhile, developed a highly favorable opinion of Villars whom he characterized as one of those few who, if born in the very depths of the provinces and in the midst of ignorance, believe that they are not born to grub in the soil. Surrounded by the beauties of nature, they not only admire them, but are driven to acquire a knowledge of them and to instruct others about them. Guettard tells us that Villars had been preparing a catalogue of the plants of Dauphiné even before the onset of their trips, and that he used the trips throughout the province to improve and extend it, teaching Guettard much about species that were new to him.[90]

The learning was not unilateral. Villars recognized in Guettard what he called a rigorous exactitude in the art of observation that pushed Villars to a higher level of difficult, precise work, encouraging his botanical ambitions. The fieldwork of 1775-1776, moreover, enlarged his scope from one small region to a large province: a particularly difficult province given the unusual variety of its terrain, and ranging in its climate and products from Sweden, Siberia, and Switzerland to Piedmont, Provence, and Languedoc.[91]

It is also apparent that the association with Guettard and Faujas stimulated an interest in mineralogy and geology. Villars subsequently never botanized without a barometer to measure elevations. As the barometer of that day required two people, one at the base and one at the summit (and he usually worked alone), he could not have obtained precise measurements of elevation, but sufficient for establishing the rough elevation of the plants he collected. He had always carried a loupe, but his new standards required greater magnification. As in time he would find the ordinary Lionnet microscope quite inadequate, he applied his mechanical skills to adding pivots, adjusting screws, and tubes, enabling him to get a magnification of 250.[92]

When Pajot de Marcheval renewed Villars' pension on a permanent basis in 1774, it carried an authorization to travel anywhere for professional advancement. At some point during their association, Guettard began urging Villars to go to Paris to compare what he had collected with the various herbaria and manuscripts available only in Paris. That trip, which would become possible in 1777, would be the culmination of Guettard's positive influence. It would enable Villars to meet and consult with the elderly Bernard de Jussieu and his nephew,

∽

[89] Villars, **Hist. Pl. Dauph.** l: xxxi-xxxiv (1786); Guettard, **Mémoires sur la minéralogie de Dauphiné**, p. 101, 131.
[90] Ibid. pp. cliv-clv.
[91] Villars, **Hist. Pl. Dauph.** 3: v-vi (1788).
[92] Bally, **Notice historique sur la vie et les travaux de docteur Villars**, pp. 29-30.

Antoine-Laurent; to become acquainted with the gardener, André Thouin; and opened up correspondence with such luminaries as Claret de La Tourrette of Lyon, Pourret of Narbonne, Deleuze of Sisteron, and Danthoine of Manosque, while sustaining correspondence already established with Séguier of Nîmes and Gouan and Cusson of Montpellier, a correspondence of which Chaix would also be a beneficiary.[93]

17

Les Baux, 2 September 1776

For more than two months I was greatly troubled, not knowing where you were and receiving nothing from you. It is not like you to treat me to such a rigorous silence. The curé de Poligny [a village between St.- Bonnet and Le Noyer], to whom I turned in my anxiety to ask if he knew anything about your situation, assured me that you had reached Le Noyer on August 16th. You can imagine the joy this news gave me. Losing no time, I am sitting down to demonstrate that fact through my letter. Receive it as the most faithful expression of my great affection.

If you can manage a few days to visit here, that will complete my good luck. But if you must put me off, please have the kindness to recompense me by writing in detail to inform me about the current state of your projects. I would go myself to your home to see you, but the exploitation of my dîme and the seeding of my land prevents me from going away now.

I have completed sketching out the native plants that I know, making all the observations now possible for me. I have met some plants that had escaped my [earlier] searches. This little work will give you pleasure. I wanted to send it to you with the curé de Poligny. But, given the rainy weather and his stops in both Gap and in La Roche-des-Arnauds, I feared that it could go astray, which would cause me great mortification.

Not having seen the famous Haller, I have often had difficulty with our alpine plants. If you can deprive yourself of [that volume] for a little time, allow me to read him.

There is a very young man in La Roche-des-Arnauds who has a strong feeling for botany, and for whom I have identified a number of plants. Would you please have the **Elémentaire** by Monsieur Rozier[94] sent to me from Grenoble for him, unless you would like to give him your own copy. I shall reimburse you for the exact amount.

Pressed to go immediately to La Roche-des-Arnauds, I terminate my epistolary conversation that I should like never to be interrupted.

P.S. I do not at the moment have any masses. If, through your offices or those of Monsieur Arnaud [curé du Noyer] to whom I address my respects, you could procure me some, I would be under great obligation to you. But do not put yourself out.

18

Les Baux, 18 October 1776

Since the honor of your visit, you are not aware to those I have made to you [in my mind] in Le Noyer. They have been very frequent I assure you. I also think that your kindly feeling for me brought you back to our slopes and to the paths of earlier times. I am

~

[93] Villars, **Hist. Pl. Dauph.** 1: xxxv (1786); Villars, "Notice historique sur Dominique Chaix," pp. 301-302.
[94] M.-A.-L. Claret de La Tourrett and François Rozier, **Démonstrations élémentaires de Botanique** (1766).

compensated for the fatigue of my field trips if, through them, I find you in the circumstances where I want you to be. . . .[95]

When you next see Monsieur Pierre Achard, vicar of St.-Julien-en-Champsaur, tell him that Dom sacristan of Durbon, brother of Monsieur Eymin, the director of the Sacré Coeur de Marseille, has been charged by said brother to extend thousands of compliments to him and to say that he will never forget him; and that Dom sacristan himself also keeps an affectionate memory of him.

When proceding on my trip to Durbon last week, planned for a long time, I went through La Cluse where I spent the night. On the walls of the church, I only found the hieracium that I know as *Hieracium amplexicaule* [L.], which grows from Les Baux up to Chaudun in the rocks along the [Petit] Buëch, quite glutinous to the touch. See if it is the one you drew at La Cluse.[96] From there I went to dig up some roots of my *Hypochaeris pontana* which I had earlier seen in alpine meadows called *lèches*, and which I planted in my garden. If God gives me life, I shall have opportunity to settle on what it is.[97]

At St.-Julien-en-Beauchêne [near Durbon], I found a variety of hieracium that is close to what we earlier called *Andryala lanata*. I believe that you have drawn the latter.[98]

I am very much obliged to you for your observation on *Carduus lanceolatus ferocior* J. Bauhin. [= *Cnicus ferox* L. = *Cirsium ferox* (L.) DC. in Lam.]. I have described the difference I have perceived between it and *Carduus eriophorus* [L.], which is quite striking. The latter likes the cold mountains as much as the former the slopes of warm hills. [= *Cirsium eriophorum* (L.) Scop.]

Hundreds and hundreds of times I have stepped on the *Aira* around Corréo about which you had often spoken to me without really examining it. It is the grassy cover in slightly humid places around Les Baux and La Roche. The leaves resemble those of *Festuca ovina* [L.]: the panicle is somewhat like that of *Agrostis capillaris* [L.], but more open, the spikelets biflora, a small awn rising from the back of the petals but barely surpassing the calyx. I believe it is *Avena* Haller, Hist., no. 1493: *Gramen palustre paniculatum minus*, Scheuchzer, **Agrostographia**.[99]

This little Cichorieae with leafless, hirsute stems is an only daughter in my garden. I would judge it to be either a *Hyoseris* or *Hieracium minimum columnae*, Tournefort, **Institutiones** 470. Do you know *Hyoseris minima* L. well? I have a specimen from you described as a variety of *Hypochaeris radicata* [L.] Have you examined its receptacle closely?

~

[95] This is the first letter addressed to Villars, not only as surgeon in Le Noyer, but as professor of botany, a position he did not yet formally hold. It may be that Chaix had just been told that the Intendant had promised Villars in confidence a medical-botanical appointment in Grenoble once his professional studies should be completed and the flora well underway.

[96] This may be *Hieracium pulmonarioides* Vill., Prosp. Pl. Dauph. 36 (1779), which is part of the *H. amplexicaule* L. group; but Chaix only recognized the latter in the Gapençais.

[97] Villars would publish it as *Andryala pontana* Vill., Prosp. Pl. Dauph. 37 (1779).

[98] This is confusing, as Chaix would later list this species as a synonym for *Hieracium andryaloides* Vill., found at St.-Julien-en-Beauchêne, Rabou, and Berthaud; while Villars would give it as a synonym for *Hieracium lanatum* Vill., Hist. Pl. Dauph. 3: 121 (1788).

[99] Published invalidly as *Avena calycina* Chaix & Vill. in Chaix, Pl. Vap. 11 (1785); *Avena calycina* Chaix & Vill. in Vill., **Hist.** Pl. Dauph. 1: 315 (1786): *in pratis udis*, at Corréo de La Roche. *Avena calicina* (sic) Vill., ibid. 2: 148 (1787): in wet meadows at Corréo de La Roche and near Sisteron.

Is it anything but *Hyoseris minima* L.? I submit my suspicions to you.

My handsome *Silene nocturna* [L.], had I not seen its coronule, I would have very easily mistaken it for *Cucubalus reflexus* [L.], as it had been labeled under that name for me. . . .[100]

I now think, following you, that my alleged *Salix cinera* [L.] is only a variety of *Salix caprea* [L.]. But I still cannot refer my *Salix amygdalina* [L.], the yellow willow of this region, to *Salix alba* [L.].[101] I can say nothing about *Salix vitellina* [L.].

I am sending with this letter a package of about forty species of seeds for the gentlemen in Paris. If they find it acceptable and have some seeds from their incomparable garden sent to me in return, I shall collect more of them next year. Please take them with you if you make the trip. If not, have them sent [to Paris] by the way available to you, making my request clear to them, which will be more acceptable thanks to your mediation and the more valuable plants from your own collection that you will be adding to the shipment.

I have sketched out a letter to have sent on to the learned and very respectable Monsieur Guettard; let me know in advance, please, when you will depart. Beyond which I shall charge you with several commissions relative to our Botany.

I have had those books I had possessed only in brochures bound in Gap and am very pleased. Casper Bauhin's **Pinax** is among them. In anticipation of your trip, send me my Gérard and Séguier so that I may have them similarly bound. When you can manage it, let me read the great Gouan while awaiting the celebrated Haller.

19

Les Baux, 15 November 1776

I am sending you observations based upon my reading of Monsieur Gouan's illustrations. Please utilize those that you judge to be sound; please criticize me about those where I have gone amiss.

1) *Hieracium scorzonaeraefolium* in my manuscript is not the one in Haller, **Enumeratio** [746, no. 14], which is surely *Hieracium sylvaticum* Gouan. . . . I think that you, too, will separate them.[102]

2) I would refer *Pilosella [majoris], seu pulmonaria luteae species angustifolia* J. Bauhin [2: 1034], Ray 240, as a variety of *Hieracium sylvaticum* Gouan. . . . We have it at Loubet, at La Grangette, with narrow leaves three inches long, subdentate, sessile, white on the edges especially at the base; the stem has leaves only at the insertion of its branches; uniflora with some bracts; the flower larger than that of [*H. s.* Gouan] with calyx white from a tissue of hairs. . . . See if it is in Haller, no. 46, as a variety of *Pulmonaria gallica tenuifolia* Tab[ernaemontanus Eicones plantarum, no. 195].[103]

3) *Hieracium conyzaefolium* Gouan. *Hieracium pyrenaicum* L. is the same as my *Hieracium blattarioides* [L.], common in our mountains. I do not sense the reason that led

∽

[100] *Silene reflexa* (L.) Aiton is a likely variant of *Silene nocturna* L. ssp. *nocturna*.

[101] *Salix amygdalina* L. [= *S. triandra* L.] is very close to *Salix alba* L.

[102] *Hieracium scorzonerifolium* Vill., **Prosp. Pl. Dauph.** 35 (1779). *H. silvaticum* Gouan, **Hist. Pl. Dauph.** 3: 125 (1788) [= *H. vulgatum* Fr. (1819)].

[103] Villars agreed with Chaix's observation, making these species synonyms of *H. silvaticum* Gouan.

Gouan to refer *Hieracium blattarioides* L. to *Hieracium amplexicaule* [L.][104]

4) Monsieur Gouan describes our *Hieracia amplexicaule* [L.] and *cerinthoides* [L.] very well; but his drawing of the latter does not correspond, as the stem has too few leaves, besides hardly resembling those on our plant.

5) I persist as to my *Hieracium intybaceum* [All.], Haller's *Enumeratia*, and as to my *Hieracium saxatile* [non Jacq.], *pilosella incana saxatilis lutea* C. Bauhin, **Pinax** 263, Ray 241, no. 18. . . .[105]

6) My *Carduus vapincensis* is *Carduus crispus* [L.] **Species Plantarum** . . . .

7) *Carduus cirsium dictus folio. lanceolato nigrius* [Magnol, **Bot.** 49], well described as *Carduum circioidem* by Séguier, is not in fact *Carduus crispus* [L.].[106]

8) Ought you not have said to me that our *ferox* thistle is *Cnicus ferox* Mantissa 1: 109? [See preceding letter]. Correct the label on the seeds by writing *Cnicus* instead of *Carduus* if you send them to Paris.

I have detected that, in the citations for your work, you are not distinguishing between the two **Mantissas** by Linnaeus and his **Observationes** [in materiam medicam (1772)]. For example, *Prunus avium* [L.], **Mantissa alterna** (1771) 397, which is **Observationes** and not **Mantissa**. According to this citation, one supposes that Linnaeus, having not mentioned said *Prunus* either in his **Species Plantarum** (1753) or in his **Systema Vegetabilium** (1774), referred to it in his supplement. And that is incorrect. . . .[107] I should very much like for you to provide at least one variety of our sweet and sour cherries within *Prunus cerasus* [L.]. Bauhin writes *Cotonaster* with you; Linnaeus, *Cotoneaster*. Check this spelling. I would prefer the latter, because the element is *cotoneum* and not *cotonum*, making the compound *cotoneaster*.

You had promised a generic character at the head of your species, and I do not see any of them on your specimens. I should like you to put it there, just as the chief virtues of useful plants. Your work would serve then for genera, species, and *materia medica*. You will sense the worth of that. If I were not speaking to a friend, I would be more reserved in my observations.

I have not put the willow from Chantoussel in the ground, it being quite dead. Given the opportunity, please get it for me. The plant to be husbanded from La Grangette suffered no alteration; it is a novelty for me.

When you are not here, your letters are my consolation. Do not deprive me of them. I am your servant heart and soul.

~

[104] Chaix was correct about the the species here, except that they all would be transferred to *Crepis*. *Crepis conyzifolia* (Gouan) A. Kerner (1872). *C. blattarioides* Vill. = *C. pyrenaica* (L.) W. Greuter (1970). *Hieracium amplexicaule* L. stands.

[105] Villars would declare *Hieracium intybaceum* All. (1773) to be a synonym of his new *H. albidum* Vill., **Prosp. Pl. Dauph.** 36 (1779) and persisted with it in *Hist. Pl. Dauph.* 3: 133 (1788); but *H. intybaceum* All. is the good species. *Hieracium saxatile* Vill., published invalidly in **Prosp. Pl. Dauph.** 35 (1779), = *H. phlomoides* Froelich in DC. (1838).

[106] *Carduus nigrescens* Vill., **Prosp. Pl. Dauph.** 30 (1779), from Gap and Les Baux.

[107] Neither Chaix nor Villars apparently knew that *Prunus avium* L. was first published in **Flora svecica**, ed. 2, 165 (1755). Beyond which Chaix was confused here, and Villars would ignore the recommendation in his flora.

20

Les Baux, 4 December 1776

From your letter of November 24th, written from Le Noyer, I learn that you had written to me from St.-Bonnet. The clarifications that you give me on the hieraciums satisfy me very much; but I regret to be without hope of receiving your letter from St.-Bonnet that no doubt responded to items in [my letter of 18 October 1776]. I beg you to reiterate the letter to the extent you can, so that I am not deprived of the insights you wanted to send to me. I [also] ask you for a more extensive reaction to my *Hieracium intybaceum* [L.]. Can it be possible that it hardly differs from *Hieracium sylvaticum* Gouan? or perhaps that is what it is? At least does it still merit the distinction Monsieur Gouan made between *Hieracium murorum* L. and it?[108]

Ray distinguishes *Carduus caule crispo* J. Bauhin very well from *Carduus palustris* C. Bauhin. . . .[109] Ray is right to say that the descriptions of thistles by most botanists have been too short and obscure, leading him to suspect what they had known or distinguished about the species of thistles.

Gouan has taught me (**Illustr.** 65) that we have *Carthamus mitissimus* L. at Les Baux [*Carduncellus mitissimus* (L.) DC.] and not *Carthamus carduncellus* [L.] [= *Carduncellus monspelliensium* All.] by the differences he establishes to separate one from the other.[110]

I see that I have poorly observed our species of violet, having relied too much on my preconceptions. I now think that we have *Viola hirta* [L.], which I had not segregated from *V. odorata* [L.] and from *V. canina* [L.]. I know *V. grandiflora* [L.] [= *Viola calcarata* L.] and perhaps also *V. calcarata* [L.]. But should I be able to refer an alpine species I have to *V. calcarata*, and which without doubt you have seen in my herbarium inappropriately labeled *pinnata*? the stem two inches tall, the flower hardly different from *V. odorata*? Does Haller make any mention of it?[111]

You improperly struck out *Rumex dentatus* [L.] in my herbarium, substituting *Rumex acutus* [L.] for it, but expressing some doubt. The plant is an annual, grown in my garden from seed sent from Montpellier with the dubious label *Rumex aegyptiacus* [L.]. You must have found *Rumex acutus* [L.] during your botanical trips. I have never seen it.

I am working to put this year's specimens in order. After which I shall devote myself a bit earnestly to preparing lectures for a mission in Champsaur after Easter, something from which I cannot exempt myself. Will you please get Monsieur Gouan's *Hortus regius monspeliensis* [1768] for me? As it is published in Lyon, it should be available. This book should provide me with the synonyms from Tournefort and from Sauvages, as well as the author's notes on various matters.[112] Your bookstores in Grenoble should be able to send it. If ever you make the trip to Paris, you will please buy at my expense the books that you know to be within my reach.

~

[108] For *Hieracium sylvaticum* Gouan, see both *H. commixtum* Jordan and *H. umbrosum* Jordan.

[109] *Carduus crispus* L. and *Cirsium palustre* (L.) Scop. respectively.

[110] Villars, correctly, placed *Carthamus carduncellus* in Les Baux.

[111] This appears as *Viola hirta* L. collected in Les Baux and around Gap, in Vill., **Hist. Pl. Dauph.** 2: 662 (1787).

[112] Joseph Pitton de Tournefort (1656-1708), and Pierre-Augustin Boissier de la Croix de Sauvages (1710-1795).

I am taking advantage of the accommodation furnished me by Joseph Chabot's sister (my nephew in Les Baux who was recently married in Le Cros-de-Saint-Laurent) to have my letter carried to you and to renew the increasingly respectful affection with which I am your very humble and very obedient servant.

21

Les Baux, 20 December 1776

Do not be troubled any longer about your letters; they have all reached me, including your last one dated the 16th, which I had requested to supplement the one I have believed myself deprived of. I have caused you a double employment; but you have doubly instructed me, and a teacher is never angry to have to reiterate his teachings for the advancement of his disciple.

You please me very much by talking to me about a *Hieracium latifolium villosum* [=*Hieracium villosum* Jacq. (1762)] and about a *Hieracium angustifolium* according to Haller [= *Tolpis staticifolia* (All.) Schultz Bip (1861)]. I also have the latter from our Loubet and La Grangette, with radical leaves like *Hieracium porrifolium* [L.], villous, narrow; the stem leaves as narrow but shorter at the time of branching; the flowers large.

I shall keep in mind that my *Hieracium scorzonerifolium* [Vill.] is a close relative of theirs. The *Pulmonaria lutea angustiori folio valde pilosa* (that I find in Ray and in Chabrey) is a variety of *Hieracium murorum* [L.] collected from the walls of St.-Joseph of Grenoble. The same *valde pilosa duplex* is quite different, as it is a variety between *Hieracium villosum* [Jacq.] and *H. cerinthoides* [L.] (according to your last letter). I do not understand this *duplex*, which troubled me also in your preceding letter. I would ask for the description in its entirety.[113]

What discovery did Gouan make to establish his *Hieracium sylvaticum*? I cannot understand it. I know the Linnaean varieties *pilosissimum* and *sylvaticum* [of *Hieracium murorum* L.], pictured in Chabrey, which Gouan excludes by his description of his own *sylvaticum*. I also know another variety that I believe to be *Hieracium murorum angustifolium*, not *laciniatum* Casper Bauhin, the leaves close to *H. porrifolium*, stems furnished with 3-4 or 5 smaller leaves: *Pulmonaria gallica tenuifolia*, **Tab.**, that Gouan refers to his *sylvaticum*. Is it the one that was his object? . . .[114]

I have not had the advantage of seeing as many monkshoods as you have. My former *Napellus* from Seyne that I was improperly calling *autumnale* should be *Napellus verus* [Lob., **Hist.** 387 = *Aconitum napellus* L.]. The stipules or bracts on the pedicels are large; the flowers are frequent on a long spike; it flowers late, toward the end of July in the garden. The other variety received from Grenoble, which you call *the small one*, I make to be *Aconitum tauricum* Clusius. The stalk is nearly glabrous above; the flowers less frequent on the spike; the stipules on the pedicels are quite small; it flowers early, about the beginning of June in the garden. Therefore, one could have been right to call it *praecox*. Since Johann Bauhin said that the monkshoods are difficult to distinguish, was Linnaeus wrong to escape

$\backsim$

[113] See *Hieracium valdepilosum* Vill. and the synonymy, Prosp. Pl. Dauph. 34 (1779). Hist. Pl. Dauph. 3: 106 (1788). The entire description was from Johann Bauhin: *Pulmonaria lutea angustiori folio valde pilosa duplex.*

[114] See the preceding letter, note 108.

the situation by merging several of them within *Aconitum cammarum*? My monkshood from Le Pleyne is *Aconitum judenbergense* Clusius. . . . The other variety received from Grenoble may well be *Aconitum thorx italica* Clusius . . . that Linnaeus refers to *A. cammarum* (with good reason I think) in his Systema [plantarum]. . . .[115]

I am undertaking ten masses for the celebration of Dominique Saint. . . .

Moved by the best wishes you address to me, I send them to you with the same spirit and the same sincerity, and can take no better model than yours to formulate them. I would almost say that [these sentiments] have established henceforth that our close association has united us as one. In this spirit, I shall always be, Monsieur and very dear friend, Your very humble and very obedient servant.[116]

22
Les Baux, 12 September 1777

I am sending you your Haller, and Gouan's two volumes and illustrations. You will need them to work on the book you have undertaken. I shall be at the summit of my joy if God permits me to see it delivered to the public. I have a thousand thanksgivings to express to you for having lent them to me during your absence. I made a compilation from them up to the hypnums [in the Mosses] and was quite delighted to meet species omitted by Linnaeus. I was no less so by being able to correct certain errors I had previously made. I do not presume, however, to have smoothed out all my difficulties or to be deflected from all my missteps. I still have not had the courage to undertake mosses and lichens, weak as I am on the more striking and less subtle botanical objects. If ever I come to studying them, I am counting on your complaisance.

My brother, whose infirmity I related to you, continues to suffer from legs and thighs swollen up to his lower stomach; but he is less fatigued by his coughing. The first doses of arnica upset his intestines very much. I do not know whether the dose was too concentrated or too copious. I had him told to continue taking it by making it weaker. He is experiencing some benefit from it. I shall keep you informed about his condition later on.[117]

I forgot to talk to you about my niece Marguerite's ailment. Some days after her first childbirth, after having been helped to lift something heavy, she had a relaxation of the uterus that inconvenienced her very much for nearly a year. At that point, she opened up

~

[115] The monkshoods that Chaix, and ultimately Villars, lumped as *Aconitum cammarum* L. are now generally grouped as *Aconitum variegatum* L. *Thore* is the common name used by shepherds in the Alps to designate the poisonous monkshoods.

[116] Although no mention is made of Villars' pending departure for Paris in this letter, one may infer from the succeeding letter that they had met before Villars' departure. Villars would later say that Chaix had offered him twenty pieces of gold as his contribution to the trip, but that he had refused to accept, recommending that such money ought to be available for loans to his relatives of little wealth. Lending money to relatives, Chaix had answered, only gets you into difficulty with them. If he had any money left over at the end of a year, he *gave* it to his needy relatives. But the gold pieces were his reserve against future accidents or illness. Villars, "Notice historique sur Dominique Chaix," pp. 301-302.

[117] Villars described *Arnica montana* L. as one of the best remedies the Vegetable Kingdom offers to medicine: as a diuretic, a stimulant, a febrifuge, and as anti-arthritic; but he recommended moderate doses to avoid severe heartburn. **Hist. Pl. Dauph.** 3: 208 (1788). It is also known to irritate the intestines and to be poisonous to some people.

on the matter to me, and I had her take mugwort soaked in wine for some time.[118] She was relieved by it; but, then pregnant for the second time, her ailment disappeared during the period of her pregnancy. Yesterday, having come here to hear holy mass in thanksgiving for her childbirth, she told me that the ailment had come back, probably after she undertook some small task. She asks you to please indicate what she might do to cure herself of this infirmity if that is possible. I have believed that this is a relaxation of the uterus rather than of the intestine. Given this limited detail, am I mistaken? I am quite upset by this matter, given my affection for her, which you know about. Permit me to recall the words of that father in the Gospel solliciting a cure for his son: *How long time is it since this hath come unto him? And he said, From a child.* **Mark** 9: 21.

I cannot write to you, my dear friend, without bringing up something about those objects whose research I have the privilege of sharing with you.

1) Have some seeds of *Carduus palustris* [L.] [= *Cirsium palustre* (L.) Scop. (1772)] gathered for me if they are still to be found. I greatly fear that those I collected at La Cou are sterile.

2) I am sending you my *Carduus* collected at Laric that I am unable to find either in Linnaeus or in Haller. Tell me if it is new or what it is. The curé de Laric has had the kindness to gather a package of seeds for me.[119]

3) I also send you the amaranth from our gardens in Les Baux. I am unable to place it anywhere but in *Amaranthus graecizans* [L.]. I am acquainted with *A. blitum* [L.] that Haller does not separate from *A. viridis* [L.], common here, its leaves emarginate. I formerly cultivated *A. graecizans* from seeds sent by Monsieur Rozier. You can assure [your readers] that the plant I am sending you is [naturalized] here.

4 The *Bromus* herein, which you also had me look at in Le Noyer, I call *intermedius*, between *B. secalinus* [L.] and *B. mollis* [L.], Haller, **Nomenclator**, no. 1503. It may be *Bromus tectorum* [L.]. *Bromus arvensis* [L.], Hall., **Nomen.**, no. 1509, is definitely something else. . . .

5) It seems to me that *Festuca elatior* [L.] and *F. fluitans* [L.] could be put with the *Poa* as Haller did with the latter and with *Poa phoenix* [L.], which approaches the former.

6) You told me that they have my *Ononis saxatilis* in Paris. Please oblige me by writing me the trivial name they give it and its description.[120]

7) *Aubletia monnieri.*[121] I definitely feel that this is not a *Buchnera* [L.], although the description of *Buchnera canadensis*, **Mantissa** 1, seems to fit it. I have always believed it to be a *Verbena tetrandra*; and you tell me that they regard it as such in Paris [that is, as *Verbena officinalis* L.]. Please give me the same information as for the preceding plant.

8) The silene found in Le Noyer is indeed *Silene noctiflora* [L.] .

∾

[118] The remedy was usually made from soaking the leaves of *Artemisia vulgaris* L. in white wine. Although reputed to have many virtues, it was a specific for irregular menstruation, probably accounting for Chaix's recommendation.

[119] Villars would find it to be new: *Carduus lycopifolius* Vill., Prosp. Pl. Dauph. 30 (1779). Hist. Pl. Dauph. 3: 23 (1788). Found at Laric and Oze near Veyne by Chaix. [= *Serratula lycopifolia* (Vill.) A. Kerner (1872)].

[120] What Chaix here called his *Ononis saxatilis*, Villars acknowledged (as *Ononis rupestris* Chaix) to be a synonym of *Ononis subocculta* Vill., **Hist. Pl. Dauph.** 3: 429 (1788). Originally published in Vill., **Prosp. Pl. Dauph.** 41 (1779). Elsewhere, **Hist. Pl. Dauph.** 1: 255, he noted its collection on his second field trip (1770) made with Chaix. [= *Ononis columnae* All. (1774) = *Ononis pusilla* L. (1759)].

9) We must not neglect our *Mentha palustris* [auct.] with stamens concealed: it differs from *M. aquatica* [L.] It flowers later, its whorls are more serrate, its corolla smaller, and its stamens always hidden in the throat. Monsieur Liottard had sent me a shoot of it from Grenoble, which has larger leaves, but which is not contradicted by the size of its stamens.[122]

You tell me of having seen the origanum with stamens inset. That may be *Origanum heracleoticum* [L.], on the surface very similar to *O. vulgare* [L.]; but which, indeed, does not show its stamens as the latter does.

10) Your hieracium, which I have called *Hieracium halleri*, you recognize as well as I to be [Haller, Hist.], no. 43, variety B in particular.[123]

11) Since my *Lactuca sylvatica* is unknown in Paris, and as you do not find it in any author, I would be very much honored for you to designate it with my name. I do not ask for that out of vanity, as God forbids me that.[124]

12) I reiterate here that I have not been able to discover any sign of a coronula in my *Cucubalus silenoides*, which would be a true silene if one ignored this Linnaean character.[125]

13) The *Euphorbia*, which you call *falcata* [L.], I believe to be *Euphorbia exigua* [L.], which I am sending you. In this season, having usually lost its leaves and involucres, and revealing itself only with its involucelles, it has often imposed itself upon me; but having for long observed it, I am no longer taken in by its disguised appearance. Here is a specimen that is quite whole (at the beginning of Gouan's volume).

Please respond as soon as possible, above all about the matters concerning the health of my relatives. I send greetings to your dear mother, spouse, and brother; my respects to Monsieur the curé [Arnaud du Noyer]. Manage a few days visit to me during October, letting me know ahead when it will be convenient for you. Beyond the duties of my position, I have nothing closer to my heart, or to please me more at the foot of my cliffs, than the association I will be able to maintain with you.

P.S. I am sending you a small packet of sand I received from Roussillon in Provence to be dusted on written pages.

23

Les Baux [ca. 30 September 1777]

Yesterday I received your letter of September 19th that I was anticipating with great eagerness. All of the critical points in it are valuable for me, and I shall benefit from them. I do hope that your presence will be salutary for my afflicted niece. I shall have my sick brother treated following your advice.

I would have many botanical matters to talk about to you here; but Neffes, the person responsible for carrying this letter to St.- Bonnet, is in a hurry.

~

[121] *Aubletia* was a genus established by Louis-Guillaume Lemonnier (sometimes Le Monnier), 1717-1799, physician to Louis XVI and professor at the Jardin du roi. [= *Verbena* L.]
[122] See *Mentha dubia* Chaix in Vill., **Hist. Pl. Dauph.** 2: 358 (1787), from Les Baux. This is now held to be a synonym of *Mentha aquatica* L., the form with stamens enclosed.
[123] See *Hieracium halleri* Vill., **Hist. Pl. Dauph.** 3: 104 (1788).
[124] Thus, *Lactuca chaixii* Vill., **Prosp. Pl. Dauph.** 33 (1779). Chaix had found this in woods around Les Baux, Rabou, and Chaudun: **Hist. Pl. Dauph.** 3: 155 (1788). [= *Lactuca quercina* L. ssp. *chaixii* (Vill.) elak.]
[125] See letter no. 12, 10 February 1775.

1) Where did Linnaeus treat *Arenaria fasciculata* that he mentions in **Observationes** 358? Did he mean his *A. tenuifolia* or the *fasciculata* in Gouan, Ill. 30?

2) I send you, or what I believe to be, *Alsine tenuifolia* J. Bauhin [3: 364], Séguier 1: 418, t. 6, fol. 2, which I found at Laric. It is an annual and early. *Alsine verna strictissimo folio,* [Haller] **Flora jenensis** 112. I am unable to count the stamens.

3) *Arenaria tenuifolia* L. appears to me to be Haller, [**Nomenclator**] no. 864 [sic = 866].

4) Please write me the synonyms for Haller no. 870 that I omitted in my compilation: *Alsine mucronata* [auct.] from which you had the kindness to collect some seeds for me. [= *Minuartia mutabilis* Schinz & Thell. (1938)].

5) I should very much like a specimen of *Arenaria saxatilis* L.

6) See if the illustration in Vaillant, t. 3, fol. 1, is for *Arenaria tenuifolia* L.[126]

7) If you can, have your *Sisymbrium monense* [L.] sent to me. I should like to cultivate it, and we shall see its difference from *S. murale* L. and from *S. erucastrum* Gouan, Ill. 42.[127]

As I am pressed, I must conclude. I am very concerned about your mother's illness. Greet her on my behalf, and your dear spouse and brother.

The attorney [Pierre-Joseph-Marie] Delafont regrets not seeing you. Do not fail to give him that satisfaction, and to Monsieur [Pompon] Gautier, the grand-vicar, as well. When you will be here in my solitude, my happiness will be complete.

P.S. If you have occasion to return to Aubesagne [in Champsaur], dig up the beautiful *Dianthus arenarius* [L.] so that I can cultivate it, or at least get some seed.[128] I am enclosing your manuscript of Séguier's three volumes. I have copied it.

24

Les Baux, 24 October 1777

I dined yesterday in Gap with Monsieur Vigne the *marchand* (small businessman), who remitted the **Flora Gallo-Provincialis** [of Gérard] enclosing your letter of October 13th, along with two bundles: *Carduus palustris* [= *Cirsium palustre* (L.) Scop.] and *Cerastium aquaticum* [= *Myosoton aquaticum* (L.) Moench] (I have never seen the latter). *Cucubalus baccifer* [L.] grows at La Roche.

Your letter was written after mine of October 11th in which I talked to you about my sedums among other things, and to which I had added fresh specimens of my small Allioni violet. If you have received it, please have the kindness to acknowledge it in your usual way.[129] La Blonde also delivered the two willow cuttings to me; but having wilted for a few days before delivery, I strongly fear for their success. The proximity of their habitat, as you indicate, consoles me. . . .

~

[126] Chaix's evident confusion about these pinks may have contributed to Villars' publication of *Arenaria hybrida* Vill., Prosp. Pl. Dauph. 48 (1779). [= *Minuartia hybrida* (Vill.) Schischkin (1936).

[127] *Rhynchosinapis monensis* (L.) Dandy ex Clapham; *Diplotaxis muralis* (L.) DC.; and *Diplotaxis erucoides* (L.) DC.

[128] Villars found this to be *Dianthus carthusianorum* L.

[129] See *Viola nummulariifolia* Vill., found by Chaix on a mountain near Gap called col de Moissière. Vill., **Hist. Pl. Dauph.** 2: 663 (1787), and note letter no. 20, 4 December 1776. Chaix's letter of 11 October 1777, while received by Villars, was not preserved.

In the [pre-Linnaean] synonyms that you are proposing, there is nothing that passes for the beautiful allium. . . .

1) *Acori radice, flore purpureo* Barrelier, **Icones Plantarum**, no. 1022. . . . *Allium petraeum umbelliforum* J. Bauhin, 2: 564 . This is our garlic of the Alps: a variety of *A. angulosum* [L.]; a variety of Haller, no. 1227. *Allium petraeum minus.* In *Icone Barrelieri radices vetustae, similes quodam modo Acori sunt, quibrus superne adhaerent radices recentes bulbosae, fibrosae. Florum capitatum hemissphaericum, pedunculis aequaliter dispersis, petalis parvis, staminibus exsertis.*

2) Ray, Hist., 1121, no. 9: "*Allii genus forte scorodopressum alterum* Lobelio, J. Bauhin; or *Allium sphaerico capite folio angustiori* C. Bauhin. . . . *Allium montaneum radic oblonga* C. Bauhin. . . ." All that is an extract from the above Ray.[130]

I had perceived Haller's mistake in attributing J. Bauhin's synonym to his *Rhagadiolus* (my *Hyoseris cichorioides* and your *H. taraxacoides*),[131] which belongs to our small *Leontodon hirtum* [Vill.] of Les Baux, and which has a long-fibrous root. I am sending you two fresh specimens that you may inform yourself about it.[132] Do you want to call the one we have at La Grangette *Leontodon crispum?* If you can take my advice about it, save this epithet for our small plant in Les Baux, unknown to Haller, which was assigned to it by Johann Bauhin, *Hieracium parvum hirsutum, caule aphyllo, crispum ubi siccatum*, with an excellent drawing. Leave the name *hirtum* for the very common one at La Grangette and in subalpine places that is *Picris*, Haller, [**Nomenclator**], no. 25, the one in our meadows and woods that differs from the alpine [plant] only from its more refined and less wild habitat. . . .[133]

The true *Leontodon hispidum* [L.] with leaves whitened by their hirsutism and by long-forked hairs is well-known to me, although rarer than the above. [Chaix goes on to argue here that his study of Haller shows there are three distinct species in question, all originally in *Picris*: *crispa, hirta*, and *hispida*] as well as other species with plumose pappus-hairs.][134]

1) *Hieracium pumilum saxatile, asperum, radice praemorsa*, C. Bauhin, **Prodr**. 66. Ray, **Syll**. no. 246. [Gérard], **Flora Gallo-Provincialis**, no. 165. This is my *Picris hirta* var. *subalpina*.[135]

2) *Hieracium narbonense caule aphyllo, folio sinuato hirsuto*, Chabrey, [**Sciagraph**], 322, with illustrations, is *Picris hispida*.[136]

~

[130] This difficult and confusing passage appears to be an attempt, beyond finding appropriate synonyms, to separate a new allium (that Villars would name *narcissiflorum* and Chaix called *grandiflorum*) from what is now *Allium senescens* L. ssp. *montanum* F. W. Schmidt.

[131] *Hyoseris taraxacoides* Vill., **Hist. Pl. Dauph.** 1: 250 (1786). [= *Leontodon taraxacoides* (Vill.) Mérat (1831)].

[132] *Leontodon hirtum* Vill., **Fl. Delph.** 84 (1786). [= *Leontodon villarsii* (Willd.) Loisel. = *Leontodon hirtus* L. (1759). Villars located it at Les Baux, Mont-Dauphin, Die, and Saint-Jullien.

[133] *Leontodon crispus* Vill., **Prosp. Pl. Dauph.** 34 (1779). **Hist. Pl. Dauph.** 3: 84 (1788). Villars located it in dry, rocky, warm places near Grenoble, Briançon, Les Baux, and La Roche. He did not follow Chaix's advice here; but the correspondence drove him to make an exceptionally extensive study of the synonymy for publication in the flora. He regarded *Leontodon hispidum* [L.] to be a possible synonym for *L. crispus*.

[134] For reasons unclear he cautions against lumping *Leontodon* with "the true dandelion," using only the name *Taraxacum*, evidently recognizing the difference in pappus; but the passage is abbreviated and confusing.

[135] Villars included this synonym for his own *Leontodon hirtum*.

[136] Villars put this under his *Leontodon crispum*.

Yesterday, I carried my botanical notes to Gap to have them bound. I put at the head a short Latin preface comprising my small botanical outings that you might read with pleasure. I have revised the description of our territory. Monsieur Delafont, the subdelegate, wants to read both parts of it. I have already had two other small Latin pieces given to him for amusement.

As I can only believe that your beautiful sage seen in Lower Dauphiné, and which I have raised in my garden, is [*Salvia*] *clandestina*, I doubt that it can be *Salvia verbenaca* L., which is Gérard, **Gallopr.** 158, *corolla labiis approximatis*. I have *corolla oro clauso* written in my notes. . . .[137]

The *carvifolia* in Vaillant [**Botanicon parisiense**] t. 5, f. 2, which Linnaeus says has affinity to both *Seseli montanum* [L.] and *Seseli glaucum* [L.], can it be the *Selinum carvifolia* of Cranz that we have from Grenoble? The latter is quite separate from the seselis. I made a selinum of it in my notes following Cranz. Perhaps it would be better placed in *Peucedanum*.[138]

I have always been sorry not to have a *Lycium* or *Rhamnus* that I saw on my trip from Avignon on the top of the ascent of Bellecombe between Le Buis-les-Baronnies and Rosans. It was a shrub, quite bristling with strong spines that appeared to be viscid. Have you found it during your botanical trips? If you cannot tell me anything about it, it will be necessary to seek greater elucidation about it.

Everything that I want you to take to Paris I will bring to you in Gap on the festival of St.-Martin. I hope to see you there. Even so, I would be pleased to have a reponse by letter from you before that time.

25

Les Baux, 19 December 1777

I have delayed having my notes on plants passed on to you on the presumption that you had made a trip to Grenoble. Your letter informs me that I was not mistaken. The curé de La Roche, my neighbor, gave me news of you. He is quite delighted to have made your acquaintance. Monsieur [François] Gaude, curé d'Oze, who is preaching the Advent sermons in Gap, spent the first week of Advent with me, pleasing me with the expectation that he would return. On Tuesday I went to Gap to get him, but fear of bad weather did not permit him to carry out that intention. During all of our discussions, he repeatedly spoke of you with regard and affection, always begging me to remind you of him. He has done that again in a letter that he wrote to me yesterday. He noted in his letter that Monsieur de Jouffroy, our bishop,[139] is nominated for the diocese of Le Mans, and that the abbé de Maillé, already provided with an abbey with 5000 livres of income, is to succeed him in the diocese of Gap.[140] This news requires confirmation for several reasons. The next mail will very likely clarify the situation.

∽

[137] A nearly-closed calyx was traditionally cited to separate *S. clandestina* from *S. verbenaca*; but the latter is now held to be a very polymorphic species with local variants, including *S. clandestina*.

[138] *Seseli carvifolia* L. (1753) = *Selinum carvifolia* (L.) L. (1762).

[139] François-Gaspar de Jouffroy-Gonssans was Bishop of Gap from 1774-1778.

[140] Jean-Baptiste-Marie de Maillé de La Tour-Landry, Bishop of Gap, 1778-1784.

When you write to Monsieur Thouin,[141] give him proof of my feelings of respect, affection, and gratitude, whether for the plants he has had the kindness to send us, or for Monsieur Sage's book on volatile alkali,[142] but especially for his having remembered me after your association [with him]. When you think it is appropriate, I shall take the liberty of writing to him for enclosure in your envelope. I am adding your Necker[143] and your catalogue of the garden in Paris to my manuscript. I have copied the plan for Monsieur [Antoine-Laurent] de Jussieu's system: the description of his orders and the species that belong to them, using only the trivial name, reserving for myself [the liberty] of characterizing them in my own way. I have drawn up a small catalogue of plants that the royal garden lacks, but which are found in our locale, in order subsequently to send Monsieur Thouin only things that can give him pleasure.

Would you be so kind as to send me the **Iter [helveticum]** by the incomparable Haller and his [**Enumeratio plantarum hortii regii et agri**] **Gottingensis** that I desire so much to read? Try to get the corrections and the authors that this celebrated author has promised you.[144] I quite fear that, by addressing them through some correspondent in Lyon, they could pass into some eager hands not willing to give them up. Take the precaution of acquainting Monsieur von Haller with the situation [as he is] so well-intentioned in your case.

Since your affairs are recalling you to Grenoble, ask Monsieur Liottard if he has either the seed or the plant of *Scheuchzeria palustris* [L.] and beg him to give you some of it. I would indeed grow it with pleasure. Monsieur Clappier told me that Monsieur Prunelle[145] had found *Laserpitium prutenicum* [L.] in his field. Ask him about it and try to get some of it. If it is possible for you, bring back some cuttings of Grenoble willows that we do not have here, or some African geraniums from one of your friends. You must not, however, let my avidity inconvenience you or drive you to commit some indiscretion.

The news reaching me from you is the subject of my complete attention and provides all the pleasure that I can feel in my gloomy solitude. . . . I wish you the most happy holidays and a new year that most conforms to your desires. . . .

Note: You will often find the same plant described and characterized in several places in my notes. You will sense the reason for it: I have gone back more than once to the same plant and at different times; and to avoid the trouble of having to recopy everything, I have allowed my initial observations to live on. Thus, the last entries are those treated last.

26

Les Baux, 8 June 1778

(Monsieur Villars' relatives are kindly asked to forward this letter to Grenoble if he has gone there, but to hold on to the accompanying package.)

◁∿▷

[141] André Thouin (1747-1824), appointed chief gardener at the Jardin du roi in Paris in 1768. He collaborated with Antoine-Laurent de Jussieu in replanting the Jardin du roi to reflect natural classification, virtually completed in 1774.

[142] Balthazar-Georges Sage (1740-1824), chemist and mineralogist.

[143] Probably N. J. Necker, **Methodus muscorum** (1771) on mosses.

[144] He probably meant Haller's **Opuscula sua botanica prius edita recensuit, retractavit, auxit** (1749) and his **Bibliotheca botanica** (1771-1772).

[145] Dr. Joseph-François Prunelle de Lière (ca. 1735-ca. 1815).

You wrote to me on May 25th from St.-Bonnet, the letter carried by the vicar of Manteyer [Jean-François Mévouilhon]. Therein, you apprise me of the death of the very ingenious and learned Mlle Motte. I am with you in regretting the loss of rare talent. May it please the Lord to raise her up for His glory and for the ornament of our century! If our obliging messenger, Monsieur Vigne, was on time, you ought to have received my letter on that same day, responding to your letter of May 10th, principally regarding questions about *Crepis tectorum* [L.] and *C. biennis* [L.].[146] I made several botanical observations, and I acknowledged reception of your notebook on the Corymbiferae on the back. I have read it with the pleasure that I always have, as you know, when reading anything from your hand. I have made notes and criticisms on it to the extent I have been capable, few in number, the work having appeared to me to be correct and accurate with the exception of the *Anthemis*. You will see my criticism. Thus, I am returning the notebook. Provide me with the reading of another [notebook] soon; for, as you are inclined to oblige me, I want to read them all in succession, one after the other, at your convenience.[147]

In this last week, I have been to Durbon. The saintly monks talked very much and very favorably about you. I came back over the mountain with your worthy relative, Monsieur Achard, as far as La Cluse. There, I looked for your *Hieracium pumilum* without finding it. In the indicated places, I saw only *Hieracium amplexicaule* [L.], *H. murorum* [L.], and your *H. andryaloides* [Vill.]. This mountain *pumilum* has always appeared to me to be *H. prunellaefolium* Gouan [= *Crepis pygmaea* [L.]. If you want to extract me from difficulty, provide me with a specimen of your *pumilum*, which you could put in the notebook that you are going to send me. Coming from La Cluse [south] to Rabioux, I found your *Sisymbrium monense* [L.] on the crumbling slopes. *Viola nummularia* Allioni is not rare in alpine pastures, nor is *Ranunculus nivalis* [L.] with its nearly double flower.

I learn that you have to go to Grenoble on the first of the month. Have a pleasant trip and, if you are to remain there for some time, write me from there. Madame de Flotte de La Roche[148] and Madame de La Roche, the nun, beg you to take an interest in their protégé as an apprentice surgeon.[149] This young man will not prove to be unworthy of any favors granted to him.

Take the trouble, at the time of your return [to Le Noyer], to bring back your *Hieracium lawsonii* [Vill.][150] from walls in Grenoble and the small willow from Petit-Chat.[151] When shall I be able to cultivate *Scheuchzeria palustris* [L.]? Should Monsieur Liottard have it? . . . If he gets it for me, I should be able to give him some plants in exchange that he does not have. At least get some seeds from him.

Note: You often make a spelling mistake when writing the verb *faire*. When it is expressed in short, as in the future of the indicative and in the imperfect of the subjunctive,

~

146 This letter is missing.
147 These are the materials that Villars was preparing for the prospectus and eventual flora of the province.
148 The comtes de Flotte were seigneurs de La Roche-des-Arnauds, dating from the 11th century. The subdelegate in Gap, Delafont, remarked in 1784 that the local nobility was not numerous, lived mostly in the country, and was "hardly rich but honorable." See Joseph-Hippolyte Roman, "Mémoire sur l'état de la subdélégation de Gap en 1784, adressé à l'Intendant du Dauphiné par Pierre-Joseph-Marie Delafont, subdélégué de Gap." Bull. Soc. Etudes Hautes-Alpes 18, 2nd ser. (1899): 169.
149 Jean-Joseph Serre of La Roche-des-Arnauds.
150 Vill., **Prosp. Pl. Dauph.** 36 (1779). Hist. Pl. Dauph. 3: 118 (1788).

it is written with a simple *e* instead of with the diphthong *ai*. In a similar case, please correct me with the same liberty I am taking in regard to you.

If Liottard fails you, young Serre, who also has a taste for botany, could go to collect said *Scheuchzeria* at Prémol by making him acquainted with it and indicating to him its locale. He would be delighted to oblige me. I am also without *Impatiens noli-tangere* [L.]

If I have not given you too much to do, try to get me a small *Geranium [Pelargonium] triste* [L.]. I could compensate the obliging gardener with one of my African geraniums that I have raised and that said gardener would not have.

27

Les Baux, undated
[Late June 1778]

I have had the pleasure of a visit from your brother. He has a plan to see the Chartreuse of Durbon along with Monsieur Mondat, his fellow-student. I gave him a letter for the Dom prior. I also entrusted your four notebooks on the plants of Dauphiné to him. I have been so happy with their contents that my critical remarks are reduced nearly to nothing. I have been surprised, however, not to find *Arctium lappa* [L.] and the [two] *Scolymus* among your Cinarocephalae; I say a few words to you about *Centaurea phrygia* [L.]; and I have struck out *Antirrhinum purpureum* [L.] that (I know not by what misunderstanding) you had introduced among our natives.

I am sending you the heads of our *Anthemis arvensis* [L.] and those produced from seed given to me by Monsieur Thouin and labeled *Anthemis nobilis* [L.], which it cannot be, being without odor and possibly an annual. It is perhaps your *Anthemis arvensis* from Paris and from Crest. See if you can find any notable difference between them.[152]

If you classify my new *Artemisia*, locate it in Les Baux, within a coppice belonging to Monsieur Mondet, occupying a small space. I have not found it elsewhere.[153]

I have lost the label or forgotten the name of two of your dried plants, so I am also sending you a sprig of each one. Determine them for me. At your leisure, please have some cuttings of the willow from Chantoussel and the Mateysine sent to me.[154] I shall read the short treatises by the celebrated Haller with pleasure, having read them previously. These coming fairs in St.-Bonnet should provide me some word from your side (at least I hope so). I wish you perfect health. . . .

~

[151] Probably *Salix aurita* L. despite Villars' choice of *S. ulmifolia* Juss.
[152] Evidently the plant from Crest turned out to be not *Anthemis*, but *Achillea nobilis* L. Vill., Hist. Pl. Dauph. 3: 257 (1788).
[153] *Artemisia insipida* Vill., Prosp. Pl. Dauph. 32 (1779). Hist. Pl. Dauph. 3: 249 (1788). A. Mutel would be unable to find it at Chaix's location, Flore du Dauphiné 2: 248 (1830). And J.-B. Verlot, Catalogue raisonné des plantes vasculaires du Dauphiné 181 (1872), remarked that evidently no specimens had been collected since Chaix's time; but that several of his specimens had reached other herbaria, notably that of Desfontaines.
[154] *Salix hastata* L. Vill., Hist. Pl. Dauph. 3: 774 (1788).

28
Les Baux, 12 September 1778

In the hope of seeing you in Gap yesterday, St. Martin's Day, I went there despite the snow and rain that made the trip very inconvenient. The dear confrere, Monsieur Gaude, curé d'Oze, who had come to my house to spend the night before, was hoping for the same pleasure. As Monsieur Garnier assured us two separate times that you had not appeared, I gave the package that was destined for you to a relative of Monsieur Joubert, who kindly assumed the responsibility to deliver it to you. By mistake, I think that I added a small packet of seeds meant for Monsieur Séguier to the one meant for Monsieur Thouin. You can tell them apart easily. The one for Monsieur Thouin is larger, and the corms of *Bulbocodium vernum* L. are attached to it. Please retain the smaller packet for Monsieur Séguier and send it back to me with Monsieur Chaix, notary of La Roche, who will be at the festival of Polligny [near Le Noyer] next Sunday. He will gladly bring back anything you might want to send me. I am eager to know what you think about the notes and additions I made in your notebooks.

~

The following letter, without address or postmark, was likely the first letter to be addressed to Monsieur Villars, Docteur en Médecine au Noyer, as would be true of letter 31. After being licensed as a surgeon in 1774, Villars had been freely prescribing treatments with plants for his patients, gradually arousing the ire of colleagues who accused him of illegal practice of medicine. To escape the danger of prosecution, he went to the Faculté de Médecine in Valence in the autumn of 1778. Enrolling, he took successively and successfully the three examinations in November and December that qualified him to be bachelier, licencié, *and, on 9 December 1778,* Docteur en Médecine.[155]

This letter offers the first hint that Chaix had become aware that Villars had become an object of professional jealousy in Grenoble as a beneficiary of official patronage, which may account for the carping about his medical practices. The more his hard work and publication brought him national recognition, the more swollen became local envy, which Chaix would seek to combat.

29
Les Baux, 21 February 1779

The Lord has granted you such agreeable qualities that you delight the hearts of all those who know you. Your merits have won you true friends from among the discerning and learned people in the best cities of the realm. You have great thanksgivings to render to the beneficent hand from which came your talents. Through the close union I have sworn to you, I, too, am obliged to bless the heavenly providence and to ask more and more for its continuing protection. Nothing could have been more kindly or more courteous than the letters from Paris that you conveyed to me. You must always do everything in your power to respond to such generous and friendly testimonials. Since you have brought me

~

[155] See B. Joyeux and A. Dejarnac, "Le Médecin Dominique Villars (1745-1814)." **Bull. Soc. Etudes Hautes-Alpes**, année 1969, p. 130.

to their attention, it belongs to you to keep me [in their minds] to the degree you believe me worthy of it. If God grants us health, we must not fail to make a trip into the Alps this coming summer to further the interests of these very kind correspondents.

Where are they on the printing of your **Prospectus**?[156] Have you resumed your botanical work? Are the notebooks for your work approaching their completion? You are thinking no doubt about writing a fair copy of all the portions that require it. Whenever my grafting will appear to be of some use to you, you may cite me in using it.

We do not know what comes next here. Beautiful June days seem to have taken the place of the February days where we actually are. The anemones, after a perfect flowering, are already fading in my garden. Along side, the *Bulbocodium vernum* [L.] and the *Scilla bifolia* [L.] are in bloom. The first narcissi are showing their stems along with the hyacinths; the fritillaries raise their heads; the liverworts of all colors, but mainly the double-flowered, border my small flowerbed with the simple and double daisies; the common cowslips, cultivated for a long time in several colors, are also waking up. The flower-buds on those pear trees we call *bittersweet* are about to open.

I must tell you, nevertheless, that my plants have suffered more this year than usual, not having been covered at all by snow. Several have even succumbed to the very cold nights of the beginning of January. I have hardly any hope that the willow brought by your brother from the Mateysine will root. Perhaps it was already dry when I received it. If you go to Grenoble, please bring some of it back again and, given the opportunity, do not forget the one from Chantoussel.

Did you receive the little poem I addressed to you at the fair for the Carnival in Gap?[157] All the news coming from you takes a great weight off my mind. You cannot write to me too frequently.

30

Chaix to Arnaud, curé du Noyer
Les Baux, 30 March 1779

I am aware how much the praise you have given to my little poem in your letter to Monsieur Villar goes beyond its merit. You would have obliged me more had you judged it with more severity. . . . But your letter will always remain precious thanks to the beauty of the style in which it was conceived and in particular because of the liberty you grant me to avail myself of your gracious condescension. Your favorable testimony has been for me (I must say it) a strong support to avoid seeing my *Florae Delphinensi carmen* pulverized and annihilated by the brutal strokes of outrageous censure from M. . . . If those subject to justice were judged in the court of first instance without right of appeal, they would often be oppressed. But raised up again by your hand, my piece is not totally sullied by the first storm it has just endured.

Soon after I wrote my little piece, I showed it to a close confrere. He, in order to oblige me, wanted to get the opinion of a neighbor on the subject. The latter, after a long delay,

---

[156] It would be published immediately before 16 April 1779, the exact date being obscure.
[157] A copy of the poem, *Florae Delphenensi carmen*, dedicated to Villars, survives in Grenoble. See Timothy Tackett, **Priest & Parish in Eighteenth-Century France** (Princeton: Princeton University Press, 1977), p. 91.

finally rendered the judgment that I am transcribing for you word for word. At your leisure, please let me have your opinion of it. If you decide it to be appropriate to draw up a criticism, I shall see to it that he reads it. I do not want to let go of the famous original. I might want to produce it after I have received the opinions of several people who have promised me to be honest in the examination of my little work. . . . It was done simply to fill some hours of leisure and recreation.

~

Villars' Prospectus for his flora of Dauphiné, having gone to press in 1778, was published no later than 16 April 1779.[158] The timing of this 49-page publication proved to be unfortunate. As early as 12 November 1773, when he was still an apprentice surgeon, Villars had notified Antoine-Laurent de Jussieu that the projected flora was underway. A second letter, 4 January 1779, indicated that the Prospectus was still in press. During the interim, in 1777, Villars had studied in Paris, becoming acquainted with both Bernard and Antoine-Laurent de Jussieu.

The chevalier de Lamarck, meanwhile, began the publication of his 3-volume Flore françoise in 1778, and the initial volume may have become known at that moment. But the work did not receive the authorization of the Academy, signed by Condorcet, until 10 February 1779, and was not presented to the king until the following month; so that the work could only have been accessible to the public after 21 March 1779.

Villars' title indicated that he was only publishing in the Prospectus a catalogue of his newly discovered plants, plus species exceptionally rare or unique to Dauphiné. He was subsequently dismayed and outraged to discover that Lamarck, who had done no fieldwork in Dauphiné, "had appropriated some of our new species," as he put it in a letter to Jussieu on 12 September 1779.[159] The expression inadvertently revealed Chaix's collaboration in the determination of species for the Prospectus that is evident in his correspondence, and which was much later confirmed when Georges de Manteyer discovered a copy of the Prospectus in proof, annotated by Villars and indicating Chaix's corrections.[160]

How had Lamarck acquired Villars' species before their actual publication? Villars later explained to Lapeyrouse that he had been obliged to provide Faujas with duplicates of his plants as part of the arrangements during their trips in 1775 and 1776; and that Faujas, whether innocently or maliciously, had given them to Lamarck.[161] Because Lamarck then published what seemed new to him as his own, without credit to the collector, the situation gave birth to legitimate controversy about priority of publication and the proper nomenclature of numerous alpine species. The first volume of Lamarck's flora seems clearly to have been available to the

~

[158] Villar, **Prospectus de l'Histoire des plantes de Dauphiné, et d'une nouvelle méthode de botanique, suivi d'un catalogue des plantes qui y ont été nouvellement découvertes, et de celles qui sont les plus rare, ou qui sont particulières à cette Province. Avec leurs caractères spécifiques, et établissement d'un nouveau genre, appellé Berardia** (Grenoble: Imprimerie royale, 1779).

[159] Gérard G. Aymonin, "1779, année faste de la floristique française." **Bull. Soc. Bot. France** 127, **Lettres botaniques** (4) 1980: 388-390.

[160] Dr. Jules Offner, **Dominique Villars, médecin et botaniste dauphinois 1745-1814** (Grenoble: F. Eymond et ses fils, 1954), p. 12.

[161] Edouard Timbal-Lagrave, "Villars et Lapeyrouse, extrait de leur correspondance." **Bull. Soc. Bot. France** 7 (1860): 681.

public several weeks before Villars' prospectus. Yet, the prospectus was available before the totality of Lamarck's three volumes. Priority of discovery certainly belonged to Villars; and, in many cases, priority in actual publication also belonged to Villars.

As his title page stated, Villars devised a new classification method for the prospectus that he called a mixed method. *That is to say, it had both natural and artifical components. It has been asserted, to account for Villars' failure to follow Jussiaean classification, that the natural method was only known through Bernard de Jussieu's disposition of the garden at the Petit Trianon and through Antoine-Laurent de Jussieu's exposition of the system before the Academy on 13 April 1774. The fundamental work was not published until 1789. As a consequence, the argument goes, Villars had no option but to rely upon the impressive system of Linnaeus.*[162]

On the contrary, Villars had been to Paris in 1777 and had consulted with both the Jussieux. Moreover, both Chaix and he owned copies of Louis Gérard's **Flora Gallo-Provincialis** *and were familiar with his similar natural classification published in 1761.*[163] *Villars says quite clearly in the preface to his prospectus that a purely natural classification, then being perfected, was the most desirable method. But he feared that it was too complex to introduce the inhabitants of Dauphiné to their flora, which was his, and the intendant's, goal. He was also aware of the liability of the Linnaean artificial or sexual method: it may offer a great facility, but the information derived about plants from its use is much less extensive. Thus, he worked out a method appropriate to his particular situation: a simple, evident, and uniform character that could encompass particular established natural families to be found in Dauphiné.*[164]

Table Showing the Divisions of this New Method[165]

Classes

    I. Plants whose flowers have only 1 stamen.

    II. Plants whose flower have 2 stamens.

        1. Orchidaea

    III. Plants whose flowers have 3 or 6 stamens attached to the petals.

        1. Gramina

        2. Junci

        3. Liliaceae

    IV. Plants whose flowers have 4 stamens.

        1. Rubiaceae

        2. Labiatae

        3. Personatae [Scrophulaceae]

---

[162] Emile Callot, **Dominique Villars: Le naturaliste philosophe, le botaniste, le professeur, étudié à travers ses manuscrits inédits** (Gap: Société d'études des Hautes-Alpes, 1982), pp. 36-37.

[163] See Roger L. Williams, "Gérard and Jaume: Two Neglected Figures in the History of Jussiaean Classification." **Taxon** 37, no. 1 (February 1988): 2-23.

[164] Villar, **Prosp. Pl. Dauph.** 9-12 (1779).

[165] Ibid. pp. 13-14.

Table Showing the Divisions of this New Method *(continued)*
> V. Plants whose flowers have 5 stamens.
>> 1. Borraginae
>> 2. Solanaceae
>> 3. Ombelliferae
>> 4. Compositae
>>> a. Chicoraceae
>>> b. Cynarocephalae
>>> c. Corymbiferae
> VI. Plants whose flowers have 6 stamens.
>> 1. Cruciferae
> VII. Plants whose filaments are united at their base into 1, 2, or several units.
>> 1. *Malva*
>> 2. Leguminosa
>> 3. Hypericaceae
> VIII. Plants whose flowers have 8 stamens.
>> 1. Onagraceae
> IX. Plants whose filaments are inserted at the internal edge of the calyx,
> and whose number exceeds 12.
>> 1. Rosaceae
> X. Plants whose flower have 10 stamens.
>> 1. Caryophilleae
> XI. Plants whose stamens are too numerous to be easily counted, inserted on the
> ovary or on the receptacle of the flower;
>> 1. Cistaceae
>> 2. Ranunculaceae or on an elongated, cylindrical receptacle without pistil.
>> 3. Amentacei
> XII. Plants whose flowers have 12 stamens.
> XIII. Plants that have no apparent stamens.
>> 1. Filices
>> 2. Musci
>> 3. Algae
>> 4. Fungi

Letters thirty-one to forty-eight were addressed to Monsieur Villars, Docteur en Médecine in Le Noyer, but sometimes in nearby St.-Bonnet. They cover the months after the publication of his prospectus in April of 1779 until his departure in late 1781 to assume a new post in Grenoble. In 1780, his stipend was doubled to 1000 livres by the intendant, no doubt in compensation for a new course in botany that Villars would inaugurate in March of 1780. That he left his wife and children in Le Noyer at the end of 1781, returning for visits on holidays, may mean that he did not anticipate a lengthy tenure in Grenoble. [1779-1781]

31
Les Baux, 24 June 1779

The fair of St. John in St.-Bonnet offers me a favorable opportunity, and I take advantage of it to greet you. . . . Monsieur Gaude, delighted by your integrity and an admirer of your talents, asked me, when leaving here yesterday, to tell you again of his affection for you. I found the plants in my little garden in quite good state with the exception of *Mimosa pudica* [L.], *Clitoria virginiana* [L.], and *Hibiscus sabdariffa* [L.] that have been chilled by the cold. [None of them being native to Europe.] Perhaps they have quite lost their lives. Their sad state distresses me; I shall not conceal the fact. . . . I found a second letter from Monsieur [Laurent] Blanc at my house, professor of philosophy in Embrun, in response to mine. He would be delighted to meet you and to join us for some botanizing in that region. I shall write as soon as possible to the illustrious Séguier, not forgetting to mention you in my letter. Many compliments to your worthy Messieurs Arnaud curé and Joubert notary, and to Madame your spouse. I wish you and your entire family a prosperous future. . . .

P.S. When you do me the honor of writing me, insure that your letter is only entrusted to Monsieur Garnier's establishment. He is so proper and so precise that, so far as his part is concerned, your news reaches me promptly.

32
Les Baux, 16 July 1799

[To Villars in care of the comte de Saix in Sigoyer][166]

Your trip to Sigoyer will probably cause you to pass through Gap either going or returning. I address my letter [to Gap] in the hope it will reach you there. I cannot come to Sigoyer as you have invited me to do. I am very pleased that you have made the acquaintance of the comte de Gruel; I know that he is attracted to botany. The excursions you will be making in that country will produce some new plants, and I am confident in participating in them given your great liberality when it comes to my interests.[167]

∾

[166] The comte de Gruel du Saix, whose seigneurie was Sigoyer, south of Freissinouse and La Roche. The family had been ennobled in the 15th century and would become extinct during the emigration. The seigneurie of Le Saix was held by the chevalier de Bimard, from an old noble family of Languedoc, residing in both Le Saix and Serres. Roman, "Mémoire sur l'état de la subdélégation de Gap en 1784," pp. 169-170.
[167] Villars, **Hist. Pl. Dauph.** 1: xxxviii (1786), dated his trip to Sigoyer, the Montagne de Céüse, the Durance near Tallard and La Saulce, and south to Sisteron, in 1778, evidently a typographical error in a work where they were frequent. It should be 1779.

On your return trip, please turn aside to visit me. Such a long time has flowed by since your last visit that I have the right to insist you come. Besides, I cannot fix either the time or the circumstances of our coming trip into the Briançonnais without a conference with you. Do come, and we shall arrange everything. Here you will see the results of my cultivation of plants from the seed sent from Paris. I know that you have seen everything in this capital, the center of the rarities in the universe; but a desert that conserves some of them is still an entertaining spectacle and will give you pleasure.. It is all the more for me who is ignorant of the brilliancy of the outside world. In this season, there are new observations to be made every day and to be recorded in my manuscript.

I have looked for your handsome myagre from Corps in Linnaeus, Haller, and Gérard without finding it. It cannot be either *Myagrum rugosum* L. or its neighbors, as it has a single silicle, very smooth and very glabrous, the flower white. It has to be placed after *M. paniculatum* [L.]. I have called it *Myagrum laevigatum, foliis lyratis, glaberrimis, silicula simplici, turbinata, laevi, an myagro similis flore albo.* *See my *post scripta*.

Our *Festuca elatior* [L.] and your *Festuca phoenix* [Vill.] appear to me to be so little different that I believe them to be the same species. The more fully-branched panicle and the more flowered spikelets in the latter, and the more open panicle and fewer flowers in the former, only derives from richer or thinner soils, more moist or drier soils. I was perhaps right to call it *Festuca elatior* L.[168] Your *Bromus perennis* [Vill.] you have not found in Linnaeus or in Haller! Can the most common grass have escaped these princes of botany?[169]

At Rabou, I have collected *Bromus tectorum* L. on old walls and in dry places.

Your sedge collected at Font-Reine is *Carex hirta* L. The hairy capsules, apices bicorniculate, permit me no doubt about it. I have always taken it for that.

You have established differences between certain hieraciums that merit them perhaps less than mine, which I have often talked about to you and that you have neglected to notice. I am very pleased that you are now examining them in their natural state.

If my letter reaches you in the time I expect, and if you can accede to my hopes, I shall have the honor to embrace you during the final week of this month. Have a good time and take care of your health.

P.S. Young Serre entered the hospital in Grenoble as a surgical pensioner on June 24th. This sucess is due in great part to your charitable representations. You will learn this news with pleasure because of the part you have taken in what concerns him, if you have not already heard about it.

* After my letter had been completed, I consulted my Barrelier [**Icones Plantarum per Galliam, Hispanium et Italiam observatae** (1714)], and I had the pleasure of recognizing our myagre there very well engraved, whereas Johann Bauhin's figure in Chabrey is

~

[168] Villars, nevertheless, published *Festuca phoenix* Vill., Hist. Pl. Dauph. 2: 108 (1787), as separate. Both species in question = *Festuca pratensis* Huds.

[169] *Bromus perennis* Vill., Hist. Pl. Dauph. 2: 122 (1787) = *Bromus pratensis* Chaix, non L., Pl. Vap. 12 (1785), or Chaix in Vill., Hist. Pl. Dauph. 1: 316 (1786). Both species = *Bromus erectus* Huds. (1762). (Chaix would use Haller's description.)

missing. When you see it in Barrelier, you will concur. I am thus now naming it as a plant unknown to our modern authors.[170]

[Synonyms]

Myagrum supinum, album erucaefoliis. Barr., **Observ.** 38, no. 358 and 1252.

Myagrum similis, flore albo. J. Bauhin 2, 895. Ray, **Hist.** 839.

Myagrum monospermum minus. C. Bauhin, **Pin.** 109, and **Prodr** . 52.

I am congratulating myself for having been able to separate this species from those of Linnaeus even before having found it in Barrelier. But do not make of this a suspicion of vain glory. I am surprised that Monsieur Gouan either did not know or forgot this plant, noted by his predessor, [Pierre] Magnol, in the wheat-grasses around Montpellier. In any case, this [recent] discovery of it in Dauphiné is due to you, as are so many others.[171]

33

Les Baux, 23 July 1779

Having no greater pleasure than to make you party to everything that concerns me, and in anticipation of the news you will have for me about Sigoyer, I write to let you know that, during the afternoon following your departure from Les Baux, I received a letter from the kindly Monsieur Séguier, along with a package enclosing about forty species of plant-seeds, brought by Monsieur [Jean-François] Nicolas, the doctor. Monsieur Séguier had at first entrusted the commission on May 24th to a person who expected to be passing through Grenoble. When that did not transpire, Séguier retrieved the material and entrusted it to a second person on June 15th, as he indicates to me at the bottom of his letter, about the same time he received your **Prospectus**, for which he asks me to extend his thanks to you pending his writing to you about it. Some of these plants are a repeat of some of those he had sent to me in his first package, the one you gave me in Le Noyer; but I have no less an obligation to this extremely kind and avid correspondent.

Among the seeds, I have found *Geranium triste* [L.] . . . and *Mimosa virgata* L. I have risked putting a few seeds [of these non-European species] into the ground, putting aside the remainder for a less advanced season in another year. I do hope that our trip in the Briançonnais will yield something worthy of this great man. Be in St.-Léger-[les-Mélèzes][172] on Monday, August 16th, the day after the Assumption of Our Lady, according to our plan. I hope to greet you there that evening. If I should meet some obstacle, I should be there the next day, God willing. Monsieur the curé de La Roche and I must go to the festival of St.-Etienne-en-Dévoluy on August 8th to greet the prior, going up there through La Grangette. If it should be convenient for you to go there, I know that your presence would give pleasure, and we could review our plans in person.[173] I am awaiting news of you from Sigoyer.

⌐

[170] He would publish it somewhat differently from the name he gave it above: *Myagrum erucaefolium* Chaix, **Pl. Vap.** 46 (1785). Chaix in Vill., **Hist. Pl. Dauph.** 1: 350 (1786): *foliis lyrato-pinnatifidis glaberrimis, silicula monosperma laevi.* From around Corps. [= *Erucastrum nasturtiifolium* (Poiret) O. E. Schulz (1916)].

[171] This accounts for the plant's appearance in Vill., **Hist. Pl. Dauph.** 3: 279 (1788) without Chaix's name.

[172] A village near the River Drac between St.-Bonnet and Orcières.

[173] St.-Etienne-en-Dévoluy is just west of Le Noyer—but over the very steep col du Noyer.

P.S. I have tried out the lenticular lens that enabled me to see the difference between *Geranium molle* [L.], received from Paris, and this small annual *Geranium* I have called *malvaefolium* Scop., which is close to *hirsutis*.[174] The seed of the former is embellished with ridges; the seed of the latter is only hairy. That of *Geranium dissectum* [L.] is pitted with tiny cavities.

Note: Our *Aster annuus* indeed belongs to Linnaeus. But, by its ray-flowers, in part entire, in part bidentate, and never tridentate, it approaches the *Erigerons*. Perhaps its calyx places it better in *Aster*.[175]

Our *Solidago canadensis*, with leaves three-veined, does belong to Linnaeus. *S. altissima* [L.] differs from it because of leaves without veins and its greater height: Barrelier, Icones 784, cited by Monsieur Gouan, Hort., renders our plant very well. I do not share Monsieur Liottard's reasons for doubting it. Perhaps the one with narrower leaves is *altissima* or one of its varieties. If you should remember about it, verify the matter when you are in Grenoble; and ask him for some seeds of *Myagrum perfoliatum* [L.] that he cultivates. I would fear to offer him any of mine, knowing him to be so obstinate.

That *Asperula* that you saw at the end of my little terrace is nothing else but *A. cynanchia* [L.]. I had asked Monsieur Thouin for *A. tinctoria* [L.], and he sent me the seed of the former species under the name of *tinctoria*. If he does not have more of the latter in the royal garden, he is as deprived of it as I am. Have you collected *A. pyrenaica* [L.] anywhere, which Monsieur Linné indicates to be near Valence?

34
Les Baux, 30 July 1779

A moment after your brother's departure, your letter of the 24th from Gap was delivered to me. My observation of the seeds from the two geraniums, repeated several times, stands: you will convince yourself about it on the basis of the material that I here enclose. The glabrous and pitted seeds of the plant I take to be *Geranium rotundifolium* L., which has a habit similar to *G. malvaefolium* [Scop.], are more like those of *G. dissectum* L. Your potentilla is perhaps not unknown to me. If it is the one I have in mind, I have not separated it from [*Potentilla*] *argentea* [L.]. But your phyteuma merits further consideration.

The compliment paid to you by Monsieur Gouan, which I believe to be sincere, gave me great pleasure. His criticism bears consideration. Take advantage of what it contributes and neglect that which appears to be unimportant or inappropriate.

Since he has compared his *Ligusticum pyrenaeum* [Gouan] to Monsieur Séguier's plant, you must believe him, cite his testimony, and abide by his desire as to his name *pyrenaeum*.[176]

I do not think that you thought wrongly about his *Carduus medius* [Gouan]. Whether

---

[174] This is possibly a reference to *Geranium foliis hirutis semi septilobis obtusis*, Haller, Hist., no. 940, now *Geranium pusillum* L.

[175] *Erigeron annuus* (L.) Pers. (1807).

[176] Villars, believing that Gouan's figure suggested *Laserpitium* rather than *Ligusticum*, published *Ligusticum sequierii* Vill., Prosp. Pl. Dauph. 25 (1779). Hist. Pl. Dauph. 2: 615 (1787). Both Gouan's and Villars' species = *Ligusticum lucidum* Miller (1768).

collected in the woods or on the bare slopes, it presents a different habit. *Lepidium iberis* L. and *L. graminifolium* L. should be lumped with some hesitation, although I agree with you.

The villous receptacle on *Crepis nemausensis* [Gouan] cannot leave it among the *Crepis*.[177]

It may be that you have made too many species of *Astragalus*. I do not believe that your *A. fetidus* [Vill.] from Queyras is different from my *A. viscosum* Vill. from [Pic de] Bure; or your *Phaca [halleri* Vill.] different from my *P. australis* [L.] from the [Pic de] Bure.[178]

You would give me pleasure by informing me what difference you see between *Bupleurum longifolium* L. and my fine [plant] from Chaudun that you call *B. vapincense* [Vill.]; for I believe it to be *B. longifolium* L., Haller [**Hist.**], no. 768.[179]

I strongly fear that your *Hypericum androsaemifolium* [Vill.] does not adequately differ from your *H. richeri* [Vill.] or my *H . alpinum* [Vill.];[180] nor your *Hypericum delphinense* [Vill.] from *H. perfoliatum* [L.] or from *H. quadrangulum* L.[181]

I cannot imagine that you would give Monsieur Gouan cause to think that you took his *Ranunculus pyrenaeus* [Gouan], which has white flowers and simple leaves, for your *Ranunculus saxatilis* [Vill.], which I believe to be a variety of *R. acris* [L.] or of *R. auricomus* [L.]![182]

Some thoughts on the pyrolas I possess: I do not have *Pyrola rotundifolia* L. I mistakenly took *P. minor* [L.] for it, which grows in our woods along with *P. secunda* [L.].

My botanizing yesterday brought me a gynandrous plant whose genus I do not find in Linnaeus. Unless Scopoli, who made new genera, has reported it, my plant will make a new genus. Try to draw it from the specimen I am sending you. It is so delicate that, despite my care, it is already drooping, because it is very nearly aqueous. If it is new, as I believe, accept it as testimony of my affection. Here is the name, the definition, and the description I am giving to it; and I am doing so in haste, because, despite keeping it in water, I see it fading from one moment to the next.

Villaria corallorhiza

Villaria scapo fugoso, radice coralloide, floribus gynandris diandris, summo petalo antice scutato, postice inverse scrotiformi.

The root fleshy, branched, composed of an aggregate of teeth, in the form of coral. The stem is a scape, very delicate, fungous, fistulose below and white, flesh-colored above, 6-10 inches in height, very smooth, round, glabrous, without leaves and without sheath, having only 2 membranous scales toward the middle. It bears 2-4 very beautiful flowers, separated from each other by about 1 inch, of conspicuous and very elegant structure. The spathes are thin membrous, lanceolate; the peduncles are 1/12-1/6-inch long; the germ

---

[177] Villars transferred it to *Andryala nemausensis* Vill., **Prosp. Pl. Dauph.** 37 (1779) [= *Crepis sancta* (L.) Babcock (1941)].

[178] *Astragalus viscosum* Vill. = *A. fetidus* Vill. = *Oxytropis fetida* (Vill.) DC. (1802). *Phaca halleri* Vill. = *P. australis* L. = *Astragalus australis* (L.) Lam. (1778).

[179] Villars kept the two separate, **Hist. Pl. Dauph.** 2: 573-574 (1787), but Chaix was correct.

[180] Chaix was right: both *H. androsaemifolium* and *H. alpinum* are the same as *H. richeri* Vill.

[181] *Hypericum delphinense* Vill. = *H. quadrangulum* L., nom. ambig. = *H. tetrapterum* Fries (1823).

[182] *Ranunculus pyrenaeus* Gouan = *Ranunculus gouani* Willd. (1800). Villars published *R. saxatilis* Vill., **Prosp. Pl. Dauph.** 50 (1779), but excluded it later from his flora. It may have become *R. adunca* Gren. (1847).

nearly globose with small purplish edges. The corolla is composed of 6 petals [3 petals and 3 petaloid sepals]; the lip in front, in the form of a shield, presents a white face spotted with red dots, with 2 white auricles at its base, terminating at the rear in a scrotiform tube. The other petals are a yellowish-white: the 2 lateral ones linear, in the form of open arms, the 3 smaller ones lanceolate and curved downward. The pistil is in the center of the flower, a whitish color, glissening, its back resting on the lowest petal, terminating in a closed calyptra containing 2 yellow stamens.

I found it in flower 29 July 1779 in the Bois des Donnes, formerly the Chartreusines de Berthaud, at a place called the *bear cave*, without being able to find it elsewhere. If you wait until tomorrow to draw it in Le Noyer, I fear that it will become disfigured through wilting because of its great delicacy. Write me a word from Gap about what you think of it and about other matters that interest me.[183]

I also collected *Milium effusum* [L.]. I am admiring a small orobanche that has come up in my garden without having been planted, which I have not yet been able to determine for want of sufficient development, so far raising four thin and simple stalks. In due course I will let you know about it

Your *Hieracium pumilum* can be found among the rocks along the Buëch. I have seen the great difference between it and *H. amplexicaule* [L.], surely meriting separation as a species.

I repeat that my *Hieracium hybridum* [Chaix in Vill.], planted in my garden, is neither *H. auricula* nor *H. cymosum* even if by some peculiarity of nature it has been procreated by one and the other.[184]

Well, quite enough for one letter. I look forward to embracing you in St.-Etienne-en-Dévoluy on August 8th.

35
Les Baux, 16 September 1779

You have had the pain of finding your mother ill upon the return [from our fieldtrip].[185] I share your trouble in every possible way. On another matter, please understand what pleasure the news that you have been made a correspondent by the Société

~

[183] It is unknown whether Villars wrote his opinion of the proposed genus or whether he waited in order to talk about it on their forthcoming fieldtrip in August. In any case, Villars would not publish it as new, giving it as *Satyrium epipogium* L., Vill., **Hist. Pl. Dauph.** 2: 44 (1787). [= *Epipogium aphyllum* Swartz (1814)] . Also see Villars, "Notice historique sur Dominique Chaix," pp. 303-304.

[184] *Hieracium* x *hybridum* Chaix in Vill., **Hist. Pl. Dauph.** 3: 100 (1788), is held to be *H. cymosum* L. x *H. peleteranum* Mérat.

[185] The trip had begun in Champsaur, presumably at St.-Léger-les-Mélèzes, and proceeded up the Drac to Champoléon on the White Drac; from there to Argentière via the Pas de la Cavale; northwestward from there into the Vallouise and on to Le Monêtier-les-Bains via the Col de l'Eychauda, finding *en route* a new *Agrostis*. They then passed through the valley of the Arsine in order to reach the Col du Lautaret. Coming down from there to Briançon, they found a superb and unknown thistle. From Briançon up to Montgenèvre; south to Gondran and Cervières to reach the Combe du Queyras. The village of Aiguilles was the last point on the itinerary noted by Villars, **Hist. Pl. Dauph.** 1: xxxix (1786). The valley of the Durance would have been their likely route home.

royale de Médecine has given me.[186] I hope that this will be only the beginning of the honors that your merit will bring to you in the future. I thank you for the two copies of the **Prospectus**. I have offered one copy of it to dear friend Monsieur Gaude in your name, which he has accepted as a testimony of your friendship. I spent two days with him at the home of our venerable prieur de Furmeyer [Jean Faure]. Both of them had many compliments for you. The trip enabled me to find, between Furmeyer and Le Mottes near the Buëch, the inula that you sent to me dried with the name *Inula britannica* [L.]. Along with you, I would gladly make *I. britannica* of it; but Linnaeus's stem-clasping leaves trouble me some. I define it as: *Inula foliis lanceolatis serratis subtus tomentosis sessilibus haud amplexicaulus.*[187]

The stem leaves on my thistle from Laric are quite similar to those of *Lycopus* [in Labiatae] in their outline, but the radical leaves are oval. The name *lycopifolius* will suffice, but one must not say that it has affinity with [*Circium*] *erisithales* [(Jacq.) Scop.].[188]

Our thistle from Lautaret with leaves resembling acanthus cannot be *Carduus tataricus* [L.] as its flowers are not triphyllous: three floral leaves do not involucrate the head; the keel on the calyx scales is not white, and so on. I nearly believe that this is *Cnicus erisithales* L. (Haller, no. 175). . . . This plant will enlighten us in the future, as it has well rooted in my garden as have the others brought back from the Briançonnais. . . .[189] You have done well to combine the genus *Cnicus* with *Carduus*, as *erisithales* with Linnaean characters should be a *Carduus*, and *Carduus tataricus* should be *Cnicus*.

Our other thistle can be either *Carduus heterophyllus* [L.] or *Carduus helenioides* [L.] following Scopoli, **It. tyrol.** 60, where he makes var. *a. caule uniflora foliis dentatis*; var. *b. caule multiflora foliis inferioribus lacinatis*. In order to take the middle course between *heterophyllus* and *helenoiodes*, one would have to call it *britannicus* with Scopoli: *Cirsium britannicum* . . . . I have not observed in our plant the bulbous roots of *heterophyllus*, Ray 305, no. 4 and 306, no. 5.[190]

The cerastium from our trip (as I said to you) is indeed the one from [Pic de] Bure, which I had incorrectly taken to be *Cerastium latifolium* L. It is *Cerastium alpinum* L., Haller, no. 888, with capsule oblong, the peduncles divaricate. The one collected in Valgaudemar at St.-Maurice, in the vale of Chasseran, *foliis subrotundis tomentosis*, is certainly *C. latifolium* L., Haller, no. 887. Ray, **Hist.**, no. 1028, renders it *capitulum oblongum semen continens*. Haller describes its fruit as *nearly round*, perhaps before its maturity. I am raising one individual plant from the seeds from Valgaudemar,

~

[186] The patent of his election as correspondent, dated Paris, 9 July 1779, can be found in the Archives of the Gauthier-Villars family in Grenoble. His name was spelled erroneously—Villars—and accounts for his adoption of it thereafter. The election recognized his work on epidemics in Champsaur and Valgaudemar. See Georges de Manteyer, **Les Origines de Dominique Villars le botaniste (1555-1814)** (Gap: L. Jean & Peyrot), pp. 148-149.

[187] See *Inula cinerea* Chaix, **Pl. Vap.** 66 (1785). Chaix in Vill., **Hist. Pl. Dauph.**. 1: 370 (1786). [= *Inula britannica* L. (1753)].

[188] See letter no. 22. *Carduus lycopifolius* Vill., **Prosp. Pl. Dauph.** 30 (1779). [= *Serratula lycopifolia* (Vill.) A. Kerner (1872)].

[189] The question about *Cnicus erisithales* will be raised again in letters 41 and 42.

[190] Both Villars and Allioni believed that *Carduus helenioides* was a variety of *Carduus heterophyllus*, but the latter is now held to be a synonym of the former within *Cirsium*.

substantially different from *C. alpinum* because of its pubescence and the rotundity of its leaves. Thus, we are settled about these two plants.

    Aretia alpina L., var. *b*. Haller, in the Col de Terre-Noire between Cervières and Queyras. Variety *a*. Haller grows on Bure where *Aretia helvetica* L. also is found in my persistent opinion.[191]

    Phyteuma scorzoneraefolia (ours) in the alpine meadows of Vallouise along the Eychauda is quite new to me.[192] The small plant in Queyras at the Col de Terre-Noire is *Phyteuma pauciflora* [L.], and the other small and frequent one in the Alps is *P. hemisphaericum* [L.] with capillaceous leaves.

    Did you inadvertently label *Cardamine bellidifolia* [L.] as *Arabis bellidifolia* [Jacq.] and vice versa? The former is about a foot tall, the leaves cauline and quite entire, growing next to streams. The latter is 1-2 inches tall, scape nude, leaves eliptic and a bit dentate, from the Col de Terre-Noire.[193] Does your *Arabis serphyllifolia* [Vill.] differ from *A. bellidifolia?*

    Our sisymbrium from Gondron cannot be *Sisymbrium pyrenaicum* [Vill.]. Perhaps it is a new species. I am calling it *Sisymbrium nanum: siliquis declinatis brevibus, caule unciali, foliis pinnatifidis.*[194]

    Ornithogalum umbellatum L. . . . *staminibus simplicibus,* contrary to Linnaeus and Haller, seems to indicate my *O. pannonicum* [Chaix in Vill.][195]

    Campanula scheuchzeri [Vill.], not *scheuczeri.*

    Bupleurum gramineum [Vill.]. I have grounds to believe that you are right in separating a new species from *B. ranunculoides* [L.].[196] I cannot say anything about your *B. angulosum* [L.]; I do not find a specimen of it.

    Artemisia insipida [Vill.]: it has no affinity with *A. tanacetifolia* [L.].

    Hedypnois: on the matter of your inverted cone. I agree with Monsieur Gouan that you are expressing your thought poorly. The upper tube of the peduncle, which is a true inverted cone, is not the calyx.[197]

    Myagrum rugosum [L.] If you mean your new myagre from Corps and Mison, it is not *M. rugosum* L. I call it *Myagrum erucaefolium.*[198]

~

[191] Transferred to *Androsace alpina* (L.) Lam. (1778) and to *Androsace helvetica* (L.) All. (1785).

[192] *Phyteuma scorzonerifolium* Vill., Hist. Pl. Dauph. 2: 518 (1787).

[193] SY= *Arabis soyeri* Reuter & Huet ssp. *jacquinii* (G. Beck) B.M.G. Jones.

[194] Villars' collection list, Hist. Pl. Dauph. 1: 307 (1786), shows *Sisymbrium pyrenaicum,* but later altered to be *Sisymbrium pusillum* Vill., ibid. 3: 341 (1788). He was seemingly uncertain about it. It may have been *Sisymbrium pinnatifidum* (Lam.) DC. = *Murbeckiella pinnatifida* (Lam.) Rothm. (1939).

[195] Linnaeus had written *filamentis emarginatis.* In altering the definition to *staminibus simplicibus,* Villars used Chaix's phrase for *Ornithogalum pannonicum* Chaix in Vill., Prosp. Pl. Dauph. 18 (1779). Later, he abandoned Chaix's species for his own *Ornithogalum lacteum* Vill., Hist. Pl. Dauph. 2: 272 (1787) = *O. narbonense* L. (1756).

[196] *Bupleurum ranunculoides* L. ssp. *gramineum* (Vill.) Hayek (1927).

[197] Villars disagreed: "The calyx of the species in this genus represents a true inverted cone, and it is contiguous with the peduncle by means of the bracts, allowing no mark of separation between them." Hist. Pl. Dauph. 3: 77 (1788).

[198] See letter no. 32.

Hedysarum chaixi. What affinity does it have with *H. saxatile* [L.]?[199]

Phaca halleri [Vill.] and *Phaca gerardi* [Vill.]. I doubt if they are different species.[200]

Ranunculus sylvaticus. I formerly used that name, but I now believe [this plant] to be *Ranunculus auricomus* [L.].

Thalictrum saxatile Chaix: I say *foliis subtrilobus argute serratis* and so on.[201]

Thalictrum simplex L., **Mant.** 78. . . from near the hospice of Lautaret, which you indicated to me as *T. angustifolium* [L.]. I do not believe you are right. If your *T. angustifolium* from Polligny to St.-Etienne-en-Dévoluy was identical, you could well be mistaken.

At Montgenèvre and Cervières, *Alyssum montanum, foliis ellipicis scabris, non spatulatis incanus. A. alpestre* [L.], *siliculis orbicularibus, oris torulosis, non vero ovalibus simplicibus.*

Our campanula from above Briançon between Le Fontenil and La Vachette must be a peculiar species, different from the varieties of *Campanula rotundifolia* [L.]. I am calling it *linariaefolia* because of the similarity of its leaves to those of *Antirrhinum monspessulanum* [L.], *campanula multiceps, caule rigido, multifloro, foliis omnibus linearibus integerrimus.*[202]

The grass from Briançon and from the Fazi plain is a Linnaean *Aira.* See if it is not Haller, no. 1470. I have not copied it; this could well be that number.[203]

The salvia you found at Valence, and brought back in seed from Montpellier, which I cultivated thereafter, which you wanted to equate with my *Salvia clandestina* [L.], and which I may have acknowledged to be *Salvia verbenaca* [L.], is surly *Salvia pyrenaica* [L.], as I originally called it. I now have some seed of *Salvia verbenaca* from Paris quite different from the above. You may take my word that *Salvia pyrenaica* can be ranked among our plants of Dauphiné.[204]

As for *Anthemis arvensis* [L.], sent from Paris in seed under that name, either they served me poorly or the plant degenerated. For the seeds produced *Matricaria chamomilla* [L.] for me, as its appearance attested even before it flowered, and about which you will be convinced by the specimen I address to you. [= *Camomilla recutita* (l.) Rauschert]. What could have become of *Anthemis arvensis* that you observed in Paris? I have certainly not confused the label.

I have collected *Trifolium filiforme* [L.] in the Garenne de Montmaur, which may well be made, following Linnaeus, a species different from *T. procumbens* [L.], common in Champsaur. We have *T. agrarium* [L.] in Les Baux on the hill that separates the woods

[199] *Hedysarum supinum* Chaix in Vill., **Prosp. Pl. Dauph.** 41 (1779). **Hist. Pl. Dauph.** 3: 394 (1788). On hillsides: Les Baux, Rabou, and from Gap to Veynes, Laric, and Serres. [= *Onobrychis supina* (Chaix) DC. (1805).

[200] Chaix was correct. Both = *Astragalus australis* (L.) Lam.

[201] *Thalictrum saxatile* Chaix in Vill., **Prosp. Pl. Dauph.** 50 (1779). **Hist. Pl. Dauph.** 3: 714 (1788). Villars noted that it would be difficult to distinguish this from *T. minus* L. were it not for its bad odor and great difference in height, i.e., shorter. He called it close to *T. foetidum* L., which is what it is.

[202] *Campanula linifolia* Chaix, non Scop., **Pl. Vap.** 76 (1785). Chaix in Vill., **Hist. Pl. Dauph.** 1: 386 (1786). [= *Campanula scheuchzeri* Vill. (1779)].

[203] *Aira brigantiaca* Chaix, **Pl. Vap.** 74 (1785). Chaix in Vill, **Hist. Pl. Dauph.** 1: 378 (1786), invalidly published. [= *Aira miliacea* Vill., ibid. 2: 81 (1787) = *Puccinellia distans* (L.) Parl. (1850)].

[204] Villars did not include *Salvia pyrenaica* L. in his flora. In 1852, Grenier specifically excluded it from his **Flore de France** 2: 715 on the grounds that it had not been refound.

belonging to the chapter of Gap from those belonging to Monsieur Mondet, and at La Grangette. We have trod upon *T. spadiceum* [L.] in the Alps.

In my parish, I have a poor old man who suffers extremely from retention of urine and, as a consequence, from a swelling of the lower stomach. The urine flows merely drop by drop, nearly continually day and night. What can be done to relieve him? I have a young nephew about twelve years old who, in the aftermath of measles this past winter, has constantly suffered from enlarged and peeling lips. Today, if a bit improved, he has two scabby openings in the mouth, eyes somewhat tender and tearing, eyelids and face somewhat puffy. What may I prescribe for him to dissipate these annoying after effects?

Please respond at your earliest convenience about these and the botanical matters. Also, let me know whether Madame de La Vallette's illness has subsided and about your dear mother's recovery. You know how much I am interested in everything that concerns you.

36

Les Baux, 21 September 1779

You will have received, I hope, my letter responding to yours touching a number of botanical matters, about which you will please tell me your sentiments. I have received a very gracious letter from Monsieur Séguier. He has expressed to me, with a kindness that honors me, his feeling about my Latin verses. Do not put yourself to any more trouble about this matter; we will talk later. Your publication will make your works and talents well known, and your preface the methods by which you have produced it. I come next to what concerns you, and I transcribe from his letter.

"I beg you to greet Monsieur Villars affectionately on my behalf and to let him know how much I have been gratified by his **Prospectus de l'histoire des plantes du Dauphiné**, which Monsieur Nicolas, physician of Grenoble, sent to me on his behalf, and for which I am infinitely obligated to him. The mixed method chosen between the artificial and the natural will be very useful to him in combining the advantage of the artificial system with the soundness of the natural families without having the inconveniences of the latter. He has, however, adopted Linnaeus by founding his divisions upon the stamens. *It is not important which method one uses*, as this celebrated botanist pointed out to me in one of his letters, *provided that it leads to the knowledge of plants, and that one attains that goal by the shortest path*. This catalogue of rare plants of Dauphiné fully announces the results of his research and makes one ardently hope to possess the [completed] work soon. It is useful to indicate all the riches to one's compatriots that they meet in the country through which they pass."

I wanted to report to you everything that this great man, whom you value so much, said to me about you in order to assure you of his approbation, in the event that he postpones writing to you because of the loss of an aunt who was extremely dear to him, which has greatly grieved him; and because of his great concern regarding the repairs of a public monument, the Maison Carrée, whose direction has been entrusted to him. He has urged me to share the collections of our [recent] trip with him. "The alpine plants," he says, "are very dear to my heart. The joy I take in them is a diversion that occupies me in my old age. This pleasure strengthens me."

Monsieur Gouan has written to me in the most civil manner. He promises me plants and seeds that are within his reach and even a copy of his observations on Belleval's copper plates in two volumes with 260 prints, that ought to have been published in Poland at the king's expense; and a copy to you, too, if the printer behaves properly.[205]

While not doubting his good heartedness, I sense that his advances are a bit self-serving. He asks me for all the alpine plants and seeds that I have available, in particular, those announced in your prospectus. He makes the same request of you. I am going to do what has been required of me, and I shall send you my package, as he recommends, which can be added to yours.

Monsieur Blanc, professor in Embrun, to whom I sent the **Philosophia botanica** [Linnaeus, 1751] and cuttings of *Geranium zonale* and *tabulare* L., will send *Gnaphalium orientale* L., or perhaps bring it here himself on the 23rd of this month, with some fruit stones from the wild white plum of Briançon for which I have asked him as I want to raise it.[206]

P.S. I found *Panicum dactylon* L. at Gap on the road near the Male-Combe bridge. You had sent it to me previously from Grenoble, and it has survived for several years in my garden.[207]

37

Les Baux, 26 September 1779

Your young patient from Rabou, Barthélemi Richier, son of Barthélemi, has taken the remedies that you prescribed for him. During the twelve days that he took your powder, he became weaker and more squeamish, nearly without appetite for eating. A spot has appeared under his left nipple. Subsequently, he took a gentian tea for six days without experiencing any greater relaxation of his stomach. During this interval, there occurred an expectoration of blood one evening and the next morning, but not a significant amount. The patient is somewhat weaker than when you saw him in St.-Bonnet, still very much labored by a dry cough, perspiring easily and at the least effort. His pulse appears to me to be regular. His arm is colder than warmer to the touch, sweaty. See what you have to prescribe for him. He came to find me this evening in order to beg you to give him your full attention. I know what you want to do in all circumstances requiring your professional services; I am persuaded, therefore, that you will do everything possible for the cure of this poor young man.

You ought to have received recently two of my letters to which I still await your response. The abbé Blanc, professor from Embrun, came to dinner and spent the night last Thursday the 23rd. He brought me two cuttings of *Gnaphalium orientale* and a number

⌁

[205] Pierre Richer de Belleval (ca. 1564-1632) was the founder of the botanical garden in Montpellier and its first professor of botany. The strange story of his unpublished plates, illustrating plants of the Alps, Pyrenees , and Cévennes, is related in Ad. Davy de Virville et al., **Histoire de la botanique en France** (Paris: Société d'Edition d'Enseignement Supérieur, 1954), p. 28. The plates were not published as anticipated here, but only in the 4th edition (1796) of La Tourrette and Rozier, **Démonstrations élémentaires de botanique** in Lyon.
[206] *Prunus brigantiaca* Chaix, Pl. Vap. 76 (1785), invalidly published. [= *Prunus brigantina* Vill. in L. (1785)].
[207] Villars did not recognize it in his flora.

of plants that I labeled for him. He has a genuine taste for botany, a good memory, and all the other aptitudes. I am fortunate to have made his acquaintance. He opens a more convenient way for me to botanize in the future in the Embrunnais. In Embrun aux Baumes, the domain of his *collège*, he has found *Orchis abortiva* L. [= *Limodorum abortivum* (L.) Swartz (1799)], a rare plant for me. In the Queyras, when he was waiting for us there: *Astragalus visicarius* L. After we left him: *Saxifraga caesia* L. at the Pic de Morgon, that is, above Savines in the forest belonging to Savines adjacent to the forest of Boscodon. There you have proof of his discernment. He will have some white-plum stones sent to me and anything else I ask for. He went on to spend the night of the 24th at Ventavon, paying a visit to the pious family of the *seigneur* of that property. I lent him your prospectus to read. He asked me to extend his greetings to you.

Please try to give me a few day's visit at a time most convenient for you. You will see that I have made quite attractive improvements in my garden, meant to accommodate a greater number of plants and with less confusion. You can bring me those notebooks on your work at that time that you want me to read. You can take back books I have from you, and we can settle the account for those you are assigning to me. You may also take home the dried plants that I am keeping for you, whatever you judge to be appropriate. Give me news of your brother and of your dear family, whom I greet with all my heart. Our association is the greatest consolation of your very humble and very obedient servant.

38

Les Baux, 17 December 1779

Yesterday, the vicar of Manteyer brought your letter of the 13th instant to me here, and I had the pleasure of adding him to my confrere from La Roche and Monsieur Chaix, the notary, who did me the honor of spending that day at my home. I regret to learn about the great number of sick who are occupying you, including some of your close relatives. I am glad to see that you are taking the designs of divine providence into account regarding your troubles and concerns. I pray that it guides your vision to its glory by granting, through your ministry, the recovery of these poor afflicted people.

Your other letter, of November 10th, had also reached me, and I had not responded to it believing that you had gone to Grenoble, consequently delaying our responses to the celebrated professor of Montpellier [Gouan]. Madame de La Roche was willing to take charge of the material I had to be sent to Monsieur Séguier, having left today for Aix and Toulon, going from there to spend the winter in Nîmes. Do not forget Monsieur de Lamarck's book for my account,[208] and let me know what resources will become available for the printing of your work. If you are to remain in Grenoble a few days, I would be very much obliged to you if you would pay a call at the home of Monsieur de Bruno in the rue Neuve, across from the back door of the Jacobins, to remember me to them and to inquire about the condition of their family. It is a house to which I have obligations, and to which I have been, and always will be, greatly attached. . . .

I have arranged the grasses and their allies from our trip in the Briançonnais in my

⁓

[208] Probably a reference to the first edition of Lamarck's **Flore françoise** (1778), first available in 1779; but there is no evidence that Chaix ever owned any of Lamarck's titles.

herbarium. I shall remind you of that for a moment by reporting my brief observations of them to you. [The following] have seemed to be new to me:

1) An *Agrostis* from Vallouise that I am calling *Agrostis villosa: flosculis muticis, villorum longiorum copia obsitis; culmi recti, bipedales. Folia 2, lineas lata. Panicula laxa. Calyces equales, majuscule, acuti, violacei. Corolla us paulo minor, e basi profert villorum manum, ipsam superantium.* It gives the misleading appearance of being *Agrostis arundinacea* [L.], but, looked at closely, if differs totally. I believe that you collected it in a meadow near the road before reaching the town of Vallouise.[209]

2) *Agrostis pumila* L., **Mant.** 31, in the Alps of Vallouise: *culmi fasciculati, bipollicares, ciati nodosi. Folia lineam lata, paniculam subaequant.*

3) The *Aira* from the Fort des Trois Têtes [outside Briançon] and from the Fazi plain could well be *Aira aquatica* L. if you do not have the latter from elsewhere. Note: I do not know, and I do not have, your *Aira festucoides* Vill., **Prosp. [Pl. Dauph.** 16 (1779)] among my plants.[210] Note: My *Poa gigantea* could well be a good species; it belongs neither to *P. compressa* [L.] nor to *P. alpina* [L.]. Note: I cannot refer my pretty *Avena* from Chaudun to any of your species or to any of Linnaeus's: *culmi pedales, folia lata, panicula erecta, spiculae pedicellatae, 5 florae cum flore imperfecto.* If you do not recognize it, I will give you a specimen of it.[211]

3) (sic) Your *Juncus spadicus* Vill., **Prosp. Pl. Dauph.** 18, that I believe to be Haller, Hist., no. 1329, I collected in Gondran and elsewhere. I would call it *helveolus*, because its spike is not chestnut-brown, but rather a pale red-wine color.[212]

4) The fine rush of our mountains, *Juncus pilosus* L., Haller, **Hist.**, no. 1327, indeed meriting specific rank as you say, I would designate it as *aggregatus*, not knowing your own designation.

5) The sedge, collected when going down from the Col de l'Eychauda toward the woods of Monêtier, believed at first to be *Carex baldensis* [Vill.], I now think is *Carex brizoides* [L.].

6) Again referring to your *Bromus perennis* [Vill.], common everywhere, I cannot be convinced that both Linnaeus and Haller overlooked it;[213] but that Linnaeus could have known it as *Festuca elatior* [L.], and that your *Festuca elatior* does not appear to me to differ from your *Festuca phoenix* [Vill.].[214]

7) I am astonished that Linnaeus did not see right away that his *Salix purpurea* [L.] is monandrous, and that he cites Arduino for support, **Mantissa** 489. I no longer doubt that

[209] *Agrostis villosa* Chaix, **Pl. Vap.** 74 (1785). Chaix in Vill., Hist. **Pl. Dauph.** 1: 378 (1786). [= *Secale villossum* L. (1753) = *Dasyphyrun villosum* (L.) P. Candargy (1901)].

[210] This is a mystery. Villars gave as a synonym Haller, no. 1488: *Avena diantha, panicula sparsa erecta, floribus auratis, basi villosis. In summis alpibus.* Verlot, **Cat. Pl. Dauph.** (1872), p. 360, calls it an uncertain plant.

[211] This may have been *Avena versicolar* Vill., **Prosp. Pl. Dauph.** 17 (1779). [= *Avenula versicolor* (Vill.) Lainz (1974)].

[212] Villars gave Haller, **Hist.**, no. 1327, as his synonym. This may well be the same as Haller, **Hist.**, no. 1326, that Chaix called *Juncus alpinopilosa*, **Pl. Vap.** 14 (1785). Chaix in Vill., **Hist. Pl. Dauph.** 1: 318 (1786). [= *Luzula alpinopilosa* (Chaix) Breistr. (1947)]. Or it may be *Luzula pilosa* (L.) Willd. (1809) as suggested in no. 4 below.

[213] See *Bromus erectus* Huds. (1762).

[214] See both *Festuca pratensis* Huds. (1762) and *Festuca arundinacea* Schreber (1771).

our osier, called *white* popularly, no doubt because of its faded bark, is *Salix purpurea* L. monandrous, whose purplish and corral-red young twigs and anthers were so vociferously noticed by Murray.[215] The diandrous willow from Lautaret would thus be *Salix helix* [L.].[216] You claim to have seen it is Valgaudemar. Keep some cuttings of it from different stalks so that, when transported to Les Baux, they will give us both sexes. . . .

Monsieur Gaude is concerned for a young man, very prepared to rear the young, who would like to find a tutorial position in Valence so that he could take his degree there. Convinced that you have good acquaintances in that city, he has asked me to approach you about it. Given the opportunity to relay this request through your brother, I hope you will be willing to oblige this worthy priest who has such fondness for you.

P.S. You are quite right to give Monsieur Charmeil's name to the phyteuma from Mont Dauphin on which he has made some very fruitful experiments. But do not call it *charmelii* in Latin; you will confuse his real name too much. Preserve it rather in its entirety with the with the spelling *charmeili*.[217] I would say the same if Monsieur [Richer de] Belleval had spelled his name *Richier*, one would have called him *Richierus* in Latin, with the *i*. And if he wrote it *Richer*, one would latinize it simply as *Richerus*, like Hallerus, thus not misleading outsiders in the knowledge of the real name.

~

On 12 March 1780, the **Affiches du Dauphiné** *published a notice that, beginning on 16 March 1780 at 2:30 p.m., Dr. Villars would begin a series of public lectures on botany, Wednesday, Thursday, and Saturday of each week. The principal goal of the course was to provide instruction for young surgical students who would soon be practicing in the region, and who were being taught and supported at government expense thanks to the intendant's benevolent concern. Villars' letters to Allioni after this date were sometimes addressed from St.-Bonnet and sometimes from Grenoble, indication that in this period before his permanent move to Grenoble he returned frequently to his home in Le Noyer.*

39

Les Baux, 27 March 1780

How benevolent you are to me! You recompense me today, thanks to your extraordinary amount of news, for the regret caused me by the too lenthy deprivation of it during the preceding months. Your valued letter of the 20th instant was delivered to me three days after its date. You can judge for yourself the service you have rendered me by obtaining for me the *stipendium* for masses to be celebrated. For the greatest part of the winter, I have celebrated them *gratis pro Deo*, not having any of them ordered, which led me to ask the curé de Poligny to cede some of his to me if possible. Given your pledge, I am therefore undertaking to offer masses at the usual rate of one louis according to the

~

[215] There are 2 stamens, but they are completely united to appear to be 1. The tree is now popularly called *red* osier. Villars made *Salix purpurea* L. a synonym for *S. monandra* Hoffm., **Hist. Pl. Dauph.** 3: 767 (1788) and called it the *white osier*.
[216] This has been treated as a form, variety, or hybrid of *Salix purpurea* L.
[217] *Phyteuma charmelii* Vill., **Hist. Pl. Dauph.** 2: 516 (1787), in Campanulaceae.

wishes of your deceased relative, who has left said sum for this purpose. When other occasions should arise, I beg you not to forget me.

Not long ago I received a very kind letter from Monsieur Gouan thanking me for my botanical shipment and allowing me to hope for the prompt reception of his notes on my plants along with some seeds, by way of the intendance under your address. He also assures me that I will be one of the beneficiaries of his publications he must give to the public within the current year, among others a sequel to the **Pinax** from [Casper] Bauhin to our time.[218] What did surprise me very much is that he asked me to urge you to have the kindness to send him your package by the parcel delivery as soon as possible, so that he might arrange your plants in his containers at the same time as mine and give us his reactions more easily. Did you not send him your package at the same time as mine? Or did yours go astray and not reach him? Please inform me about this matter.

When you have additional news in regard to the printing of your work, do not keep me in ignorance of it. But I need not spur your punctuality to have sent whatever you want to get back from me, and I know only too well how much remains for me to do. I have no pleasure comparable to repeating to you often the great affection that I have for you, Monsieur and very dear friend. . . .

P.S. The poor blacksmith in Rabou died a few days ago of a cancer of the upper lip, perhaps because of the stress from the plasters administered to him by the alleged specialist on such infirmities from Gap. This poor wretch had consulted you, and you had forewarned me what is fate would be. The young man with the chest ailment is not better, so far as I now know.

40

Les Baux, 30 April 1780

Stricken (as I, too, have been) by the illness of the illustrious and amiable Monsieur Séguier, you have in addition feared for his life. I hasten, therefore, to remove your concern. I have just received a letter from him, by way of Madame de Flotte de La Roche, which informs me (thanks be to the Lord) of his recovery from a troublesome and stubborn cold that confined him to his room and bed for two whole months. He has been very pleased with my botanical shipment, mentioning a number of my "fine plants;" and he adds that, when his health is fully restored, he will be able to examine them all at his leisure and will let me know which of them appear to be doubtful. He hopes to have the third volume of his **Plantae Veronenses** soon from Verona. One of his friends, whom he has commissioned to bring the book to him, must pass through [Gap], and he wants me to have a copy.[219]

As for what concerns you, here are his words: "Greet Monsieur Villars fondly and tell him how much I admire him." Two boxes were sent with his letter, one full of precious

---

[218] Chaix's personal library contained only one title by Gouan, **Hortus regius monspeliensis** (1762), nothing published after that date. "Catalogue de livres de Botanique avec les prix qu'ils ont coutés." Archives H.-A., Série L 1007.

[219] Jean-François Séguier had published **Plantae Veronenses** in 2 vols. in Verona in 1745-1754. Chaix indicated ownership of 3 vols., ibid.

packets of seeds, and the other containing four cuttings of [African] geraniums: *odoratissimum, acetosum, gibbosum,* and *fulgidum* [all L.], but nearly dead from dryness, having been enveloped only in *Zostera marina* [L.], and having been cut for a month. I can hardly count on their taking root, but rather on the seeds that this honorable man will have the kindness to send me, God willing.

I have read [C. L.] Willich's **Observationes botanaicae** [1762] with pleasure, and I have transcribed the essence of some of his observations, the remainder appearing to me to be included in our 3rd edition of the **Species Plantarum** or in the **Mantissa** Linnaeus. This leads me to speak a moment about botany. Should not Haller have our *Dianthus alpinus* [L.] from the Briançonnais (which is indeed *alpinus* L. with scales as long as the calyx)? I had believed it to be detailed in . . . [Scheuchzer], **Itinera alpina** 310.[220] But Willich does not think so, assured that Haller's plant has a calyx with "short exterior;" and I find nothing written by [Haller] in his **History** to indicate that to me.[221]

Willich judged wisely that the *carvifolia* of J. Bauhin belonged to something other than what Linnaeus and Haller cited. You call it *Peucedanum carvifolia* [Vill., **Prosp. Pl. Dauph.** 25 (1779)] quite appropriately. But then why should you name Linnaeus' selinum *carvifolia,* making it an angelica?[222] . . . I would prefer that you call your angelica *tenuifolia* with Rivinus, Ruppius, and Dillenius. Its leaves do not resemble those of caraway at any time in their development; they surely fall short of being decussate. Such a labeling will always offend the botanist. I like your *Peucedanum* and your *Sesseli carvifolium* under that name.[223] You have promised me your notebook on umbels when it is finished. I shall read it with great eagerness as well as the others I have not yet seen.

We have learned here that your protégé, young Serre, a student at the hospital [in Grenoble], has enlisted as a surgical assistant, and that he left Grenoble at the beginning of April to take ship with his company. This news has greatly surprised me. I wish him a great success in his own right and for the support of his family in the future.

Take care of your health, and give me your news as often as possible.

P.S. In about two weeks, I shall make a response to Monsieur Séguier. I am proposing to go down to Oze on May 10th to spend several days with my very dear confrere, Monsieur Gaude. You will certainly not be forgotten in our conversations. If you go into Valgaudemar, bring back some cuttings for me of that water-willow that puzzles us, especially of the male sex. And dry two or three specimens of *Geranium lucidum* [L.] properly for me, which, according to you, grows at St.-Jacques-en-Valgaudemar. Take a close look at the characters of *Cerastium aquaticum* L. growing at Villeneuve, which is perhaps only a variety of *Stellaria nemorum* [L.].[224] My bearded and pubescent cerastium from St.-Maurice-en-Valgaudemar is adorned these fine days with a corolla white as snow, quite deeply notched. Is it *Cerastium*

~

[220] Villar interlined here that, to judge by the description, the species was *Dianthus deltoides* [L.].
[221] Villar adds, "It is not there."
[222] *Angelica cardifolia* Vill., **Prosp. Pl. Dauph.** 25 (1779). It does not reappear in the flora.
[223] See *Sesseli annuum* L. ssp. *cardifolium* (Vill.) P. Fourn. (1937).
[224] Villars published *Cerastium aquaticum* L. as *Stellaria aquatica* Vill. because "of its resemblance to *Stellaria nemorum* L." [=*Myosoton aquaticum* (L.) Moench (1794)].

latifolium L. or a new species from among our discoveries?[225] *Oportet sapere ad sobrietatem. Viola nummulariifolia nostra frequens obvia est Bauxii ab amniculo Buech ad Corrie in vallibus nostio pago obversis, juxta viam Vapincum ducentum. Centies cam pede trivisti. Specimina hic subjunxi.*[226]

41

Les Baux, 2 June 1780

I owe you a response to two letters: of May 8th from Le Noyer that accompanied Monsieur Gouan's notes; and of May 25th from Grenoble bearing [Dr. Henri] Gagnon's seeds from Egypt. I respond to the latter first: I am quite pleased that Monsieur Gagnon wants to share with me the riches that have been sent to him from Egypt, a country whose productions are so little known to me. If the occasion arises, extend to him my most sincere thanks. The spring season being already advanced, and all my pots being occupied, I shall not sow these precious seeds. I shall reserve them until next year, if God permits, in order to give them a longer time of continuous vegetation.

You have addressed a good source for the correction of your manuscript on the maladies of the Valgaudemar etc.[227] Knowing your good character, I sense how much you will benefit from the opinions that will be given to you. I do know that you [already] have obligations to Monsieur Guettard; but he, also, has not forgotten the services you have rendered him [in 1775 and 1776]. I am delighted that he has made his gratitude to you public.[228] Your telling me of his work has given me real pleasure.

I turn now to your first letter that gives me the opportunity to speak about botany to you. Do not become annoyed; try to put up with my scribbling. In any case you know, even without my request, what you must do with the notes from Monsieur Gouan. Although his judgment does not appear to us to be quite sound on various points, we are obliged to believe him to be sincere, and we are equally obliged to him for have troubled to give us his judgment. I employ here the advice of the great apostle [Paul] that I have perhaps cited to you on other occasion: *Always follow after that which is good, one toward another, and toward all.* I ·Thessalonians 5. 15.

1) Monsieur Gouan appears to me to have gone a bit astray in referring your [*Hieracium*] *lawsonii* and my [*Hieracium*] *saxatile* to his *Hieracium sylvaticum*, **Illustrationes** [p. 56].[229]

He is not right to merge your *Hieracium prenanthoides* [Vill., **Prosp. Pl. Dauph. 35** (1779] with *H. cerinthoides latifolium* after your verification based upon the Tournefort

∾

[225] The description in Vill., Hist. Pl. Dauph. 3: 647 (1788), would suggest *Cerastium alpinum* L. ssp. *lanatum* (Lam.) Ascherson & Graebner (1917).

[226] *Viola nummulariifolia* Vill. had been published the previous year in the prospectus; but in Hist. Pl. Dauph. 2: 663 (1787), Villars would say "I owe this species to Monsieur Chaix."

[227] Villars, **Observations sur une fièvre épidémique qui a régné dans le Champsaur et le Val Gaudemar** (1781).

[228] In the long preface to Guettard, **Mémoires sur la minéralogie du Dauphiné**, 2 vols. (Paris: Clousier, 1779).

[229] *Hieracium lawsonii* Vill., **Prosp. Pl. Dauph. 36** (1779), is a good species. *H. saxatile* Vill., ibid. 35 (1779), has been regarded both as a synonym of *H. lawsonii* and as = *H. phlomoides* Froelich in DC. (1838). Gouan's *H. sylvaticum* (1773) may have become *H. sylvaticum* Lam. (1786) and be = to *H. vulgatum* Fr. (1819).

herbarium. But as he is assured that your *H. prenanthoides* is nothing other than his *H. cerinthoides*, ascribe to him his *Cerinthoides rami corymbosi; folia caulina semi-amplexicaulia, oblongo-ovata, basi latiora*, **Illustrationes**, p. 58, which should reasonably suit him. I believe you are well grounded in your verification.[230] (Your *H. scoronerifolium* [Vill., ibid. 35 (1779)] is only a pure variety of the *cerinthoides*. I am sure of that.)

I do not have your *Hieracium valdepilosum* [Vill., ibid. 34 (1779)] nor your *H. pulmonarioides* [Vill., ibid. 36 (1779)]. I doubt if they are true species. (Your *H. cydoniaefolium* [Vill., ibid. 34 (1779)] differs very little from *H. prenanthoides*: the difference in its habit perhaps only derives from a wilder, more exposed natural habitat.) According to Monsieur Gouan, my *Hieracium spurium* is *H. dubium* L. . . .[231] I am not far from believing that your *Leontodon alpinum* [Vill., **Prosp. Pl. Dauph.** 34 (1779)], which, according to Gouan, is Haller, Hist., no. 25 [*Picris caule nudo, unifloro, foliis asperis, dentatis*], is only a variety of *Leontodon hispidum* L. . . .[232]

2) Our professor [Gouan] supports my feeling about the *Cnicus* from Lautaret by making it *Cnicus erisithales* L. (1763). [= *Cirsium erisithales* (Jacq. 1762) Scop. (1769)]. I am congratulating myself about it. It is doing well in my garden. Do not forget yours from Lemps. Bring me some stalks of it for cultivation. That being the case, you have been wrong to say that *Carduus lycopidolius* [Vill., **Prosp. Pl. Dauph.** 30 (1779)] has affinity with *Cnicus erisithales!*[233] (The *Centaurea seusana* [Chaix] is only a variety of *C. montana* [L.] according to Gouan,[234] as is another taller centaurea, very branching and with narrow leaves, from Monsieur Liottard, which I am cultivating.

The professor [Gouan] has made a change in the genus of our *Hedysarum supinum* [Chaix] that may well be his *Onobrychis italica perennis* from his **Illustrationes**, cited from Tilli [**Cat. pl. horti pisani** (1723)]. Check on this if you have this author. What he ought to have taken immediately into comparison was *Hedysarum saxatile* L.! I am astonished that he does not know the latter.[235] It may be, as the doctor says, that your *Turritis raii* is *Cardamine petraea* L., Haller, [Hist., no.] 453.[236]

If, as I believe, your *Brassica cheiranthos* is *Brassica erucastrum* L., in Gouan, **Illustrationes** [43-44],[237] then certainly the plant we know as *Brassica erucastrum* L. is the one in Haller, Hist., no. 459, [*Eruca caule hirto, foliis hirsutis, semipinnatis,*] *pinnis*

<hr>

[230] Villars, **Hist. Pl. Dauph.** 3: 110 (1788), would publish his own version of *Hieracium cerinthoides* and discuss the disputed interpretations.

[231] *Hieracium* x *spurium* Chaix ex Froelich in DC. (1838). *H. cymosum* x *H. pilosella*. See *H. cymosum* L. var. *c*, Vill., **Hist. Pl. Dauph.** 3: 102 (1788). From Les Baux.

[232] Verlot, **Cat. Pl. Dauph.** (1872), indicates that no specimens of *Leontodon alpinum* survived in either of the Villars or the Chaix herbaria, and that the species was probably a form of *L. hispidum* L. as Chaix says here.

[233] Villars persisted in putting *erisithales* in *Carduus*, thus accounting for the affinity. **Hist. Pl. Dauph.** 3: 20 (1788).

[234] *Centaurea seusana* Chaix, **Pl. Vap.** 61 (1785). Chaix in Vill., **Hist. Pl. Dauph.** 1: 365 (1786), from near Manteyer on Mt. Seuse. [= *Centaurea triumfetti* All. ssp. *triumfetti* (1773)].

[235] *Hedysarum supinum* Chaix in Vill., **Prosp. Pl. Dauph.** 41 (1779), on hillsides at Les Baux and Rabou; from Gap to Veynes, Laric, and Serres. [= *Onobrychis supina* (Chaix in Vill.) DC. in Lam. & DC. (1805)].

[236] *Turritis raii* Vill. = *Arabis stricta* Huds. (1778).

[237] *Brassica cheiranthos* Vill., **Prosp. Pl. Dauph.** 40 (1779), stands as *Rhynchosinapis cheiranthos* (Vill.) Dandy (1957).

subrotunde dentatis, petalis longo ungue, ochroleuces. . . .[238] It thus appears that [Gouan] does not know our *Brassica erucastrum. . . .*

3) *Cerastium* (I am referring to my *lanuginosum* from St.-Maurice [in Valgaudemar]) *foliis ovatis lanuginosia, capsulis oblongis,* differs from *C. alpinum* but has affinity to it: *foliis lanuginosis ovatis, non ovato-lanceolatis dumtaxat subhirsutis; a repente pedunculis 1-2 floris, non ramosis; capsules oblongis non globis.*[239]

4) *Primula queyrasensis foliis ovalis serratis, margine farinosis, calycibus glabris brevibus,* is variety *f* of *Primula auricula* L.; Ray, Hist., 1083, no. 3. Our *Primula viscosa* [Vill.] is Haller, Hist., no. 613, *foliis carnosis, hirsutis, odoratis etc.* Your *"calycum laciniis capsula longioribus"* in **Prosp. Pl. Dauph.** [21] is not a good characterization, as several varieties of *P. auricula* L. have such a calyx. I would define this species as follows: *Primula (viscosa) foliis subserratis, subhirsutis, glutinosis, item calycibus.*[240]

5) . . . We must not let our *Herniaria alpina* from the Briançonnais fall into oblivion.[241]

6) My *Festuca pumila,* Haller, Hist., no. 1439: *foliis setaceis, duriusculis, panicula collecta, spiculis trifloris subaristatis. Culmi 3-4 uncialis, glumae splendentes, viridi et violaceo variae.* In the Alps of Chaudun and Valgaudemar.[242]

7) Your *Festuca elatior* and *phoenix* [Vill., **Prosp. Pl. Dauph.** 17 (1779), are affinitive to the highest degree.[243] Your *Carex hordeistichos* [Vill., **Prosp. Pl. Dauph.** 18 (1779), is surely nothing other than *Carex hirta* [L.] and ought to be suppressed. . . . Your *Carex gynobasis* is Haller, Hist., no. 1386.[244] . . . My *Carex argentea* is Haller, Hist., no. 1388.[245]

8) I am enclosing a specimen of sedge collected in my meadow that I have not previously had, and which I do not know how to find in Linnaeus. If it is known to you, tell me its name. I call it *Carex spicis femineis oblongis, sessilibus, bracteatis, distinctis, capsulis muticis, glaucescentibus* or Haller, Hist., no. 1374.[246]

9) This [enclosed] leaf collected in Laric I believe to be *Crataegus torminalis* [L.]. Am I mistaken? I do not remember whether you have reported this tree among our plants of Dauphiné. I want to collect its fruit or to get some cuttings at a suitable time.[247]

So there you have much scribbling, dear friend; about which, nevertheless, I want your clarifications, which you will please give me when you have the leisure. Find out about

---

[238] *Brassica erucastrum* L. stands as *Erucastrum nasturiifolium* (Poiret) O. E. Schulz (1916).

[239] In Vill., Hist. Pl. Dauph. 1: 332 (1786), Chaix listed his *Cerastium alpinum* L. from St.-Maurice as *totum valde tomentosum est.* The plant, raised from seed in his garden, had retained this native character. It appears that he was describing *Cerastium lanatum* Lam. [= *Cerastium alpinum* L. ssp. *lanatum* (Lam.) Ascherson & Graebner (1917)], the mat-forming, lanate variation. He believed it to be different from *C. latifolium* L.

[240] Villars did not alter his description in the flora. SY= *Primula hirsuta* All. (1773).

[241] *Herniaria alpina* Chaix, Pl. Vap. 75 (1785). Chaix in Vill., Hist. Pl. Dauph. 1: 379 (1786), a good species from Vallouise.

[242] *Festuca pumila* Chaix, Pl. Vap. 75 (1785). Chaix in Vill., Hist. Pl. Dauph. 1: 316 (1786). Villars somewhat altered the description in ibid. 2: 102 (1787) to make the species *spiculis teretibus subquadrifloris.* [= *Festuca quadriflora* Honckeny (1782)].

[243] Villars retained only his *Festuca phoenix* in Hist. Pl. Dauph. 2: 108 (1787).

[244] Vill., Hist. Pl. Dauph. 2: 206 (1787). [= *Carex hallerana* Asso (1779)].

[245] Chaix in Vill., Hist. Pl. Dauph. 2: 206 (1787). [= *Carex alba* Scop. (1772)].

[246] Villars did not cite Haller's no. 1374 as a synonym for any of his sedges.

[247] Villars would report the species to be around Tullins and between Crest and Montélimar. It is now held to be *Sorbus torminalis* (L.) Crantz.

Monsieur and Madame de Bruno and about their family, who will be close to my heart until the day of my death. I know what I owe to them. If your stay in Grenoble is long enough, send me your own news from there, something I always receive with great eagerness. I shall take the trip that you have proposed to me into the Valbonnais and Valgaudemar with great pleasure at the most favorable time for us to find *Ligusticum gmelini* [Vill.] and so on if God deigns to favor us.[248] Take care of a health that is as dear to me as my own.

42

Les Baux, 22 June 1780

The thistle from Lemps along with the *Aira aquatica* [L.] reached Les Baux in quite good condition. Each plant, put in the location you said would be suitable for it, has lost no time, thanks to its vegetation, in giving me happy prospects for a successful rooting. Your *erisithales* will soon show me the form of its leaves, of which you entirely deprived me, no doubt to prolong my gratification. Although Monsieur Gouan has called our thistle from Lautaret *Cnicus erisithales*, I retain a suspicion about his opinion, sensing how much this thistle differs from the one in Daléchamps, and will incline to your verdict without difficulty. But since we are on the subject of thistles, would it not be both useful and quite natural to make one genus of all those that have a pappus of simple hairs and refer to the genus *Cnicus* those that have a plumose pappus, such as your from Lemps, ours from Lautaret, *C. eriophorus, C. tuberosus,* and so on?[249]

As you have not responded to any of the items in my [last] letter sent to Grenoble, either you did not receive it, or you are postponing in order to reply in person. I infer the latter from the ending of your letter, allowing me to look forward to the pleasure of greeting you.

I come to botany as usual. I found a vetch in the Mother of Charity's garden[250] that was unknown to me, which I shall call *Vicia parviflora*. I cannot refer it to any of my authors: elongated peduncles with 5-6 flowers; the corolla very small, the standard light blue, the wings and keel bluish-white; the pods oblong and pubescent; stipules small and linear; leaves oblong and truncate; the tendrils triphyllous. If you go through Gap, take a look at it. I have asked the Mother Superior to allow it to mature as I want to collect seeds from it.

At La Grangette, I found *Juncus filiformis* [L.]. I observed *Carex vulpina* [L.] there, sometimes having only female spikes. The enclosed specimen of *Astragalus* I believe to be *arenarius* L., collected from our subalpine turf, which I had [previously] regarded to be a variety of *Astragalus glaux* [L.]. I believe I perceived a difference between the two.[251]

Anthemis cota [L.] grows at Laric. I am finding proof through shoots of it I dug up and that are vegetating in my garden. I brought back another anthemis with it that I have not

<hr>

[248] *Ligusticum gmelini* Vill., Prosp. Pl. Dauph. 24 (1779) [= *Pleurospermum austriacum* (L.) Hoffm. (1814).

[249] The principle for segregation was sound enough; but thistles with the plumose pappus should have been referred to Miller's *Cirsium*, which neither Chaix nor Villars recognized, not to *Cnicus*.

[250] Marguerite de Colvin (ca. 1717-1786), superior of the Maison de Charité de la ville de Gap from 1750 until her death. See Georges de Manteyer, **Les Origines de Dominique Villars le botaniste** (Gap: L. Jean & Peyrot, 1922), pp. 168-170.

[251] Recognizing only *Astragalus glaux* L., Villars discussed the problem in **Hist. Pl. Dauph.** 3: 459-460 (1787).

yet determined. Monsieur Thouin sent me a species with the name *Anthemis nobilis* [L.] that is neither the latter plant nor our *Anthemis arvensis* [L.]. It could be the one I asked them for in Paris, the one you talked about to me. It appears to be a biennial.

Before responding to the pressing needs of your patients, who will no doubt be demanding of you after your return, divert yourself for two days here in my solitude. The pleasure of your conversation will make me forget the tedium caused by your long absence.[252]

Are you aware that an article strongly criticizing Monsieur Guettard's **Minéralogie du Dauphiné** has already appeared? Monsieur de Sainte Croix, who is here in La Roche, has spread the word. He mentioned the critic's name, but no one was able to repeat it to me. You will certainly not ignore anything concerning this matter.

43
Les Baux, 22 September 1780

If, as I presume, you have made a trip to Marseille, you will have many important things to tell me. I am savoring the pleasure of it in advance; but while awaiting the actual satisfaction of [that pleasure], permit me to open the conversation by letter.

I was in Durbon. They would like very much to see you at that pious house, talking extensively to me about you. The Dom coadjutor had written to you but has not received your response. Knowing your punctuality, it was easy for me to exonerate you by attributing the lapse to the infidelity of the letter carriers. He should have some chemical and medical procedures to communicate that could be of subsequent use to you in your practice. Write to him and give them some hope that you will go to see them at first opportunity.

I found *Serapias longifolia* [L.] [= *Epipactis palustris* (L.) Crantz] that I had not previously seen, and which I recognized from one single flower remaining on a half-eaten plant. I shall go to collect it next year in its season if God so permits. This trip also provided me with *Scirpus acicularis* [L.] [= *Eleocharis acicularis* (L.) Roemer & Schultes (1817)]. From Durbon I went to spend the rest of the week with dear friend, Monsieur Gaude, [at Oze]. He is as fond of you as one could be and has charged me to remember him to you.

Last week, I went to the [Plateau du] Bure to collect seeds for Monsieur Thouin. I was also on Mont Aurouze with a plan to collect *Ranunculus rutaefolius* [L.], either in seed or in live plants to have them go to seed in my garden, so that it can be distributed to our correspondents. But I found it impossible to find even one stalk of it, either because I could not find its location in the mist, or because it had totally disappeared in this season, as I

~

[252] In letter 41, Chaix accepted the proposal for a collecting trip into the Valbonnais and the Valgaudemar. It is probable that they made the trip in the period between 22 June and 22 September 1780 (letters 42 and 43). Villars, **Hist. Pl. Dauph.** 1: xxxix (1786), reports that they brought back, alive, a very rare *Geranium argenteum* L. from Vieux Chaillol that they had never before seen. Later in 1780, he was in the Dévoluy west of Le Noyer and as high as the Grande Tête de l'Obiou. As he regarded his collections as very advanced by then, he subsequently turned his attention increasingly to grasses, mosses, and the lichens of the Alps. He would publish some observation on cryptogams in the **Affiches du Dauphiné**, 16 February 1781.

covered the entire mountain. Next year I shall resume the search after the snow melts.[253] Despite this frustration, I did bring back [*Ranunculus*] *glacialis* [L.] in quantity. If it seeds, I shall make note of its cotyledon.

I am sending you the seeds of *Anthemis*, as they alone mark the difference between species: *Anthemis arvensis* [L.], *semina coronata, striata, laevia, pallida*. *Anthemis cotula* [L.], *semina scabra, nigra*. *Anthemis mixta* [L.] [= *Chamaemelum mixtum* (L.) All. (1785)], *receptaculum maturescens longe conicum, semina minima ovalia*. *Anthemis cota* [L.] [= *Anthemis altissima* L. (1753)], *semina compressa, nigra, magna*. Perhaps in Paris you took *Anthemis cota* for an *A. arvensis* different from ours.

I have been pondering our willows again. I cannot refer your willow from Laffrey either to [*Salix*] *hastata* [L.] or to [*Salix*] *aurita* L. I am temporarily calling it *S. crispa*. The one from Chantaussel does not appear to me distant from *S. cinerea* [L.]. *Salix fragilis* L. is the same as my *S. amerina*, the peasants' *sweet willow*, with 2, 3, or 4 stamens, that I first suspected of being *S. phylicifolia* [L.]. It is quite possible that this is not the *fragilis* of the English. It may be the *amygdalina* of Haller, but not of Linnaeus, as it is quite distant from the *triandra* [L.].

When you come to retouch your work before giving it to the printer, I should like you to have my manuscript before you. You could make use of certain observations that can have escaped you, whether for species or for characters.

Note: [J. A.] Murray, **Systema vegetabilium**, 121 [1774], excludes variety B of *Scabiosa columbaria* L. in **Species Plantarum** and transfers it to *Scabiosa lucidam*. Villars' Dauphiné?[254]

[From the margin at the top of the page]: *Anemone myrrhidifolia* [Vill., **Prosp. Pl. Dauph.** 50 (1779)], planted from seed, has only produced this cotyledon all year without other leaves or stems; and, at the top of its small root, one can see a well-formed hibernaculum for developing the plant next year. I have made this observation on five or six shoots that have come up for me.[255] Draw this cotyledon. It remains to find out whether other anemones have a similar germination. This reminds me of the [germination] of *Berardia* [Vill.].

44

Les Baux, 23 October 1780

For nearly a month, I have been tormented by a chest cold that has greatly weakened me, perhaps brought on by a checked perspiration or by some blast of air during excessive perspiration. I have no appetite and take little but bouillon and herbal teas. My cough has steadily been dry and deep. At night in particular, I have a heaviness on my stomach or chest that gives me a feeling of great uneasiness. I do not believe that I have had any fever, but I fear that all this will degenerate finally into serious consequences.

~

[253] The exceptional interest in finding more specimens of *Ranunculus rutaefolius* L. was explained by Villars, Hist. Pl. Dauph. 3: 741 (1788), who suspected it merited separation from *Ranunculus*. [= *Callianthemum coriandrifolium* Reichenb. (1832)].

[254] *Scabiosa lucida* Vill., **Prosp. Pl. Dauph.** 18 (1779).

[255] Villars, Hist. Pl. Dauph. 3: 727 (1788), said he had segregated this species from *Anemone alpina* L. because of its considerably larger size. The separation did not stand.

Perhaps I need a mild purgative. Please come to me, as you quite know my needs. Your letter reached me last Saturday. I had been without news of you since your departure from here in mid-August. *Veni et vide. Addictissimus tuus famulus.*

45

Les Baux, 10 November 1780

Thanks to the Lord and to the care that you gave me, I am today completely recovered. I have regained my usual appetite and my normal strength. The celery tea managed to get rid of the phlegm that had inconvenienced me for so long, and I did not need the rhubarb that you had also prescribed for me. I was in Gap on Monday to buy some sheep to consume my hay. I picked up your pleasant letter there, proof again of your friendship, and at the same time enclosing the welcome untangling of the difficulty about *Arenaria fasciculata* [L.] and *Alsine mucronata* [L.].[256]

My illness had no effect on my mind [to account for] my gross error. But how to extricate myself from the difficulty, seeing on the one hand the synonyms of Haller and Séguier applied by Linnaeus to *Alsine mucronata*, which I recognized very well from the engraved figure in Séguier's first volume and through Linnaeus' **Mantissa**; and on the other hand not having Haller's description in my manuscript and being unable to recall his figure? This point is now clarified. And, at the same time, I understand that *Alsine tenuifolia* Vaill., in Séguier I, t. 6, f. 2, if seen as pentandrous, is *Alsine mucronata* L.; and, if seen as decandrous, is his *Arenaria tenuifolia*. But take care not to refer *Alsine pentstemon* in Séguier III, 175, no. 5 to it. This synonym relates to *Alsine segetalis* L. as [Séguier] describes the veils cited by Linnaeus from Séguier. This latter plant could also sometimes well be decandrous, as Linnaeus, **Mantissa** 359, is not far from putting it among the *Arenaria*. All that is confirmed by your autopsy of the pentandria and decandria in *Arenaria fasciculata*.[257]

I do not have *Carex capillaris* [L.] from anywhere except [Col de] l'Eychauda. I had left it undetermined in my herbarium; but, on the basis of your thought about it, I do not doubt at all that that is what it is. I am enclosing one example of it with this letter, but do you not have it among the dried plants brought back from there? Its thinness makes the name *capillaris* even better, unlike the drawing in Séguier that represents it as being quite thick. The disposition of the spikes is quite the same, and Séguier's description is more suitable to it than is Haller's no. 1394.

After having compared several species of *Salvia* to yours found in Valence and to the one, received from Monsieur Gouan, that I still cultivate without having been able to settle on it, I finally believe to have found its rightful place: *Salvia pinnata* [L.]. The synonym from Linnaeus, **Goëtt.**, p. 298, certainly describes my plant; it quite relates to the description I had made of it two years ago. The synonym in Morison [**Plantarum historiae universalis**] is *Horminum virginianum* that I have also cultivated and is only a variety of

[256] In Vill., **Hist. Pl. Dauph.** 1: 333 (1786), Chaix will cite Villars about the variable number of stamens in *Arenaria fasciculata* L. [=*Minuartia rubra* (Scop.) McNeill (1963)]. And he would list *Alsine mucronata* L., giving *Alsine tenuifolia* in J. Bauhin, Haller, Séguier, and Linnaeus as possible synonyms. [= *Minuartia mutabilis* Schinz & Thell. ex Becherer (1938)]. Villars did not recognize the genus *Alsine*.
[257] Villars, **Hist. Pl. Dauph.** 3: 629 (1788), asserts that there may be either five or ten stamens.

Salvia pinnata according to Linnaeus, **Mantissa** 318. How is this plant growing in Valence? It can have been brought there and become naturalized: other plants furnish us proof of that possibility. I am also enclosing two leaves of it under this cover.[258]

Here is the synonym from Linnaeus, **Goëtt.**, p. 298, for the true *Salvia clandestina* [L.]: *Salvia foliis pinnato-dentalis aspiris hirsutis verticillis pilosis, floribus apetalis. Herba Salvia verbenacae, foliis Salvia certophyllae pumila*, Murray, **Syst. veget.** 65. None better. Perhaps you do not have this excellent synonym. How different from the [species] above![259]

I had greatly desired to come to see you within your family before the onset of winter, and I should have already done so had it not been for my illness. I still have not given up hope, if God wills. In that case, I shall bring you the plants that I have dried for you and the seeds for Monsieur Thouin with a list of those I would like from him for the coming year. Write to me when you can. I have no greater pleasure than to converse with you by letter, only rarely being able to do it in person.

P.S. I am not counting on going to Gap tomorrow, St. Martin's Day. If your mother can find a dozen pounds of good cheese for me, I would be very grateful to her. I will reimburse her for the expense and could pick it up when I come to see you.

46

Les Baux, 2 January 1781

The favorable news that you have finally received from your brother after awaiting it for so long has given me a share of your joy. Waiting for it like you, I had long since had cause to be troubled. Now our minds can be at ease.[260] Let us bless the Lord who disposes all things at His will. All events that concern us will be for our good if, as they are disposed in the order of His sovereign providence, we receive them accordingly. That God, in His goodness, protects you, favors you, I am assured that it is to His glory and that my wish is the most reasonable.

As you inform me that you are going to Grenoble soon after the first of the year, I hasten to send you my commission. Please have [the following] bound for me as you have it done for yourself:

1) My parish missal that is coming apart: with gilt edges or simply tinted in red, including the ends of the ribbon page-markers, taking care that it is well wrought and well sewn to be sturdy enough for frequent and long use.

2) Allioni, by adding it at the end of my manuscript of the **Hortus parisiensis**.[261]

3) The three volumes by Séguier, taking care to put the plates of the figures in their proper places.

~

[258] Villars lined out *pinnata* and wrote in *lyrata* [L.], a species from eastern North America. *S. pinnata* is native to the eastern Mediterranean region. Neither species appeared in the flora.

[259] *Salvia clandestina* L. = *Salvia verbenaca* L., a polymorphic species with many variants.

[260] This suggests that they had learned belatedly that Jean-François Villar had received the degree of *maître ès arts* from the University of Valence on 21 July 1780. See Manteyer, **Les Origines de Dominique Villars**, p. 138.

[261] This was not a title in Chaix's personal library, indicating, as is suggested here, that he had copied Vaillant's **Botanicon parisiense** (1727).

4) Bergen's **Flora francofurtana** [1750], to be covered with basan.

5) The parish **Hours.**

I have had the shipping cost paid to the messenger. Agree on a price for the whole lot with the binder whom you may pay on my account if you get back the completed work before your departure. If not, you may send me word of the total I owe, and I shall have him paid. The messenger from Gap will bring back my books to me, and I shall pay him again on recovering them. Urge the binder to hurry the work for me because of my need for the missal, and give him the enclosed instructions on a separate sheet of paper. Please give me his address in the event I have to have said work picked up from him.

In Grenoble, please buy for me, or get for me in some other way, Weis' [**Plantae**] **cryptogamicae** [**florae gottengensis** (1770)]. This author pleases me very much, and he should be greatly useful and sufficient to enable me to start learning those plants that are his subject, as I have neither Haller's descriptions nor the figures in Dillenius, [**Historia muscorum**]. It is up to you to decide whether you want your own copy of Weis back that I now have.

Monsieur Céalis has had a journal of [P.-J.] Buchoz sent to me in which my poor piece on belladona was inserted.[262] The announcement of works by the Abbé Soulavie is curious and interesting.[263]

A bit of botany, a matter more to my taste:

1) *Orchis morio* [L.]: discovered in the alpine meadows at La Grangette. *Orchis vero mascula* [L.], *in paludibus ubique, petalis magnis patulis disjunctis, purpureis.*

2) *Periploca monsp.* Gouan *in plantes siccit est Cynanchum monspeliacum* [L.].

3) Your *Arenaria biflora* [L.] . . . is *Arenaria ciliata* [L.]. . . .

4) *Ephedra maritima* Gouan *in notis* is *Ephedra distichya* L. . . .

6) Your *Arabis serpillifolia* [Vill.] is not *Arabis bellidifolia* Cranz, t. 3, f. 3.

7) The *Thlaspi* that Cranz sculps in t. 3 for *alpestre* is *montanum* L. . . .

8) What does Linnaeus do with his *Brassica erucastrum* in **Hortus Cliffortianus?**

P.S. If you have any news concerning the printing of your work, or any new plans about that matter, do not keep me in ignorance about them. Many people of a certain rank always ask me when your work will appear. The comte de La Suche very much wishes to read your dissertation on Champsaur printed in the **Mémoires** of the Société royale de médecine de Paris.[264] I fear that Monsieur de Jussieu is less favorable to your edition than you think. I hope I am mistaken.

47

Les Baux, 30 April 1781

Our epistolary circulation, a bit numbed by the winter frosts, revives with the encouragement of the spring zephyrs. Letters on the 12th and the 17th of this month have given me your valued news. Both one and the other are very precious, but the latter is

~

[262] See Chaix's observation on *Atropa bella-donna* L. in Vill., **Hist. Pl. Dauph.** 2: 498 (1787).

[263] Jean-Louis Giraud, called Soulavie, **Prospectus de l'Histoire naturelle de la France méridionale** (1780). The work, published in Nîmes in 7 volumes, would follow: 1780-1784.

[264] See letter no. 41.

especially so for its contents and for the items that accompanied it.

The Alsatian histories by Mappi [**Historia plantarum alsaticarum**] and by Lindern [**Hortus Alsaticus**], which you have been pleased to present to me, have given me a real pleasure. I add them to the other testimonials of your affection for me. The Murray that you have promised me for reading will immensely please me.[265] Please do not deprive yourself of Weis in my favor inasmuch as you would suffer through not having it. I am having it sent to you herewith in order to complete your section on cryptogams, along with the botanical treatises by Ray, Tournefort, etc., and the [**Classis Cruciformium emendata**] by Crantz [1769] that you had lent to me. I am also sending Monsieur Thouin's list back to you, from which I have written down the things that I can provide for him so that you may do your best to satisfy this ardent correspondent. I have planted his seeds with great care and am watching for results with great anticipation. In addition, the letter from Monsieur Allioni in which I do not see that this famous professor opposes your **Prospectus** very much, although he has given some other names to your plants. I do not believe that he is rejecting very much by calling certain of your plants varieties. It is quite unfortunate that his **Auctarium taurinense** [1762] did not reach you. No doubt you will have to ask him for it again. And, finally, the abbé Pourret's extensive, cordial, and instructive letter! I am overwhelmed that he is paying so much attention to me. In his last letter, he tells me of a shipment of seeds that he has entrusted to Monsieur Séguier to have sent to me by way of the intendance. If you receive them, I am depending upon your usual exactitude.

The abbé [Jean-Joseph] Rossignol, professor at the Collège d'Embrun, has had the kindness to have sent me some printed lessons that he has just published under the title **Botanique élémentaire**. As this small work is only an abridged compilation taken from abbé Rozier's **Elémentaire vétérinaire** on Tournefort's system, I have been surprised that this professor, whom I believe to be a former Jesuit, has assumed the style of author on the frontispiece of this compilation and, in particular, by the final observation in his conclusion where he anticipates those who might criticize his **Botany**, as they had done with his **Geography**, calling them Aristarchus [i.e., severe critics]. In my letter of thanks, I said to him quite frankly that the shafts hurled aginst his piece would not strike him, as it contains nothing of his own. I made some critical observations to him about Tournefort's system that he promotes as an infallible and adequate key for every beginner.[266]

I turn now to our proposed botanical items:

1) *Onopyxis* Barrelier, **Icones** 1116, that [Etienne] Danthoine [of Manosque] wants to refer to our *Carduus nigrescens* [Vill., **Prosp. Pl. Dauph.** 30 (1779)], is nothing other than *Carduus nutans* [L.], given the cited figure, the synonymy, and the description in Barr., **Obs.**, p. 80, no. 911.

2) I confess to my superficiality and my error in treating *Laserpitium pseudonoides* [L.]. The figure and the description attributed to this plant do not conform to our *Peucedanum*

---

[265] Chaix's personal library included the first volume of the 13th edition of Linnaeus' **Systema plantarum**, published by John A. Murray as **Systema vegetabilium** (1774). A.H.-A., Ser. L. l007.

[266] Rossignol was actually headmaster of the Collège d'Embrun, well on his way to being a prolific publisher on many subjects. Later, as an émigré in Turin, he would republish this title in 1805 as **Botanique élémentaire où l'on apprend à connoître les plantes sans le secours d'aucum maître**, evidence that he was untouched by either hostility or irony.

carvifolia [Vill., **Prosp. Pl. Dauph.** 25 (1779)], in particular because of its seeds and its leaves *trifariam divisa, foliola longe petiolata, pinnata.*

3) Take good care to rework the two bitter vetches in question. Linnaeus was wrong, in his **Systema**, to refer Gérard's fourth species with branched stalk and blue flowers, which you have collected in Lower Dauphiné, to his *Orobus angustifolia* [L.]. You have not read Gérard correctly when you tell me that this author suggests *that* species is *Orobus angustifolia* L. He says quite the contrary on p. 493, Observation 1.[267] In his **Species**, Linnaeus characterizes his own plant as having a simple stalk, with yellow flowers, and with subulate stipules: thus different from that of Gérard.[268]

Now I also doubt whether the [*Orobus*] from our meadows around Gap and Les Baux can belong to the Linnaean species, as it differs from it by its stipules that are lanceolate and semi-sagittate; by its numerous fruits, up to 12 and more; by its very white flowers without a hint of any other color, at least when in full bloom. It differs from Gérard's by its simple stalk, .. by stipules that cannot be called subulate, by its nascent spikes on the upper wings, and by its numerous flowers that have no hint of blue. It should be compared with the figure in Gmelin, [**Flora**] **sibirica** 4, t. 5, p. 14. Until then, I shall call it *Orobus pannonicus* Jacq. or *austriacus* Crantz. The flower has a yellowish color only before opening or in fading. I do not know the reason that led Monsieur Gouan to call it *asphodeloides.*[269]

4) I was mistaken regarding the synonym I wanted to refer to our *Polypodium fragrans* [L.] [= *Dryopteris fragrans* (L.) Schott (1834)]. But I hold to what I said about our *Polypodium cristatum* [L.] [= *Dryopteris cristata* (L.) A. Gray (1848)], and I claim that you criticize Weis inappropriately about his synonymy: *frondibus subbipinnatis* in order to describe the lower pinnas, truly tripinnate, only the most upper being pinnatifid. He cites Mappus, **Hist[oria plantarum] Als[aticarum]**. Compare them.[270] I agree that he has not examined the points of fructification in *P. filix-femina* closely enough, which, following you, I observe give the impression of little dolphins, for which reason one cannot be a variety of the other.[271] I have no specimen of *P. cristatum* with the very black points that you have put into its character. Take a look at the enclosed shoots of both plants. Am I mistaken on the matter of *P. cristatum* by not understanding your plant?

5) You are telling me the best things in the world on the question of *Brassica erucastrum* [L.] [*Erucastrum nasturtiifolium* (Poiret) O. E. Schulz (1916)]. I think, as you do, that Linnaeus has confused two species in his **Species**, having in front of me here

∿

[267] Gérard, **Flora gallo-provincialis**, p. 493, no. 4: *Orobus caulo ramoso, foliis quaterno-pinnatis, linearibus, stipulis, semi-sagittatis, subulatis.* Obs. l: *Licet hanc orobo angustifolio* L., **Sp. Plant.** 729, *nullatenus differre agnoverit Celeb. Linnaeus, noluimus tamen eam conjungere, cum haec caulem gerat ramosum, flores caeruleas.* . . .
[268] Originally *Orobus vicioides* Vill., **Prosp. Pl. Dauph.** 41 (1779), Villars would revise the name to be *Orobus angustifolius* Vill., (non L.), **Hist. Pl. Dauph.** 3: 435 (1788), using the Gérard species as a synonym, and declaring them to be affinitive to the Linnaean species of that name. He noted that "Monsieur Chaix thinks it is a *Lathyrus* on the basis of the pistil." [= *Lathyrus filiformis* (Lam.) Gay (1857)].
[269] *Orobus albus* L. fil., Vill., **Hist. Pl. Dauph.** 3:436. [= *Lathyrus pannonicus* (Jacq.) Garcke (1863). A ssp. has been recognized: *asphodeloides* (Gouan) Bässler (1966).
[270] *Polypodium cristatum* L. had appeared in Vill., **Prosp. Pl. Dauph.** 51 (1779), but would become merely a synonym for *Polypodium aristatum* Vill. in **Hist. Pl. Dauph.** 3: 844 (1789). [= *Dryopteris carthusiana* (Vill.) H. P. Fuchs (1958)].
[271] Villars, **Hist. Pl. Dauph.** 3: 845 (1789), would say that the points of fructification are oblong, not round or reniform as in other ferns.

Brassica cheiranthos [Vill.], a true *Brassica*, [= *Rhynchosinapis cheiranthos* (Vill.) Dandy (1957)], and the other one in question in [**Hortus**] **cliffortianus** (a true *Sisymbrium*). Since you agree with me that the latter is a *Sisymbrium*, why do you hesitate to call it *Sisymbrium erucastrum* [Poll.] as I am? That way the chaos is rectified. The one with white flowers does not contradict itself when cultivated. It is always an annual of the smallest constitution. I have called it *Sisymbrium dubium*. It merits the specific rank better than many others.

6) I ask you what are the Linnaean equivalents of *Alsine saxatilis* Lindern t. 2, *Alyssum serpyllifolium ejusd.* t. 5, and *Hieracium pratense ejusd.* t. 7?

7) If you have occasion to see *Orobus tuberosus* L. [= *Lathyrus montanus* Bernh. (1800)] in your woods, transplant several roots of it in your garden for me and for the seeds Monsieur Thouin requests. He is never in our region, and I have only one wretched specimen in my herbarium.

8) Thanks to the convenience of the fair, I am sending two cuttings of *Geranium tabulare* L. to Monsieur Lavallette.

Doctor d'Héralde[272] told me last Thursday in Gap that you would come to see me in Les Baux before your departure for Grenoble. I am counting on you to keep your word. I shall deviate from my solitude as little as possible in order to await you with the exception of next week when I hope to spend a few days with dear Monsieur Gaude. You will not be forgotten in our conversation.

The storekeeper Martin came here to give me a commission for an herbarium. He wanted to give me earnest money, but his word is sufficient for me. I shall talk to him when I meet him again in Gap, as he will have seen the person from Provence who wants the herbarium.

We are on the way to marrying my niece Magdelon to a young resident of my parish. The match pleases me very much. I hope that God will bless our plan. In that case, she will be replaced in my domicile by one of her younger sisters. Thus, it is not only on stage that the scenes are changed. I have to accustom myself to such revolutions. *Tua, pater, providentia gubernat.* . . .

48

Les Baux, 5 October 1781.

Monsieur Gaude, curé d'Oze, whom you esteem, and who is as fond of you as of me, has great confidence in you regarding his infirmities. During the winter of 1780, at the time he was conducting the services for the parish of St.-Marcellin, he was out on a trip in very deep cold, but even so in a sweat because of his movement. A need obliged him to remove the glove from his right hand, damp from perspiration. He felt a quite sharp pain on the back of his hand almost immediately that grew worse when he was saying mass, as he was obliged to keep the hand bare. For about the past eighteen months, this pain has

~

[272] Delafont, the last subdelegate in Gap, called Dr. d'Héralde the most distinguished of the three physicians then residing in Gap and attached to the general hospital, known both as the Hôpital Sainte-Claire and the Hôpital de Charité. D'Héralde was well educated and left his fine library to the City of Gap for its library. See Joseph-Hippolyte Roman, "Mémoire sur l'état de la subdélégation de Gap en 1784, adressé à l'Intendant du Dauphiné [baron de la Bove] par Pierre-Joseph Delafont, subdélégué de Gap. **Bull. Soc. Etudes Hautes-Alpes** 18, 2nd ser. (1899): 257.

not ceased, sometimes less, sometimes more severe. I have just seen him this week in Furmeyer, very tormented at intervals, nearly unable to use that hand, or sleep, or remain at rest, even at table, without becoming greatly stressed. There is a bit of swelling and a warmth somewhat above normal. He begs you (and I join with him) to indicate to him what he can do to cure himself. It is probably a blockage of perspiration on that [bodily] part. We await the earliest possible response from you.

Monsieur [Augustin-Bernard-Hyacinthe] Escallier, curé de Gap, returned the three texts by [Jean-Joseph] Rossignol to me last Monday: **Trigonomètrie, Plan d'étude,** and **Vûes sur l'Eucharíste,** which had been given to him by [Urbain] Rougier, curé de Romette. As I did not find [the text] on **Sensations** included, which you had also taken from my house, I am a little concerned about it. You must see if you still have it, or remember to whom you lent it initially, so that this text does not go astray. I can hardly believe that you would have sent off the three without the latter, as you knew that I was in no hurry to get them back. Monsieur Blanc of Embrun has just promised me two texts that I lack in order to have the entire output of said Monsieur Rossignol.

Monsieur Buchoz, when informing me that he had discontinued his journal two months ago, has sent me the printed list of the different works that he has published or soon will. As he offers me whatever would give me pleasure, in exchange for some of our plants, I shall ask for his **Plantes nouvellement découvertes représentés en gravures** in which he has established some new genera: *Lieutautia* [= *Miconia* Ruiz & Pav.]; *Lassonia* [= *Magnolia* L.]; *Ronnowia* [= *Omphalea* L.]; *Sparmannia* [= *Rehmannia* Libosch.]; and *Trochera* [not recognized]. He will not ask me to pay for the work, of course, as I am not in a position to meet the announced price.[273]

You will think that I have become purely a scribbler! I still hold today to what I wanted to uphold previously, which I repeated to you in my last letter in July.[274] *Iberis umbellata* L., Barr., **Icones** 893, no. 1, seen now in its maturity, does not seem to differ from my plant from Mont Aurouze: *fructus corymbosus persistit.* Take a look at this plant-top collected at Gap in a pot. The following year I stripped seeds from both plants. The fruit on mine[275] also does not take a racemose form. Thus, we must get *Iberis amara* [L.] from elsewhere, a native of Switzerland.

I am also sending you a leaf from an umbel sent by Monsieur Pourret in seed under the dubious name *Athamanta sibirica* [L.] that he had recently received from [P. S.] Pallas of St. Petersburg. We shall see it bear fruit next year. . . .[276] Monsieur Pallas' plant is not *Athamanta condensata* L. [= *Seseli condensatum* (L.) Reichenb. fil. (1867)] that I grew from seeds received from Monsieur Gouan, but that I do not refer to *Libanotis major* Haller, no. 744, as he does in Gouan, **Ill.** 83, t. 26. I shall speak to Monsieur Pourret about it. . . .

I hope that you will come to visit me in my poor solitude before your departure for

<hr>

[273] Pierre-Joseph Buchoz, a physician originally from Metz, published over 300 titles, a fecundity alleged to have been based upon his skill in finding the manuscripts of others that he published as his own. None of his titles appear in the catalogue of Chaix's library.

[274] This suggests a missing letter, perhaps one that never reached Villars.

[275] *Iberis aurosica* Chaix, Pl. Vap. 45 (1785). Chaix in Vill., Hist. Pl. Dauph. 1: 349 (1786), from Mont Aurouze. *I. umbellata* is a Mediterranean species, not alpine.

[276] This Russian species, having also been put in *Libanotis* Hill, is now *Seseli sibericum* (L.) Garcke (1849).

Grenoble, as you promised, and can prolong this advantage for me. Please respond to the items I have set forth here, and do not let anything you know of interest to pass in silence.

P.S. The mercury in the barometer at sea level reads at the 28th degree, and it comes down on the summit of Vieux Chaillol to the 19th degree. You have calculated that said summit is 1834 fathoms above sea level. According to the authors I have, each line represents only 12 fathoms of height or depth. Now, each degree being composed of 12 lines, 9 degrees make 108 lines, which, when multiplied by 12 fathoms, I find comes only to a total of 1296 fathoms. Please tell what error causes my calculation to be so much lower than yours. I shall be very much obliged.[277]

---

[277] A *toise* or fathom = 1.949 meters. Villars' calculation would make the mountain to be 3,578 meters in elevation, Chaix's 2,526 meters. It is calculated today to be 3,163 meters.

*Letters forty-nine to eighty-eight cover four years: beginning in late 1781 with Villars'
move from Le Noyer to Grenoble to accept appointment as Médecin titulaire de l'hôpital
militaire breveté du Roi, ending in late 1785 when the initial volume of* Histoire des
plantes du Dauphiné *was in press.*

*The new position brought Villars an additional 800 livres a year, as he was
privileged to retain the pension of 1000 livres granted by the intendant. This recognized both his continuing work in botany and his need to pay for a second domicile in
Grenoble. Even though Villars continued to give his annual series of botanical lectures,
which would seem to have been a justification for additional remuneration, his relative
wealth seems to have sparked jealousy within the small medical staff. To make matters
more uncomfortable for him, he found a disagreeable rivalry between the physicians and
surgeons of the city and the Brothers of Charity who administered the hospital for the
crown, led by Father Dominique Durand, a surgeon.*[278] *One can only surmise that,
under such circumstances, the continuing peaceful and congenial association with
Father Chaix provided Villars with more than a valued botanical collaboration.
[1781-1785]*

49

Les Baux, 20 November 1781

Having wanted to send my shipments for Paris and for Narbonne with this letter,
which have taken me several days to prepare; and having been occupied moreover by parish
duties and domestic cares, contrary to my real inclinations, I have greatly delayed writing
to you. You have my genuine apologies. I have received the loupe that you have had the
kindness to buy for me for 1 livre 10 sols, the amount I have deducted from your
promissory note.

I have sown *Sesamun orientale* [L.] several times without it ever having come up. I shall
try again with what you received from Monsieur de Saive; but try to get the clover *sulla* of
Malta from him, or *Trifolium coronarium*.

I am sending you my parcels for Monsieur Thouin along with a list of seeds I want
from him for the coming year. Add to it whatever you judge to be appropriate; and have
[the shipment] reach him as fast as possible, by the means that you know, so that the roots
of the *Isopyrum thalictroides* [L.], which he may not have, do not perish from excessive
dryness. In addition, I am sending you my shipment for Monsieur Pourret. Take the same
care with it, entrusting it to the surest and fastest route. You well know the value of these
small items. Finally, I send you the letter from dear friend Monsieur Gaude. Please do not
overlook the testimonial regarding his case for Monsieur Pascal, secretary to the
prosecuting attorney. This matter is so close to his heart that, on St. Martin's Day in Gap,
he repeatedly begged me to approach you.

These commissions out of the way, I am at leisure to talk about botany to you,
succumbing to the attractions this amiable science inspires in me. Thanks to the **Flora
lapponica**, I have taken a good crack at our willows. Perhaps you will protest at some of

∼

[278] "Notice biographique sur Dominique Villars rédigée par lui-même [en 1805]," in Manteyer, **Les Origines
de Dominique Villars**, pp. 212-213.

the results, but I have done nothing without close consideration. Do not condemn me in haste or from prejudice.

1) *Salix triandra* [L.]. *Duritate habitus, foliis minoribus, subtus incanis ab [S.] amygdalina differt.* Haller, no. 1637. Rare in Le Noyer, abundant at La Roche and at Veyne.

2) *Salix amygdalina* [L.].[279] *Ad vimen feritue ut non sit fragilis.* Haller, no. 1636. *Est fragilis anglorum.* Ray, 1420, no. 3. Rare in Le Noyer and Les Baux. Our sweet willow. *Ad [S.] alba proceritatem excrescit. Affinitus ejus cum triandra a Linnaeo.* . . .

3) *Salix fragilis* [L.]. *Non fragilis Anglorum.* Certainly *Salix daphnoides* Vill., [**Prosp. Pl. Dauph.** 51 (1779)], Haller, no. 1638. Fl. lapp. no. 349, t. 8, fig. 6. The black willow in Champsaur and Dévoluy.

4) *Salix purpurea* [L.] (*melius monandra*). Haller, no. 1640. Our white osier.

5) *Salix myrsinites* [L.]. Haller, no. 1645. Fl. lapp. no. 353, t. 8, fig. b, and t. 7, fig. 6. . . . At [Mont] Bayard, in Dévoluy, around alpine springs.

6) *Salix myrtilloides* [L.]. Haller, no. 1646. Fl. lapp. no. 357, t. 8, fig. l and fig. k. Dried specimen from D. Vill. *Ego nunquam legi.*

7) *Salix glauca* [L.]. Fl. lapp. no. 363, t. 8, fig. p and t. 7, fig. 5. Found in 1779 at Lautaret below the hospice along small streams in the Oisans valley. . . .

8) *Salix aurita* [L.]. *S. ulmifolia* Juss. . . . Fl. lapp. no. 369, t. 8, fig. y. Haller, no. 1652. Gouan, Ill. 78. Around Laffrey according to D. Vill.

9) *Salix lanata* [L.]. Haller, no. 1651. Fl. lapp. no. 368, t. 8, fig. x and t. 7, fig. 7; also no. 361, t. 8, fig. n. At Les Baux, La Roche, and Veyne, in swampy areas. *Hybrida* Scop. . . . is a variety.

10) *Salix lapponum* [L.]. Haller, no. 1643. Fl. lapp. no. 366, t. 8, fig. 1. In the valley of Monêtier, around col d'Arsine, in northern rocky places near perpetual snow.

11) *Salix arenaria* [L.]. Haller, no. 1642. Fl. lapp. no. 362, t. 8, fig. o and fig. 9. . . . At Lautaret, col d'Arsine. D. Vill. called it *lapponum*.

12) *Salix rosmarinifolia* [L.]. Haller, no. 1644. On Mont Bayard.

13) My *Salix oleacea*. Bauxiensis Vill. mss. *varietas capreae foliis amygdalinis.* Haller, no. 1653. *Salix oleae sylvestris folio alpina* Rudb. Lapp. [99]. Fl. lapp. no 367, t. 8, fig. u. Varietas 3, *capreae*, **Species plantarum**. Found in the Loubet woods at Les Baux. . . . I believe this to be a hybrid born from [*S.] capreae* [L.] and [*S.] viminalis* [L.].[280]

14) *Salix cinerea* [L.]. Fl. lapp. no. 358. Haller, no. 1655. Found at Chabottes, at Chantoussel. This is not *S. repens* L. . . .

Monsieur Allioni has informed you regarding [his] naming of some plants hitherto unknown. But I think that you can retain your own trivial names that are as good as his by merely citing his as synonyms.

I am reading the **Flora lapponica** closely and with much pleasure. I should like to turn next to **Analecta transalpina**. I would like very much to learn in detail about the duties you have to fulfill on behalf of the intendant. Keep me informed about them and, in your letter, improve upon these items that have been my subject. Do not leave me in ignorance of anything worthy of consideration.

~

[279] Now generally held to be a synonym of *S. triandra* L., which has several subspecies that form hybrids.
[280] This was published as *Salix oleaefolia* Vill., Hist. Pl. Dauph. 3: 784 (1788).

~

In late 1781 and in 1782, Villars was greatly occupied in the establishment of a botanic garden in Grenoble, reducing his field work to brief trips around the city.[281] *He engaged Pierre Liottard to be the head gardener, the position he held for the remainder of his life. Liottard's achievement was remarkable. Barely able to read and write, ignorant of proper spelling, without a knowledge of Latin, he learned his Linnaeus thoroughly, knowing the botanical descriptions virtually by heart. In his eyes, the only people who counted were botanists; and he regarded as criminal anything that could impede the culture of plants. He became furious one day when a student crossed one of his flatbeds. Berriat-Saint-Prix tried to calm him, noting that no damage had been done. Liottard replied that when Rousseau had come into a garden he had stayed in the middle of the walks: that he was a man who respected plants! A simple, natural, independent man, Liottard sought nothing more than the satisfaction of his garden. He was fatally injured in 1796 when a piece of ornamental stone over the entry to the Botanic Garden fell on him.*

50

Les Baux, 21 December 1781

[Addressed in care of Madame Amar, merchant in the Grand Rue, St.-Bonnet.]

If you have kept your promise and returned from Grenoble to spend the Christmas holy days with your family, my letter will find you in your native country. I am ready to remind you of [your] duties there, as permitted me by your absence. And, if time and your affairs should allow you to come as far as here, I would indicate an even greater satisfaction to you. In anticipation, I am rejoicing over your presence with your beloved relatives.

You have said more than once that you are a Thomas (in that regard, today is your festival); and you are right to be so in regard to me who would have misled you so frequently were it not for your prudent, habitual doubt. But your merit [in that quality] is even more striking; for, according to Saint Gregory the Great: *plus nobis Thomae infidelitas ad fidem, quam fides credentium discipulorum profuit.* Even so, I come again to try your patience.[282]

[*Salix*] *folia elliptica-lanceolate, integerrima, utrinque tomentosa,* Haller, no. 1643. *Folia lanceolate, integerrima, utrinque hirsuta, superius villis albis, confertis adspersa, subtus crassissimo vellere tecta,* **Fl. lapp.**, no. 366, t. 8, fig. t. *Salix lapponum* [L.] , **Sp. pl.**, 1447, is indeed our cottony willow in the northern High Alps, which I could name *bombycina* because of the thick down on its leaves; and which Linnaeus called *lapponum*, because the Lapps make their fires entirely of it when they are in the mountain pastures during summers. **Fl. lapp.**, p. 293. If Haller does not see it in Linnaeus, Linnaeus saw it in Haller.

. . .

I know that you want to call my *Salix arenaria* L. [by the name] *Salix lapponum* L.; but, in doing so, you overlook the definitions and the illustrations. *Folia superna glabre, inferne sericea; id sericum matura amittunt,* Haller, no. 1642. *Folia ovata, acuta, supra subvillosa, subtus tomentosa,* **Sp. pl.** 1447, no. 23. **Fl. lapp.**, no. 362, t. 8, fig. 1019. Its leaves are

~

---

[281] Villars, **Hist. Pl. Dauph.** 3: xiii (1788).

[282] On this date, Chaix had not yet received Villars' reaction to his long list of willows, but is here anticipating disagreement as he returns to that genus.

uniformly oval, pointed at their extremity, nearly glabrous above. Whereas those of my *Salix lapponum* are always lanceolate, never oval, and always covered on both sides.

I am equally unconvinced that your diandrous willow from Lautaret, close to *Salix purpurea* L. in leaf, is *Salix glauca* L. The description and the figures in **Fl. lapp.**, t. 1, fig. 5, and t. 18, fig. p., show me leaves *ovato-oblongis*; yet, yours from the royal garden has leaves *acuminato-subrotundis.*[283]

Like you, I would believe that *Salix arabuscula* L. is only a low variety of *S. fragilis* from the Alps; but *folia subdiaphona* in **Sp. pl.**, and *fere pellucida* in **Fl. lapp.**, no. 352, t. 8, fig. 2, and no. 356, troubles me a bit.

If you are not to visit me, honor me with your response by letter, apprising me of everything new and interesting. The worthies of La Roche would be delighted to meet you. They have often spoken to me about you, most recently the curé, my respectable confrere and neighbor.

51

Les Baux, 19 February 1782

The silence has been long! Since the 28th of September, date of your last letter, I have had no news of you other than what I have had the consolation to read in the public notices.[284] At the moment I am living for your epistle, always a livelier expression of your sentiments. The complications of the greater world, which you have depicted for me, are not unknown to me, although without having had much experience in it. I remember from my Horace seeking the delights of the retreat and satirizing urban manners. In our time, Linnaeus congratulated the Lapp sleeping on the branches of *Betula nana* or between cushions of *Polytrichum vulgare*, living in the woods, near the snow, content in the possession of a few reindeer. *O felix Lappo - - -* (**Fl. lapp.**, p. 269). *O sancta innocentia - -* -(ibid., p. 270). If I cannot claim my hearth to be the asylum of innocence, at least I can call it the seat of tranquility in comparison to the turmoil, the dissimulation, the intrigues of the outside world. . . .

Your thoughts on the duties of physicians, occasioned by the abbé Blanc's brochure, are excellent.[285] Your reasoning about mucilage is learned, but it has appeared to me a bit abstract. You ought to have been able to develop your thought more clearly. . . . I have noticed several mistakes by Monsieur Blanc, in particular regarding his genip.[286] He has thanked me, saying that he will add a page of corrections to the copies still under his control—most of them. You will find herein a letter from him that I have retained for some time being uncertain where to find you.[287]

⁓

[283] See *Salix caesia* Vill., **Hist. Pl. Dauph.** 3: 768 (1788), a new species from Lautaret.

[284] Probably **Les Affiches du Dauphiné**.

[285] Abbé Laurent Blanc, **Essai sur les propriétés des plantes** (Embrun: Moyse, 1781); and Villars, "Analyse de l'Essai sur les propriété des plantes," **Affiches du Dauphiné**, 28 December 1781.

[286] *Genipa* L. (Rubiaceae), trees yielding the honeyberry.

[287] Laurent Blanc, born in Caléyère near Embrun, is now obscure. He taught philosophy at the Collège d'Embrun and devoted his leisure to botany. His 30-page herbal was meant to be a serious contribution to the public by providing information on both the medicinal and nutritional properties of plants. It reads delightfully today and would be a challenge to translate faithfully to retain the charm of its genre.

Monsieur Thouin, who has written to me, applauded very much the founding of your botanical garden. He charges me to mention him to you and to tell you that he does not forget you in regard to Monsieur de Jussieu to whom he has remitted your letter. He will send me some seeds, very likely through you. Do you not have any news from the abbé Pourret? I am quite concerned, for, at the time of his last letter, both he and his closest relatives were convalescent. Tell me something about that situation.

I am sending you a little essay on my rustic amusements to be given to Monsieur Giroud if he will consider its publication. You may withhold the short notice written separately from him . . . if you think it might offend him.

Jean Thomé, one of my parishioners, qualified as poor, has a case pending before the court for prosecution by the attorney-general, Monsieur Jacquemet being the prosecutor. Please have the kindness to mention the matter to Monsieur Pascal, secretary to the attorney-general, and to said Monsieur Jacquemet, prosecutor in Grenoble.

You argue. That you have always seen the leaves of your *Salix lapponum* consistently covered on both sides. You are led to believe that Linnaeus, the most precise of botanists, did not see [that character], as he did not mention it.

I answer. You are right if you only consult **Species plantarum** where the *utrinque* is omitted, without my knowing from where it comes. But, if you recall the descriptions in **Flora Lapponica**, where the author treats this willow, having it before him, and which I copied for you in my last letter, and [assuming] you have not suspected me of deception, you will be convinced to the contrary. In the meantime, while you read in that very book, and while you examine the figure with your own eyes, I am here transcribing from [Fl. lapp.], no. 366, tab. 8, fig. t., for a second time. . . . By supposing that Haller did not see *Salix lapponum* L. in nature, but rendered his judgment based upon **Species plantarum** without consulting **Flora Lapponica**, he would have had reason to make his [Hist.], no. 1643 inclusive of our willow; to render it, I am saying, as foreign to Linnaeus. Moreover, the *Salix foliis integris utrinque hirsutis lanceolatis* Haller, **Helv.**[288] 155, t. 5, f. 2, which Linnaeus excludes from *Salix arenaria* in **Mantissa** p. 499, is it really separate from the one in Haller, **Helv.** 155, t.5, cited by Linnaeus in *Salix lapponum*, since the terms in Haller are the same, *verbo ad verbum* as those in **Flora lapponica**? *Quid ad haec, Thoma.*[289]

I thank you indeed for sending me the intendant's ordinance. I shall make it known to my friends.

The month of February here is making us pay with usury for the indulgences of the months of December and January. You endeavor to heal the body; the care of that part of man is not to be neglected; but there comes a moment when it must succumb. The treatment of the soul has eternal consequences—but should I want to moralize at the end of my letter? With the fondest and firmest devotion . . . .

∼

[288] Haller's earlier title, **Enumerato methodica stirpium Helvetiae indigenarum** (1742).
[289] Villars, believing that Linnaeus had never seen the willow in question, would publish his own *Salix lapponum* Vill., **Hist. Pl. Dauph.** 3: 780 (1788). Verlot, **Cat. Pl. Dauph.** (1872), p. 308, believed that Villars had one of two varieties of *S. lapponum* L., the other being *S. helvetica* Vill., **Hist. Pl. Dauph.** 3: 783 (1788).

52

Les Baux, 12 March 1782

The parcel of seeds from Monsieur Pourret is not as copious as this zealous correspondent had intended had he been permitted more time. He is now quite recovered as are his relatives; but he is very much occupied by a matter concerning his family interest and is obliged to go to Toulouse in their defense. I do not know whether he approves all the names of the dried plants I sent to him; but he said to me that, having only been able to give them a quick glance, he finds in the lot "a matter for discussion." You quite believe that the names I gave to them are not different from yours. Having sent both the plant and the seeds of *Cnicus erisithales* [L.], I find among his packets one containing his *Cnicus erisithales*. I shall take good care of it. I have just sown his *Illecebrum paronychia* [L.] [= *Paronychia argentea* Lam. (1778)]. In addition, he promised to send me dried specimens along with *I. capitatum* [L.] [= *Paronychia capitatum* (L.) Lam. (1778)] and *I. verticillatum* [L.] upon his return to Narbonne after Easter when he will send me everything he had intended for me. You will have any part of it you wish.

Under this cover you will find 115 small packets of seeds for your garden in Grenoble. Some of them are a bit old and perhaps will rebel against coming up; others will respond to your expectations. But Monsieur Liottard has such an aversion to me that, if he learns that they came from me, he may prefer to drown them in the pond of the Porte de Bonne than to put them in their place in the garden. Since poor Gay, I have not sown white poppy.[290] I no longer have any of it, and I doubt whether Monsieur Delafont has any of it. Have some sent from Lyon and reserve a bit of it for me.

The abbé Blanc charges me to send you this copy of his essay in which he has added some *corrigenda* and an observation not in the first copies.

We already knew about the capture of Fort St.-Philippe. The confirmation in your printed leaflet allows us no more doubt about it.

My *Salix lapponum*. I have not convinced you. But you have also not made any progress with me. So, I renew the debate.

You argue: You believe, therefore, that *utrique hirsutis* is the same thing as *utrique tomentosis*. I do not believe that.

I answer: I, too, am quite far from believing that, and you wrongly attribute that imputation to me. It is you who are not taking *tomentosum* in the Linnaean sense: *folium tomentosum villis intertextis vix conspicuis tegitur, ergo saepius albidum,* **Philosophia botanica,** p. 49. Linnaeus, who gives this epithet to *Salix arenaria, rosmarinifolia, caprea, viminalis,* . . has refrained from imposing it on *Salix lapponum*, which exceeds all other willows in its long down. As the one most covered with hair, it is the only one that fits the epithet *hirsutum velu,* covered with hair, with bristling hair. Please understand the terms of our author and do not quibble.

You argue: You have always seen the leaves of your *Salix lapponum* consistently covered on both sides.

I respond: I have never seen them as such if, by *covered,* you mean *tomentosa*: simply

⤳

---

[290] Probably *Papaver somniferum* L. Chaix's estate, AH-A, Série F 676, indicates that Pierre Gay had married Chaix's niece, Marianne Chaix; and that he borrowed money from Chaix on 11 February 1787.

tight-cottony. But I have seen them bristling with hair on both sides, *utrique hirsuta.*

You argue: **Flora lapponica**, by giving a difference between the top and the underside of the leaf, condemns instead of justifies you.

I answer: The slowest scholar in the fifth form will explain that *utrinque hirsuta* means villous on both sides. But I believe it means *superius villis albis confertis adspersa, subtus crassissimo vellere albo tecta*, **Flora lapponica**, p. 292. Let us speak no more about the texture of the leaves. You agree that your *Salix lapponum* has smaller leaves than mine, more obtuse (*ovatis acutis Salix arenaria*). But, on the contrary, mine has lanceolate [leaves], (*lanceolatis S. lapponum*) *utrinque sensim versus extremitatem attenuatum*, **Philosophia botanica**, p. 46, and half again larger. The figures in **Flora lapponica** back me up. I am sending you in this regard one leaf from each of these willows. See which one merits being called *hirsuta, lanceolata*, and which one *ovata, supra subvillosa*. Figure *t* in **Flora lapponica** gives a length of two inches and a quarter for *Salix lapponum*. Does your alleged willow have leaves of that length? *Quam bene salix lapponum mea!* Consequently, I am holding to my own opinion.

The truth is sometimes so enveloped in clouds that one finds it only after much research. *Every man is given to lying.* **Psalms** 115. If I took pleasure in having found [the truth] regarding the subject of our dispute, I would allow myself to be guilty of the greatest rashness. I sense the weight of my ignorance; I weep in the depths of my affliction. . . . Take the above refutation in good spirit. Only good friends, such as we are, hide nothing from one another, as is our practice. My heart will always be open to you, however sterile and unworthy it may be. With this pledge, I am, Monsieur and very dear friend, your very humble and very obedient servant.

P.S. The world is transitory: an eternal dwelling is permanent. *For we have not here an abiding city, but we seek after the city which is to come.* **Hebrews** 13. 14.

53

Les Baux, 15 March 1782

I foresee that the reasons I recently cited in my last letter to you, included in a parcel of seeds I sent to you, in regard to my *Salix lapponum*, will have made no more impression upon your mind than those I exposed to you previously. Perhaps your reasons will also not convince me. Since you now have in your hands a leaf from each of the willows in contention, which I sent with said letter, you would do well to send them in a letter to Monsieur de Jussieu, asking him to pronounce on our disagreement. I hope that he would give us his opinion. That is the way the two shepherds in an Eclogue of Vergil would choose a judge for their bucolic songs. The matter is quite close to my heart.

I presume that the holy days of Easter will bring you back to [Le Noyer], and that perhaps you might be willing to extend your steps up to my poor abode. I beg you to do so if at all possible. If, beforehand, you should have a word of response for me, write to me at once through the post. I shall take care to ask the postal director about letters. . . .

P.S. If you come, do not forget to bring me two or three cuttings of *Salix aurita* [L.] from Laffrey. I am nearly out of it. I would ask Monsieur Liottard for *Myagrum perfoliatum* L. and for his *Laserpitium prutenicum* L., but ———. Perhaps you will find something in my garden to transport to the one in Grenoble.

54

Les Baux, 19 April 1782

Everything that came from the hand of Monsieur de Linné has such an influence on me that, not content merely to have read it, I should always like to own his work. But the modesty of my means is a barrier. In order to respond to your opinions and to satisfy myself in my desire to be useful to you in some way, I am sending you not only the treatise on the culture of plants by this celebrated author, but the one on [vegetal] economy drawn from the three kingdoms.[291] The former will be of use to you and to your gardener for general principles; the latter can provide you very interesting material quite suitable for presentation during your introductory lectures at the beginning of your botanical courses. You will find my translation very simple, as I would have to fault myself had I deviated from the wording in the original whose strength and brevity you know.

Please tell me whether you have found my shipment of seeds for your garden and whether Monsieur Liottard has not rejected them all. I am adding another seven or eight packets to this mailing. I am already yearning [for the seeds] promised to me by Messieurs Thouin and Pourret even though the season in this country is hardly favorable for germination.

Since Monsieur [J.-L.-Alexandre] Giroud sends you his notices, give me the pleasure of letting me read the one in which he has published my observation on the cattle fly.[292] Several of my friends have asked me for it, but I have not yet received it from anyone. I shall take care to keep it for you.

A girl from Manteyer, about 24 years of age, having consulted Messieurs Jean and Patras about an ailment afflicting her but without getting any relief, has sought me out on the matter. As the condition is beyond my powers and knowledge, I have not prescribed anything for her, but have assured her that I would ask you to grant her your help. For about twenty months she has been afflicted by a hoarseness, which cuts her speech such that she can only speak in a low voice, and only with effort. When she wants to cough, she can only wheeze through her windpipe; and a bit of fatigue caused, for example, by a climb a bit steep, impedes her respiration very much. She appears to me to be sound of lung, her stomach also being good. I do not believe that this is an effect of the vapors, as there is no respite from her discomfort.[293] I presume it is a contraction of the glottis or an inflammation of the epiglottis, brought on by a sudden chill after warmth. She has not been able to tell me any cause for it except for having drunk from a spring at harvest time, the moment when the ailment began. She does not remember having strained her voice nor having experienced any violent fear. See what you can tell me for her relief. The cure that you could obtain for her would lend honor to your knowledge and attest to your charity.

Try to get back your copy of **Systema naturae** from Monsieur Donnet. I should like to spend a little time on the animal kingdom, in particular that which concerns our province.

~

[291] See letter no. 55.

[292] Dominique Chaix, "Observation d'insectologie, sur les tumeurs ou varus des bêtes à cornes, occasionnées par l'insecte appelé par Linnaeus, *Aestrum Bovinum,*" **Affiches du Dauphiné,** 8 March 1782.

[293] Both hysteria and hypochondria were traditionally attributed to vapors emanating from the uterus and rising to the brain, a *periodic* phenomenon.

Have *Uvalaria [amplexifolia* L.], *Scheuchzeria [palustris* L.], *Seseli elatum* [L.], *Salix pentandra* [L.], and *Impatiens noli-tangere* [L.] grown in your garden? I offer you everthing I can for its variety. If you do me the honor of coming to see me around mid-May, you will possibly be somewhat more satisfied with my garden then during your last trip here.

Remember my commission for two silver servies.

55

Les Baux, 14 June 1782

Monsieur Garnier, apparently waiting to give me the deposit and letter personally you had addressed to him, so as not to entrust them to persons unknown, did not deliver your letter until the 5th of this month; and I was unable to have the two silver services retrieved until last Saturday. That explains my tardy response. I am gratefully crediting you with an expenditure of 60 livres. Please accept my thanks for it until our next meeting. The willows from Laffrey were brought to me Trinity Sunday, but very sick from dessication. I fear for their recovery. Earlier, I had also received your seeds of *Aconitum pyrenaicum* [L.], *A. lycoctonum* [L.], and *A. variegatum* [L.]. You no doubt believe that the last two of these species are different from the ones we take by those names. God permitting, I shall inform myself on the subject.

Should you not have received [by now] two shipments I made to you toward the end of April under the intendant's cover? You have not yet uttered a single word to me about them. I have grounds for supposing they have gone astray. The first shipment contained a good number of seeds for your garden. The second was bringing you my translation into French of two articles extracted from the Acts of Stockholm[294] concerning the culture of plants and the utility of botany that you asked for. I also asked you in the latter to indicate a remedy to me for relieving a virtuous girl from Manteyer who has suffered a loss of voice for nearly two years. . . . She often asks me for your response.

Your brother has promised, upon leaving his place [at the University], to cede it to a student in Valence who is from Oze, and in whom Monsieur Gaude is very much interested. This worthy curé, my very close friend, who is extremely loyal to you, and who I visited during the week of Pentecost, has asked me to beg you very earnestly to write to your brother not to forget his promise. That is to say, to turn over his position to said student whom he knows, and who he believes to be prepared to replace him, since he marked him for the position both to favor him and to give Monsieur Gaude pleasure. Assure him of my affection. When I next go to Gap, I will see Monsieur de St. Genis [vibailli de Gap] or one of the grand vicars.

I did a little botany this week, given the opportunity when Monsieur Deleuze, who arrived in Manteyer on Sunday, came here to see me three or four times.[295] Besides Gérard's *Geranium pyrenaicum* (that we have everywhere here),[296] he knows another *Geranium pyrenaicum*, pentandrous, around Sisteron, in Murray, [Linnaeus] **Systema** [**vegetabilium**]. Do you have it in Grenoble?

~

[294] Presumably the **Acta societatis regiae scientiarum Upsaliensis**, published in Stockholm from 1744-1751.
[295] Jean-Philippe-François Deleuze (1753-1835) of Valernes, just north of Sisteron.
[296] Louis Gérard's *Geranium pyrenaicum*, **Flora gallo-provincialis**, p. 434, no. 12, referred to N. L. Burman, **Spec. Bot. Geran. 27 (1759)**, which is still the correct reference.

For several years I had *Viola mirabilis* L. in my garden but never examined it until now. It fills the preserve of Manteyer from where I had perhaps brought it back. Without doubt it is not a new find for you.

Having reviewed the descriptions by Linnaeus and Haller, as well as the figures of Chabrey,[297] with Monsieur Deleuze in regard to the two *Crepis, tectorum* and *biennis*, we do not doubt (in accord with my earlier opinion) that the one with amplexicaul cauline leaves, very elongate at their extremity, corollas entirely yellow, the root shallow, is *Crepis tectorum* [L.]; and the one with a stem purpurescent at base, covered with tiny purplish prickles rendering it scabrous; with leaves lyrate, pinnate, or half-pinnate, the terminal pinna short and triangular; whose exterior florets are red, and whose root descends quite far, is *Crepis biennis* [L.]. Please transcribe for me, from Monsieur Liottard's copy of Linnaeus, **Flora Suecica**, those species that approach these two plants when you do me the honor of responding to me. I notice every day in our meadows that the latter plant, being the first to open, closes by noon; and that the former, opening a little later, closes between three and four o'clock. It only remains for me to sow them separately to see which is annual and which is biennial. We would be criticized with good reason if, through our mistake, we gave a false lead on two plants as common as these and that our authors have described for us with care and at length.

If you have a leaf of *Salvia verbenaca* [L.], send it to me in your letter. I sense that I do not have this species among the number I am cultivating, even though some seeds with that label have been sent to me. Do not fail to plant the following in your garden: *Scheuchzeria palustris* [L.], *Impatiens noli-tangere* [L.], *Uvularia amplexi[folia]* [L.] [= *Streptopus amplexifolius* (L.) DC.], and the female *Rhodiola rosea* [L.]. Please collect for me the seeds of *Drosera* [L.], *Antirrhinum cymbalaria* [L.], and *Antirrhinum bellidifolium* [L.] [= *Anarrhinum bellidifolium* (L.) Willd.] etc.

Your work on the plants in the region around Grenoble must be such that it enhances your history of all the plants in the province. But you must take care not to allow the improvement of the former to work to the detriment of the latter; that is to say, that one does not reduce the value of the other. If you have to have Latin synonyms put into your composition, have me see them in advance. You are not always correct in Latin diction, not having learned Latin fundamentals. In giving you this opinion, I am more than convinced that any other grammarian, who is a bit of a botanist, can substitute for me very satisfactorily. I only mean to offer you what little I can.

P.S. It is clearer than the day that I am consistent with Haller and Linnaeus in regard to the two *Crepis* in question, although the style is less brown in my *Crepis tectorum*, the only character favoring me less. The latter must be an annual, as I see it in our fields being worked and sown with *transailles*.[298] The other one never appears, because it would have been smothered in the furrows; its root is fusiform; and Haller relates it to *Hieracium montanum rapifolium* C. Bauhin, **Prodr.**, p. 65. Are you possibly reverting to your first opinion?

~

297 D. Chabrey [Chabraeus], **Stirpium icones** (1677), the figures for J. Bauhin, **Historia plantarum universalis** (1650-1651).

298 The name used in the Midi for the sowing done between the spring harvest and the fall planting.

About my *Crepis tectorum*: *Cichorium pratense luteum laevins*, C. Bauhin [**Pinax**], no. 126. *Hujus flores, capitula et sementa omnium quae in Anglia sponte provenuit hieraciorum minima sunt, si hyoserim excipias, caulis autem major et firmior reliquit. In pascuis frequens occurrit et aestate floret*, Ray, **Hist.** 234.

About my *Crepis biennis*: *Hieracium erucaefolium hirsutum*, J. Bauhin, [**Pinax**], no. 127. I am quite sorry not to have this synonym in my Ray . . . . Read your J. Bauhin.

Having *Carduus medius* Gouan before me in all its vigor, I must ask you to tell me between which thistles Monsieur puts the latter species. I have not copied his description in my *adversaria*.

I am boring you to death with my verbiage. Running out of paper, I am obliged to close down. Adieu.

56

Les Baux, 4 July 1782

Monsieur de St. Genis, with the best will in the world, is disposed to do--for you and your brother--everything to please you. When your brother wants to submit to the tonsure, he has only to present himself to Monsieur de St. Genis. The bishop will preside on whatever day that may be; and, if you desire, he would confer the four minor orders upon him at the same time, a day of double offices. But we believe that it would be more appropriate to separate the tonsure from the four minor orders by issuing dimissory letters so that he can receive [the minor orders] immediately along with the subdeaconship in the seminary on the day of ordination. In that way, his advancement will be more rapid. Monsieur de St. Genis is quite delighted that [your brother] is going to the seminary in Viviers where he, himself, went. He will recommend him to the superior with whom he has ties by letter. He also told me that if you have not preferred the Séminaire de Viviers for very plausible reasons, he would grant you a half-pension for the seminary in Gap; or even a full pension if necessary, and if the bishop, who requires two years of seminary for the priesthood, should make an exception to his rule in your brother's case upon reception of attestations from the superiors. . . .

I should like very much to believe all the reasons you cite for why my alleged *Crepis biennis* is really *Crepis dioscoridis* L., and that my alleged *Crepis tectorum* is really *Crepis biennis* L.; but I must point out to you that both of them are biennials. I see the plants, born in our meadows this year, that will only flower next year. They are nearly equal in height: 2 feet. The first one, more glabrous, more fistular ; root straighter and deeper; stem always red below and up to the nodes; calices larger, mealy; petals red below; pappus stipitate on the seeds.[299] The second one never has the height here that Linnaeus and Haller give it; the calices smaller; petals entirely yellow without the least hint of red; styles all yellow; pappus sessile. How can Monsieur Gouan only make varieties out of our *Crepis dioscoridis* above (so large in all its parts) and of our *Crepis virens* [L.] from Valgaudemar (so small in all its parts)? I cannot conceive of it if he has seen these two plants. Following you, I have to admit that *Crepis tectorum* is unknown to me. . . .

∾

[299] In the previous letter, Chaix had described *Crepis tectorum* as an annual. It is possible that he describes here *Crepis taraxacifolia* Thuill.

The small willow from the marshes of Bayard (where I was yesterday) can only be referred to *Salix rosmarinifolia* L., as was our earlier opinion, a variety somewhat different from Haller's no. 1644. It is *Salix pumila angustifolia prona parte cinerea* J. Bauhin. . . . *Salix pumila linifolia incana* C. Bauhin. . . . *Salix humilis repens angustifolia* Lobel. Ray 1422, no. 10. I am sending two branches of it from here.

Do you not take that species one sees on the eastern slope of Bayard, between Chauvet and La Croix, for *Ononis antiquorum* [L.]? Quite different from the [*Ononis*] *spinosa* [L.] of our fields and along their edges, it is--even during its tender growing period--quite bristling with long spines, so that cattle do not browse on it.[300] Yesterday, it was not yet in flower, whereas the other species had already flowered very well. When you come to see me, remember to bring me some of it in flower.

Monsieur Deleuze well knows our *Geranium pyrenaicum* in Gérard with tufts of hair at the base of the petals, which is also Hudson's plant according to Murray, **Systema vegitabilium**, p. 514. In order for us to see its difference from the first *Geranium pyrenaicum*, pentandre, Murray [ibid.], p. 513, no. 59, which he has also seen in Sisteron, he has promised me some seeds of it.

Have you *Androsace villosa* L. in your collections? I do not have it, as I took *A. carnea* [L.] for it, which I recently collected at La Grangette. I have *A. lactea* [L.] from Orcières.

I shall always receive your news with renewed pleasure along with whatever you give me concerning affairs of state. You know that my position hardly permits me to learn about them.

57
Les Baux, 2 August 1782

I had a visit last week from our student in Valence. I do not yet know what arrangement he will have made with Monsieur de Saint Genis, but I do not doubt that he has been welcomed in accord with his hopes. He will probably indicate to you what has been settled between them. I have been surprised that the hope held out to you for a half-scholarship at the seminary in Viviers has gone up in smoke. Your candidate hinted to me that he would prefer the seminary in Valence, because he knows it already. As for me, I would prefer him to be at Viviers, and I would presume that Monsieur de St. Genis will also have suggested that [choice] for him. Let him take care not to sign any public document or to submit to tonsure before the end the year's study at his university. That could have negative effects upon his degrees. *Omnia secundum ordinem fiant.* I **Corinthians** 14.

I cannot discuss theology with you; you would waste your time talking to me about medicine. I turn to botany that is our common ground and to which we are equally attracted.

1) You request of me in your last letter of July 12th, after having given your opinion regarding the dwarf willow from Bayard, still more information about it. I give it very willingly. That willow is, in truth, *Salix repens* L. *Salix rosmarinifolia* [L.] is not on that mountain. A dried specimen, which I had previously collected with you, with narrower

~

[300] The popular name for this genus was *l'arrête-boeuf.* But in England it is called the rest-harrow, because its thickly-matted roots resist the harrow. See *Ononis spinosa* L. ssp. *antiquorum* (L.) Arcangeli, a legume.

leaves, led me to take it for *rosmarinifolia*. But I am now better informed and believe it is quite the same willow as *S. repens*. That is sure.[301]

2) Monsieur Liottard's mallow, which you have indicated to me as *unknown*, is without doubt *Lavatera trimestris* L. *Malva stillata* J. Bauhin. Ray, **Hist.** 598, no. 2. Three years ago I had cultivated the white-flowered variety. The one from the said Liottard is more commonly flesh-colored, but has also produced the white flower for me. Since then, you would have determined it as well as I had you taken the trouble to examine it.

3) Misled by Monsieur Gouan, you have made *Crepis dioscoridis* L. a variety of *Crepis virens* [L.] in your manuscript. Pay attention to correcting it according to your recent reflections.[302]

4) Our catmint [or catnip] with leaves lanceolate, cordate at base, notably dentate, common in our mountains and called *chatuegne* by our countrymen, I raised this year from seed sent by Monsieur Thouin labeled *Nepeta nepetella* L., and I believe it to be such; without distinguishing it from the one in Gérard, **Fl. gallopr.** 274, no. 2, it is *Cataria tenuifolia* Clus.; *Mentha cataria minor* C. Bauhin; Ray 548, no. 3; Haller, no. 247; Allioni, **Auct. Pedem.** I do not know why Linnaeus did not give any synonyms. The "red when young" in Linnaeus, the violet in Haller, must not deflect us. I have seen it fading to violet. Clusius gives it "glistening-white flowers."

5) Some seeds from Paris labeled *Anthemis altissima* L. have produced only *Anthemis arvensis* L. for me, the seeds ribbed: proof that the latter plant is in the Paris garden, perhaps without being well known there. I had already received it from Paris once before.

6) Now that I am growing *Iberis umbellata* [L.], I am very confident that the one from Mont Aurouze is *Iberis amara* L.

7) *Illecebrum paronychia* L. [= *Paronychia argentea* Lam. (1778)] has been born to me from seeds from Monsieur Pourret. The figure in Barrelier, **Icones** 726, depicts this species very well. Our plant from the Alps at Champolléon and from the Briançonnais, not being referable to any [species] in Linnaeus, I am calling *Illecebrum alpinum*. Here is its description along with the difference I see with *I. paronychia: Perenne, fruticosum, procumbens; folia elliptica, glabra, congesta, sessilia; flores in axillis, foliorum crebri, congeste* etc.[303] [*I.*] *paronychia vero an perennis? herbacea; folia oblongo-linearia, laxa, distantia, subpetiolata. Caetera sempus dabit.*

8) The affinity reminds me of our *Herniaria* from the Briançonnais. I still believe it to be *H. fruticosa* L. Gérard, **Gallo-prov.**, p. 336, var. 2. Do not let this pass in silence.[304]

9) *Valantia granulata* Juss., *V. hispida* L., *V. muralis* L. are born to me from seeds sent from Paris. *V. glabra* [Vill., non L.] we collected near Montgenèvre [= *Cruciata glabra* (L.) Ehrend. (1958)]. Try to get me, either in seed or plant, the *Valantia aparina* [L.] with *Androsace villosa* [L.] that you promised me.

10) Among the great number of sages I have already cultivated, I have not yet been

∽

---

[301] Both species remain good, but are related within the *Salix repens* group. Villars, **Hist. Pl. Dauph.** 3: 767 (1788), would give Bayard as a location for *S. repens* L.

[302] The error was not corrected in Villars, **Hist. Pl. Dauph.** 3: 142 (1788).

[303] *Illecebrum alpinum* Chaix, **Pl. Vap.** 20 (1785); Chaix in Vill., **Hist. Pl. Dauph.** 1: 324 (1786). [Probably = *Paronychia kapela* (Hacq.) Kerner ssp. *kapela* (1869)].

[304] This became *Herniaria alpina* Chaix, **Pl. Vap.** 75 (1785); Chaix in Vill., **Hist. Pl. Dauph.** 1: 379 (1786).

able to distinguish *Salvia verbenaca* [L.]. Do you have it in Grenoble? I would ask you for its seed, but Monsieur Deleuze has promised me some of it. Here are the species I have: *Salvia pyrenaica* [L.], whose leaf you earlier drew for me and that is completely clear. Three times I have been sent a species from Paris, labeled *S. haematodes* L, that I believe to be *Sclarea syricae, fore albo* Tourn., [**Institutiones rei herbariae**], 179. *Horminum silvestre majus, flore albo, integris foliis* Barrelier, **Icones** 187, which I am calling *Salvia roseo-alba*, and which certainly is not *S. haematodes*.[305]

Pardon, my dear doctor, so much verbiage! The womb of a solitude such as mine can only be fruitful in reveries. If any one of them appears judicious to you, your suffrage will be all the prize I anticipate. May the palaces of the great that you frequent, the conversations of a world well-born that you share, not let you forget the hearth of your poor villager, the racks of his solitude, or the gloomy silence of his abode. *In propria pelle quiesque quiescat. . . .*

When will we see you? And when in person shall I be able to congratulate you on your title *médecin militaire royal* that you forbid me giving you by letter?[306] I await your response, and I am perhaps your most affectionate servant.

P.S. Thanks to the Almighty, we have rain today here, of which we have been deprived for the past excessively hot and dry five weeks.

58
Les Baux, 10 October 1782
[Chaix's hope for a visit from Villars was frustrated when Villars became sidetracked by a series of duties.]

In due course, I did receive the parcel of dried plants from Monsieur Pourret, but I have not yet received the other package of live plants he mentioned to me in his letter. I quite fear that some unreliable hand has deprived me of it forever. Please ask [Dr. Jean-François] Nicolas about it as he delivered the other parcel to you. At the same time, please make an accounting of all you are spending on my behalf. The shipment contained about 120 plants, nearly one-third of which are not new to me. The remainder are giving me great pleasure. I was putting them aside for you in particular so that you could see them during the visit I was anticipating. I am now sending you some fragments of them in the opening pages of my Mappi, unable to share with you the others without depriving myself of them, the plants being singles. I indicate those that appear to me to be the most valuable and could most merit your attention.

Plantago pilosa Pourr. [1788]. *Media inter albicans* [L.] *et P. loeflingii* [L.]. [= *Plantago bellardii* All. (1785)].

Campanula grandiflora Pourr. *Non est C. grandiflora* Lamarck *seu C. medium* [L].

$\sim$

[305] It appears that Vill., **Hist. Pl. Dauph.** 2: 402 (1787), published this as a new species, *Salvia agrestis* Vill. But that it, like *Salvia pyrenaica* L., were never again found and were rejected by Grenier, **Flore de France** 2: 715 (1856). The great variability of *Salvia verbenaca* L. and *S. pratensis* L. may account for the confusion.
[306] His actual title was Médecin titulaire de l'hôpital militaire breveté du Roi. On 4 August 1782, two days after this letter was written, the religious order that administered the hospital gave him the additional title Médecin des Pauvres de l'hôpital de la Charité de Grenoble. Manteyer, **Les Origines de Dominique Villars**, p. 155.

Habitus C. spicatae [L.], *sed folia non hirta, glabra autem, oris retrorsum ciliatis, integerrima, confera; flores axillares, magni etc. Narbonae.*[307]

> *Galium maritimum* [L.]

> *Galium divaricatum* Pourr. [ex Lam. (1778)] *rami plurimi, divaricati, divisi, filiformes.*

> *Phyteuma pauciflora* L. ex Pourr. *an bene?* non. *Cauliculi longiores, foliosi; capitula conferta secus nostra, lecta* at Queyras.

> *Diapensia pyrenaica* Pourr. My *Aretia alpina* [L.]. Haller, **Hist.** no. 618 [= *Androsace alpina* (L.) Lam. (1778)].

> *Primula integerrima* [= *P. vitaliana* L. or *P. integrifolia* L.].

> *Veronica ponae* Gouan.

> *Salvia liliaefolia.* Ortega *in Hisp.*

> *Salvia triphyllos* Pourr. My *S. triloba* L. fil. [1781].

> *Lamium grandiflorum* Pourr. *an laevigatum* L.? *Folia cordato-lanceolata, serrula, verticilli conferti, approximati; calyx calicisque dentes quam [Lamium] albi ac muculati duplo majores. Pulchra planta, non est L. orvala* [L.]. [= *Lamium maculatum* L. (1763)].

> *Bromus sylvaticus* Pourr. [?]

> *Melica amethystina* Pourr. [= *Melica bauhini* All. (1789)].

> *Arundo arenaria* [L.] [= *Ammophilia arenaria* (L.) Link (1827)].

> *Salicornia herbaceae* [(L.) L.] [= *Salicornia europaea* L. (1753)].

> *Scabiosa cretica* [L.].

> *Dianthus pungens* Pourr. *Non est pungens* L. **Mantissa.** . . . [= *Dianthus subacaulis* Vill. ssp. *brachyanthus* (Boiss.) P. Fourn. (1936)].

> *Allium triquetrum* [L.]

> *Alyssum spinosum* [L.] [= *Ptilotrichum spinosum* (L.) Boiss. (1839)].

> *Saxifraga groenlandica* [L.]. *Mihi certe S. geranioides* [L.].

> *Saxifraga retusa* Gouan. *Mihi Saxifraga oppositifolia* [L.].

> *Saxifraga hedypnois* [Pourr.?]. *Mihi caespitosae varietas Saxifraga caespitosa* [L.] *nostra videtur S. groenlandica consulatur* L., **Mantissa** 383.

> *Cistus glutinosis* [L.], **Mantissa** [246]. [= *Fumana thymifolia* (L.) Spach ex Webb (1838)].

> *Cytisus argenteus* [L.] [= *Argyrolobium zanonii* (Turra) P. W. Ball (1968)].

> *Astragalus stella* [Gouan].

> *Bupleurum pyrenaeum* Gouan [= *Bupleurum angulosum* L.]

> *Athamanta sicula* [L.].

> *Scandix australis* [L.].

> *Rosa pyrenaica* Gouan [= *Rosa pendulina* L.]

> *Daphne dioica* Gouan [= *Thymelaea dioica* (Gouan) All. (1789)].

Note. Examine this *Saxifraga stolonifera* or *sibirica?* I do not know if this is haste on the part of Monsieur Pourret, but I see a fructification quite different from the saxifrages. Give me your opinion and send it back to me in my books. I only have this one individual,

~

---

[307] Apparently *Campanula grandiflora* Pourr. was not published. See both *C. lanceolata* Lapeyr. and *C. precatoria* Timb.-Lagr. as close to the description.

and I will call it to the attention of our agreeable correspondent.

I have read the supplement by Monsieur Linné's son with much pleasure.[308] He does not appear to me to be as reserved in the establishment of species as his father was; and, for the first time, I see some women's names in botany: *Pommereulla carolinea.*[309] I have taken extracts about the plants that could be in my range. Therefore, I will return your book when you want it. Not only do the Allionis, the Scopolis, the Jacquins, the Lamarcks and so on impoverish our **Flore du Dauphiné** in regard to its rare plants, but even the living Linné!

For instance: *Arabis hispida* L. **Suppl.** appears to me to be your *Turritis raii* [Vill., **Prosp. Pl. Dauph.** 39 (1779) [= *Arabis stricta* Huds. (1778)]. *Primula glutinosa* [Wulfen in Jacq. (1778)] is certainly not your *Primula viscosa* Vill., [**Prosp. Pl. Dauph.** 21 (1779)] [= *Primula hirsuta* All. (1773)], Haller no. 613. *Serapias xylophyllum* L. **Suppl.** appears to me to be Haller no. 1298, var. B. [= *Cephalanthera longifolia* (L.) Fritsch (1888)]. You must have it in Grenoble, as I have a dried specimen of it from Monsieur Liottard that I have not been able to refer to any older species.[310]

You will find toward the end of my said Mappi some plants from my collecting this year. My *Carduus aurosicus* (from Mont Aurouze) appears to be new.[311] I cannot refer it to any of the species of the authors I have. *Iberis aurosica* (from the same place) either is only a variety of [*Iberis*] *umbellata* [L.] or an unknown species, as this is certainly not [*I.*] *amara* [L.] or [*I. odorata* [L.].[312]

Myagrum rugosum L. [= *Rapistrum rugosum* (L.) All. (1785)] is common in Les Baux. Please do not make it into *Myagrum perenne* L. [= *Rapistrum perenne* (L.) All. (1785)] whose seed Monsieur Thouin sent me. Our annual is clearly *M. rugosum* L.

· *Galium uliginosum* L. is found along the Buëch. I think that is what it is. I do not know if you cite it in your manuscript. I am not confusing it with your *Galium jussiei* [Vill.], which I know very well.

I retract what I said to you in a letter [no. 57] in regard to *Nepeta* no. 2 Gerardi. It is not *Nepeta nepetella* L., but *Nepeta nuda* L., that grows at Loubet and La Grangette, and which I cannot distinguish from *Nepeta violacea* L. [= *Napeta nuda* L.]. I will stand on my other observations about which, having still had no response, I am waiting for a word from you, as well as about those I have just made to you here.

In Monsieur Rossignol's exercise book, I am sending you some specimens of cryptogams. Please have the kindness to determine them for me and to send them back, each one with a label, separated by leaves in my books. I had expected to get you to do this here, if you had come.

Please have the kindness to have the enclosed books bound for me according to the instructions that I am addressing to the binder. Pay the cost and indicate to me what it is, when sending them back to me by the messenger, so that I can make a grateful payment.

~

[308] C. von Linnaeus, filius, **Supplementum plantarum systematis vegetabilium** (1781).
[309] *Pommereulla* L. f., an Asian genus in Gramineae.
[310] Villars, **Hist. Pl. Dauph.** 2: 51 (1787), correctly selected Haller no. 1296 as the right synonym, but objected to the trivial name *longifolia.*
[311] *Carduus aurosicus* Chaix, **Pl. Vap.** 60 (1785); Chaix in Vill., **Hist. Pl. Dauph.** 1: 364 (1786).
[312] *Iberis aurosica* Chaix, **Pl. Vap.** 45 (1785); Chaix in Vill., **Hist. Pl. Dauph.** l: 349 (1786).

If you have some new subject to send to me, say *Androsace villosa, Impatiens noli-tangere,* or some products from your garden, said books can serve as their carriers.

I have two species of *Cheiranthus* here that I cannot find in Linnaeus. They came up in my garden without my knowing from where they came. I shall describe them carefully and send them to you. I have another plant, pentandrous, monogynous, petals united, regular, capsular, whose genus I do not know. The Lord limits our insights in order to keep us on our toes.

When will our anticipated flora see the light of day! Try to raise the financial backing for it. I fear being gone before its appearance, and, moreover, that the works that continue to be published on this material will not leave you many new plants.

I pray the Lord every day to keep you in health midst the dangers you are obliged to face. You are the dearest person I have in this world. I embrace you with all the affection of which I am capable.

P.S. Open the tied-up books with plants therein with care. I have had my niece, who went to Gap, pay the shipping charges to the messenger. She sends you her respects.

59

Les Baux, 3 January 1783

Despite the long delay you made me suffer, I shall never suspect your heart of cooling in my regard. I am pleased to believe myself too high in your esteem to be neglected. During that long silence, I could only blame your pen for not having exercised itself in my favor since August 27th, so far as I know. Thanks to the neglect of the postal director in Gap, your letter of August 9th only reached me on November 5th, three months after its date. In it, you spoke of a collection of 400 or 500 dried plants for a herbarium wanted by a gentleman in the intendance. At that time, I could not undertake the commission. But, as it has not been fulfilled by someone else, and you raise the matter again in your last letter of December 24th, I accept it and take on the responsibility. I shall have executed it by the end of August, please God. Please send me the bound notebook [for the collection] here in the spring.

I am happy with my bound books and thank you for them. I have made note of the 4 livres that you have paid. You have left many of my mosses in doubt, perhaps because several of them lacked fruit. I doubt if I shall be fortunate enough to meet them again in fruit. My *Hypnum squarrosum* [L.] is quite different from *Hypnum filicinum* [L.]. My *repens* near the rocks of our fountain in Les Baux is not *Hypnum triquetrum* [L.]. I doubt whether you have labeled *H. crista-castrensis* [L.] well. Do not be angry with a childish criticism founded upon my ignorance. You have not had time to make a considered examination of my uncertain plants. Please give me satisfaction when it be possible for you. But I had hoped you would have at least said a word to me about my *Carduus aurosicus* that appears to me to be a new species.

We have at Sigottier, near Serres, *Jasminum fruticans* [L.] growing naturally; *Cistus laevipes* [L.] [= *Fumana laevipes* (L.) Spach (1936)]; and *Pistacia terebinthus* [L.]; and at the headsprings of the Charence, along the road, *Scirpus holoschoenus* [L.], which I have not seen before this year.

Myosotis scorpioides montana radice perenni, corollis majoribus, Haller no. 591, is truly a separate species from *M. arvensis*, an annual.[313] I inappropriately took a variety of *Valeriana montana* [L.], which I now call *Valeriana appendiculata* on account of its two leafy wings at the base of the petiole, for *Valeriana tripteris* [L.], which I believe only has radical leaves.[314] (Collected from the base of the Valgaudemar.) If you have more than one specimen, spare me one. Your letter, so long in the post, gave me two excellent specimens of *Androsace villosa* [L.]. I thank you for them.

Do not put into your public garden, at least without great precaution, the plants from the following short list, of which I am the unfortunate victim:

Rejetatrices importunae[315]

Antirrhinum linaria	*Frageria vesca*
Oegopodium podograria	*Physalis alkekengi*
Artemisia pontica	*Potentilla anserina*
Asclepias syriaca	*Stachys palustris*
Achillea ptarmica	*Triticum repens*
Asperula taurica	*Scutellaria galericulata*
Lepidium latifolium	*Saponaria officinalis*
Lysimachia lutea	*Panicum dacytlon*
Oxalis corniculata	

In regard to Monsieur de St. Genis, I shall not forget your brother's concerns. But nothing is pressing to hasten him to take holy orders. He has only now just left the world; he must now learn to bear the ecclesiastical yoke. Through an examination, you can adequately demonstrate your knowledge or your aptitude to acquire it; but possession of the ecclesiastical spirit is quite another matter and essential in our profession for one's own salvation and for the good of the faithful. Two years in seminary is not too long a noviciate. A dispensation of this very wise rule sometimes gives rise to much groaning. Holy orders should come in due course when one is convinced of one's suitability. I speak to you before God and according to my conscience. I know, moreover, the extent of your interest in being relieved of his responsibility. But God, who gave you the *beginning*, will not refuse you the *successful outcome*, if that is His glory.

Nothing surprises me when it comes to Messieurs Clappier and Liottard. Their education, based upon the principles of their master [Rousseau], could only have been a monstrosity in regard to religion and humanity.

I am sending you one example from each of my aconites that you requested for Monsieur Vitet. In the future, I should be able to serve him better. The shoots you had cut for your extract all perished. Last year, some wretched large rats devoured them in bud along with the *Allium ursinum* [L.], without touching any other plants, before I caught seven or eight of them in a rattrap. I am not surprised that Linnaeus, in **Flora Lapponica**, reported women putting this plant in their soup.

Very sensible of your courtesy, I am returning to you your two volumes of

∾

[313] Chaix was criticizing Linnaeus here for have made *Myosotis arvensis* a variety of *Myosotis scorpioides*. *M. arvensis* (L.) Hill (1764) is regarded as a biennial.

[314] *Valeriana tripteris* L. has middle cauline leaves, 3-foliolate or with a pair of small basal lobes.

[315] Most of these were Linnaean species, but the source of this list is unclear.

Supplementum Systema Reg. Analecta transalpina by Monsieur de La Tour d'Aigues [of Aix-en-Provence]. Along with said aconites, I am enclosing one shipment for Monsieur Thouin and one for Monsieur Pourret that I beg you to have sent on to them by whatever you judge to be the safest route.[316] If you find it possible, have a copy of [F. G.] Weis, [Planta] **cryptogamicae [florae gottingensis]** bought for me. This work would help me to examine the members of this family about which I know little. I trust the expenditure will not give you any difficulty. Apprise me of interesting news about affairs of state.

We are in mourning for the death of Monsieur de La Roche, who died in Gap of a sudden and bloody apoplexy while asleep in bed at the residence of the abbé de la Villette, on the night of December 25-26, after having dined as a guest of the bishop on Christmas Day. When his bed-curtain was opened in the morning, he was cold, dead for about four hours. The chapter and the city of Gap have celebrated his funeral according to his dignity and rank. His body was buried on the 28th at La Roche. Yesterday, at Les Baux, we held a funeral service for him. *Watch therefore, for ye know not the day nor the hour.* **Matthew 25. 13.**

To wish you simply a happy new year would be inadequate to express what is in my heart. I shall say to you, therefore, in brief: *fauste valeas in annum et faustissime in longissimas annos.*

P.S. In my letter to Monsieur Thouin, I have congratulated him on *Thouinia* L. f. in **Supplementum [plantarum] systematis vegetabilium.** [= *Linaciera* Sw. in Oleaceae.]

60

Les Baux, 12 February 1783

When I was told in Gap last Saturday that you were in Le Noyer, the news at first rejoiced me, thinking that you had gone there for the festival of St. Agatha; but my joy was shortlived when it was added that your wife's illness had called you home. I shall be anxious about that until you inform me that she is better. I urged Monsieur Garnier to get you to come as far as our neighborhoods if possible for you. My friends from La Roche and Manteyer, who dined here today, and to whom I talked about you, would have been quite delighted to have had you for a table companion.

I owe you two responses:

1) To your letter of last December 31st enclosed with the long-awaited shipment from Monsieur Pourret: Despite having been en route for a period of five months, as the shipment nearly entirely comprised bulbous plants, I am very hopeful that at least the greatest part of them will take root. This shipment gives me all the more pleasure as I had come to despair for news of it. In his letter of last January 15th, Monsieur Pourret complained forcefully about your silence. I transcribe what concerns you here so that you can calm his anxiety: "Monsieur Villars has not written to me for a century. I no longer dare to write to him. I fear to intrude upon him, so I am addressing my letter directly to

---

[316] Claret de La Tourrette to Jean-François Séguier of Nîmes, 14 April 1783: "Here is a package that came from Grenoble, sent by Monsieur Villars on behalf of Monsieur Chaix, two very distinguished botanists of Dauphiné, to be passed on by you to the Abbé Pourret." H. Duval, "Lettres inédites de Claret de Latourrette." **Revue du Siècle**, no. 163 (December 1900): 665.

you. Please tell me whether I can continue to use the route through Monsieur Villars and ask him what can be the reasons for his silence in my regard. I sense all that I am losing in being unable to indulge my taste for our congenial botany to the fullest extent."

He made some critical notes for me on our plants, and he promises me more in greater detail on others. I will send them on to you later. You will no doubt have had the kindness to send him my package that was included with your books and the one for Monsieur Thouin that I sent you by messenger.

2) And to your letter of last January 20th that I retrieved from the mail last Saturday: Since peace is now assured to us,[317] I shall not dwell upon the dispositions of the states concerned, the details of which you kindly furnished me, but will go directly to botanical matters.

1. How can my *Valeriana appendiculata* be *Valeriana tripteris* [L.]?[318] Its radical leaves are never cordate, but round or oval-oblong. Its leaves are often entire without appendages. I believe I have reason to make it only a variety of *Valeriana montana* [L.]. Do me the pleasure of sending me *V. tripteris* of which I believe I have only one leaf collected in the Valgaudemar.

2. How are the authors to be reconciled? *Iberis linifolia: Folia caulina linearis integerrima; corymbi hemisphaerici* L., Sp. pl. *Folia caulina serrata* Murray, Syst. Veg. *Flores primum corymbosi dein racemosi* Gouan, Ill. 41. The one I had previously grown under this name has all the characters noted by Monsieur Gouan: cauline leaves linear, long, narrow, quite entire, terminating in a yellow glandula. Such leaves separate it from *I. amara* [L.], even if the racemous flowers approach it. My *Iberis aurosica* [Chaix] always has a hemispheric corymb, as does the *linifolia* in Linnaeus; but it is certainly an entirely different species than the one I have just spoken about, and I concur with you that, its leaves not being lanceolate, it differs from *I. umbellata* [L.]. I have it in the garden and will continue to observe it.

3. My two mints, *latifolia* and *subspicata*, cannot be referred to [*Mentha*] *hirsuta* [Hudson], **Mantissa**, because their stamens are always enclosed within the throat of the corolla.[319]

4. As for my *Carduus aurosicus* [Chaix], you are not thinking about it when you claim to see it only as a variety of the *defloratus* reported by Haller [no. 164]. Have you forgotten that *Carduus defloratus* [L.] has very long, nude peduncles; its heads very small; it scales lanceolate, imbricated, with very small terminal spine? Can you say that of my above said thistle? On the contrary, its peduncles are sessile;[320] its heads twice as large; its scales open, obvious, very long, subulate, very spiny. I am sending you a head of it again. Take the trouble to compare it with those of *Carduus defloratus* that you must have in Grenoble. Do you believe me so limited as not to be able to distinguish them? I have only seen the former

~

[317] The Preliminary Treaty of Paris between Britain and the United States, 30 November 1782, had forecast the end of the American Revolution. The general peace would be worked out in 1783.
[318] See letter 59.
[319] This appears to become *Mentha dubia* Chaix in Vill., Hist. Pl. Dauph. 2: 358 (1787): *foliis hirsutis acuminatis serratis ovato lanceolatis: floribus spicato capitatis, staminibus corolla brevioribus.* [= *Mentha aquatica* L. (1753), the form with stamens enclosed.]
[320] He would say *pedunculis brevissimus* when publishing the species as new.

along the base of Mont Aurouze, whereas the latter never stops skinning my legs in our mountains. When you can prove to me that the magpie is not different from the blackbird, you will be able to convince me that these two thistles are the same.

5. I shall hold back on the name of *Galium hercynicum* Weigel. *G. uliginosum* L., *G. saxatile* L., and *G. pusillum* L. will come along when they can.

I am instructing the man carrying my letter to bring back your response to me if you have one to make. He is going on to Lyon to take possession of a small inheritance and will remain there only a few days. Nothing feeds my zeal more than a conversation with you by letter, since Providence prevents us having it in person. It seems to me that adding [G. H.] Weber [**Spicilegium florae gottingensis**] to the [**Plantae**] **cryptogamicae**] **flora gottingensis**] by Weis could be a great help to me for this part of the Vegetable Kingdom. I beg you to do everything to obtain a personal copy for me.[321]

Keep me informed on the state of your family. I am not forgetting your brother's interests in regard to our superiors. I shall try to have him granted the letters dismissory for the four minor orders, effected during the four offices of Pentecost, before the completion of his first year in seminary. . . . He could then be made a subdeacon at Christmas, a deacon during Lent, and a priest at the Pentecost of his second year in seminary. That is, provided he meets expectations. Remember me to him when you write to him. I am entirely devoted to him. It would be superfluous to repeat to you the inviolable affection I shall have for you for life.

P.S. Monsieur Buchoz has just decorated an exotic plant with the name *Villaria* in his prospectus for 1783. I congratulate you on it.[322]

61
Les Baux, 12 March 1783

I need not beseech you when it is a matter of rendering a service to me or to those for whom I am concerned. Here is the situation: Joseph Thomé, son of my old friend Jean, called *Bras d'Or*, a soldier in the La Fère regiment, in garrison at Calais, and from the city of Gap, had obtained a semester of leave to return to his father's house for a change of air. He has just received an order to rejoin his company at once. His health does not allow him to set out on a trip, as you will see from the certificate signed by his doctor in Gap and from your own examination of his state. As a consequence, he is obliged to enter the Royal Hospital in Grenoble as a patient. Please have the kindness to expedite your medical certificate to be sent to his captain so that he will not be suspected of disobedience by his command!

I must tell you in confidence that this young man, having already served for eight years, must have his leave confirmed until next St. Martin's Day [November 11th]. You will understand the expense for him to go from here to the most distant part of the kingdom and then to return. You can render him a great service if, for reasons of poor health, he can be spared this long journey. I am counting on your prudence and on the friendship I am pleased to enjoy with you.

∿

[321] Both titles were in Chaix's library at the time of his death. A.H.-A., Série L 1007.
[322] The name was never recognized.

On last February 24th, the date of Carnival in Gap, I saw Monsieur de St. Genis. He asked me expressly to extend his compliments to you and to tell you not to be concerned about the orders to be conferred upon your brother. You believe that the letters dimissory can only be sent on the request of the superior in the seminary who has the subject under his supervision and can judge his qualifications. But Monsieur de St. Genis assures me that it is common practice in that seminary to have the seminarians take the minor orders during their first year at the four offices of Pentecost; and that, if your brother should not be among that group, it would be a sign that they are not pleased with him. The major orders will then be conferred upon him subsequently, *omnibus consonantibus.* Your letter, sent during his absence, was discussed with him, and he talked to me about it himself. In a word, do not worry. Let the candidate, taking advantage of a very valuable period, be concerned for nothing but his piety and his knowledge.

... Perhaps my botanical correspondents will soon send something to your address for me. I am confident of your promptitude, but the ground in this country is hardly suitable. It has rained or snowed continually for nearly three weeks. We did not have so much snow during the entire winter. May God, in his mercy, give us better days. The people wail in misery and for their families and their livestock.

62

Les Baux, 17 June 1783

Pleased by the sweet anticipation of your visit during Pentecost, I did not respond to your letter of May 16th, sent with the two bound notebooks meant for the herbaria in question. I accepted your commission last year; and, beginning with the first spring plants, I set my hand to the task and will try to see it through to the finish. I do not know whether I am permitted to put any grasses in the collection. I foresee some difficulty in putting any cynareae in it on account of the thickness of their involucres. I shall be obliged to account for only a few of them.

The abbé Blanc, professor at Embrun, spent two days here during the week of Pentecost. I am counting on going to join him in Embrun in mid-July after our hay harvest to see the plants of that country. Monsieur Pourret had notified me of his intention to make a trip through these regions this spring (I believe I made that known to you). In case he passes through Grenoble and stops there for a few days, please inform me of it as quickly as possible, so that I do not leave for the Embrunnais and can await him at my cottage. I let Monsieur Blanc, the vicar of La Roche, know about the step you took concerning his purchase of a watch. He will abide by all the decisions you kindly noted in your last letter to me, June 10th, from St.-Leger [-les-Mélèzes]. I am quite delighted by the news you give me from time to time about your brother: his health, his progress, and his decided taste for the ecclesiastical life. I am convinced that virtue will come along with the favorable talents God has given him. Remember me to him when the occasion arises. Father Cornille, professor at the seminary in Gap, recently told us something I did not know about the conduct of Monsieur Christophe. Heavy work in the fields is improper for someone in our state.[323] The Levites had no portion in the division of the promised land. God provided

⁓

[323] Pierre Christophe, from a laboring family, was serving as a priest in Le Noyer at that time.

otherwise for their subsistence. It appears to me, from what you have told me about the situation, that his temperament would benefit from greater supervision. I shall speak to Monsieur de St. Genis about it, even to Father Cornille, who often sees the diocesan superiors, so that they became aware of his situation. *Who is weak, and I am not weak?* 2 **Corinthians** 11.

Your silence in my regard on matters botanical limits what I can take up with you. The observation you make to me about *Hypericum coris* [L.] allows me to speak.

1. I feel, and I have always felt, that our *Hypericum* with hyssop-like leaves is different from *Hypericum coris* according to the authors' descriptions and illustrations. I do not know whether Linnaeus misstates by indicating stipules on *H. coris* in **Species Plantarum**, for the stipules are *structurae ad basin foliorum*. At least the small leaves, which appear in the axils of our *Hypericum*, are merely ornaments on the young axillary branches. Haller, Hist., no. 1040, gives his *coris* punctate anthers. I have had in my notes for a long time: *Antherae nudae; caulis teres; folia oblonga, pellucido-punctata; calices et petala pallide lutea, nigris glandulis secundum oras exornata*, as a description of our plant. It is not, therefore, *Hypericum coris*, putting us in agreement; but I have been unable to refer it to anything in the authors I have. If the trivial name in [Gérard] **Flora Gallo-Provincialis** [-------],[324] you could perhaps give it the name *hyssopifolium* with the description *Hypericum* [-------] *tereti; foliis lineari-oblongis, pellucido-punctatus; calicibus petalisque, serrato, glandulosis* [-------] *in collibus apricis*.[325]

2. It is true that my *Iberis aurosica* [Chaix] is nothing but *Iberis amara* L. The racemose fructification, which had not developed on the mountain, now appears quite clearly in my garden. The question about the plant is quite decided. The victory is yours, as usual.[326]

3. I have never run across *Thlaspi montanum* L. We only have *Thlaspi alpestre* L. here. Perhaps Monsieur Liottard had sent the former to me earlier. If you have it, I would like an example.

4. My *Allium palustre* [Chaix], that I am now calling *Allium raii*, **Hist.** 1118, no. 5, is surely a species different from *A. schoenoprasum* [L.], the spring onion, or chives. I will stand by it.[327]

5. *Ornithogalum umbellatum* [L.]: *filamenta emarginata* L. has always bewildered me.[328] I have been [-------] that you put *Ornithogalum pannonicum* Chaix in your prospectus, as it is only a smaller variety of the above not meriting this name from Clusius. You did well, therefore, to write *filamenta omnia simplicia*, as Murray, **Systema**

◁

[324] The manuscript was damaged by rodents several places in this passage. The probable reference in Gérard is to p. 399, no. 7, the equivalent of *Hypericum coris* L., from which Chaix wished to segregate a new species.

[325] *Hypericum hyssopifolium* Chaix, **Pl. Vap.** 25 (1785). Chaix in **Vill., Hist. Pl. Dauph.** 1: 329 (1786). From Les Baux and Rabou, in open woods.

[326] See letter 58. Villars' victory was short-lived. He would accept *Iberis aurosica* Chaix as a good species, as initially published in 1785-1786, in **Hist. Pl. Dauph.** 3: 289 (1788).

[327] *Allium palustre* Chaix, **Pl. Vap.** 17 (1785). Chaix in **Vill., Hist. Pl. Dauph.** 1: 321 (1786). [= *Allium schoenoprasum* L., a quite variable species found in damp and wet ground.]

[328] Villars commented in **Hist. Pl. Dauph.** 2: 272 (1787) that "its stamens are simple in our region, the filaments merely dilated at their base."

vegetabilium 271, has written *base dilatatis* instead of *emarginata* without making an issue over it with his master [Linnaeus]. . . .[329]

6. My *Scirpus ruppi*, Haller, **Enum.** 249, **Hist.**, no. 1225, which is found in all ditches, is quite different from *Scirpus palustris* [L.] that I only find here at Gap.[330]

7. I believe that a rush found around Gap is Haller's no. 1318, Ray's no. 1307: *Juncus bulbosus* L. But I do not see any bulb with roots. Moreover, Ray says only *radix obliqua juncea fibrata*, which accords with my plant.[331]

8. My *Geranium malvaefolium* Scop. is pentandrous with five rudimentary sterile filaments. Murray, in **Systema vegetabilium** 515, has observed it well within *Geranium pusillum* [L.], Linnaeus having perhaps regarded it as a variety [of *pusillum*]. Monsieur Deleuze thought it to be *Geranium pyrenaicum* [Burm. fil.], Murray 513. Said Monsieur Deleuze gave me another annual *Geranium, foliis reniformibus, palmatibus, linearibus acutis*, that I believe to be the true *Geranium pusillum* L. You must have it in Grenoble.[332]

9. Have no doubt that *Myosotis montana* Chaix [**Pl. Vap.** 50 (1785); Chaix in Vill., **Hist. Pl. Dauph.** 1: 354 (1786)], in Barrelier, **Icones** 404 and Haller, **Hist.**, no. 591, a perennial,[333] differs from *Myosotis arvensis* [(L.) Hill], an annual, Haller, **Hist.**, no. 590. I have also found a white-flowered variety of it.

10. If you are going to take, as *Valeriana tripteris* [L.], the plant with radical leaves round, the cauline lanceolate with one of two appendages on the petiole, how are you going to square that with *foliis radicalibus cordatis* L., *caulinis ternatis* L. and Haller? I have found it *foliis caulinis simplicissimis*. Moreover, what are you going to do with the figure in J. Bauhin? I believe that your plant is *Valeriana montana* [L.] *foliis ovato-oblongis*. As for the true *V. tripteris*, I have some leaves of it from Valgaudemar. As for our small *foliis* [------], Monsieur Deleuze thinks so, too, unless it is a very distinctive variety of *V. montana*. (I am enclosing two leaves of it here.) Let me have your *V. tripteris*.[334]

But this scribbling is hardly in tune with the seriousness of your concerns, so let us stop it and turn to more interesting matters.

Please, Monsieur, have the kindness to find out if a recent work is available in Grenoble: **Pensées théologiques relatives aux erreurs du tems** by Reverend Father Nicolas Jamin, a monk of the Congrégation de Saint-Maur and formerly prior of the Abbaye royale de St. Germain-des-Près. A 5th edition was published in Brussels in 1774. If you find it, I would be pleased to have you send me two bound copies by way of the Gap messenger, indicating their price so that I can send you the correct amount. If you do not find it, let me know that, too. I can have it ordered from Avignon. I should recommend that you add

~

[329] *Ornithogalum pannonicum* Chaix in Vill., **Prosp. Pl. Dauph.** 18 (1779). [= *Ornithogalum lacteum* Vill. (1787) = *O. narbonense* L. (1756)].

[330] See *Scirpus halleri* Vill., **Hist. Pl. Dauph.** 2: 188 (1787). [= *Eleocharis quinqueflora* (F. X. Hartmann) O. Schwarz (1949)].

[331] Haller's no. 1318 is *Juncus bulbosus* L.

[332] See *Geranium dubium* Chaix, **Pl. Vap.** 23 (1785). Chaix in Vill., **Hist. Pl. Dauph.** 1: 327 (1786). In Champsaur near St.-Bonnet. Verlot, **Cat. Pl. Dauph.** (1872), said that Chaix's species was smaller in every part than *G. pusillum*, but nevertheless the same species.

[333] Villars believed, correctly, that Haller no. 591 was *Myosotis scorpioides* L.

[334] See *Valeriana rotundifolia* Vill., **Hist. Pl. Dauph.** 2: 283 (1787), published as a new species but with the proviso that it might only be a variety of *V. montana* L. In the 19th century, it was variously treated as either a form or a synonym of *V. montana*.

this little book to your library, sensitive as you are to the poisoned shafts that rash and antichristian minds launch against the verities of our holy religion, either without knowing them or feigning to misunderstand them. You will be able to draw from the book the reasons that reduce them to silence. The principles and the fundamentals of our faith are depicted with the greatest clarity and with the greatest precision. The philosophes, with the ornament of eloquence, put forth sophistries; but their profound reasoning proves nothing. *Mentita est inequitas sibi.* **Psalms 26.** The man without religion is no man. He cannot reason and merely holds the highest rank among the brutes. He respects no law. He cannot control himself, but is a slave to his passions. . . . A pitiable state! Forgive me this digression into morality. Speaking to a friend as sincere and as religious as you are, I do not have to conceal my manner of thinking. The more I reflect upon the irreligion spreading in the world outside, the more I have an affection for the poor dwelling of my solitude. *I shall die in my nest.* **Job** 29. 18. For the rest of my life, you will be the dearest of the friends I have on earth, and my love for you will extinguish only with my last breath.

63
Les Baux, 2 October 1783

I have put off writing to you too long (perhaps I have not done so since the month of June), rather presuming that some happy circumstance would cast you into our [ecclesiastical] cantons and give me the pleasure of greeting you, a pleasant hope later confirmed by your brother in person and more definitely by your last letter of August 28th. Some obstacle to carrying out your intention must have arisen since you did not appear here in the course of your intended trip to Le Noyer in order to establish the patrimony in support of the clerical title of our worthy ecclesiastic upon his entry into the first holy order, which occurred, I believe, at the last ordination.[335] I had imagined that the fairs at St.-Bonnet had been the cause of your delay, but now I feel sure your affairs required a return to Grenoble. In any case, I pray the Lord to guide your efforts and to cover you with the wings of His protection. I would have had many things to discuss with you, and I would have drawn much enlightenment from your presence. As I can recompense myself only through epistolary connection, you will have to put up with considerable detail.

In due course, I received Father Jamin's **Pensées théologiques.** I do not tire from reading this excellent compilation that I would like to know from memory as my *credo*, given the unfortunate unbelief that infects most of those who represent literature and good taste. I have made a record of the price on your note, as well as the sum of 78 livres 12 sols that the vicar Blanc of La Roche has credited me for the purchase of his watch. By my calculation, when these figures are added to the other items that you have paid for on my behalf, you owe me no more than 54 livres 13 sols today.

The drying of the plants for the herbaria in question is completed. Please have the kindness to indicate to me, as you promised you would, the order you want me to follow

~

[335] Through an attorney, Dominique Villars granted his brother, Jean-François, a lifetime pension of 100 livres a year to enable him to enter the priesthood. The document was dated 1 September 1783 from Grenoble. Manteyer, "Les Origines de Dominique Villars, le botaniste (1555-1814). **Bull. Soc. Etud. Hautes-Alpes** 40, sér. 4 (1921): 138.

in their arrangement in said herbaria. After your response, I shall go to work on them immediately.

Monsieur Danthoine sent me a small shipment of very unusual plants, most of which are not in Linnaeus. I have been very touched by his courtesy, but he is badly advised to ask my judgment of them. Monsieur Deleuze is working intently on botany and is a very good observer. I took a brief trip to Embrun at the beginning of August. The lakes of Siguret at St.-André [d'Embrun] provided me three or four new species [--------]. The [-------] plant gives me great difficulty as I was not able to observe it adequately because of the smallness of its fructification and the rainfall during my botanizing.[336] I am sending you herein a specimen of it; please give me your opinion of it. Here is what I have been able to observe about it :

Radix dura, lignosa: an annua? Caules plures, dodantales, ascendentes; rami divaricati; folia linearia, glabra, alterna; flores axillares, sessiles; calices 1-phylli, filiformes, angulosi, persistentes, ore 4-5 dentato; petala 4-5 minima, emarginato-dentata, purpurescentia; in fauce haerentia stamina 4-5; stylus unicus, filiformis; stigma globosum; capsula cylindrica, tecta, unilocularis; semina plurima, minima. Somewhat resembles *Velezia,* but is not *Velezia rigida* [L.]. Note that *Velezia* is 5-androus digyna. Murray, **Systema,** 220. My unknown plant grows along the upper small lake.

Myriophyllum verticillatum [L.] was found in the lower small lake.

The new species of epilobium I am calling *Epilobium ebredunense: caules simplices, tenues, decumbentes, glabri; folia breviter, petiolata, ovato-lanceolata, dentata; flores parvi, ex albo purpurascentes.* A specimen for you enclosed.[337]

Recently found at La Grangette: *Valeriana tripteris* L. *foliis radicalibus omnino insignitur dentalis, caulinis constater ternalis, dentalis. Valeriana montana* [L.] *variat foliis uno alterovo foliolo appendiculatis. . . .*

Iberis aurosica [Chaix] *floribus corymbosis, non racemosis.* It is not *I. amara* [L.], as correctly noted: *foliis subdentatis, integerimis.* Nor is it *I. odorata* [L.]: *foliis superne dilatatis serratis. . . .*

Geranium pratense L. . . is not native to Dauphiné. The geranium grown in my garden, from seed sent from Paris and labeled *Geranium sylvaticum* [L.], is certainly what we have in Dauphiné. . . . Nor do I believe that *Aconitum variegatum* L. grows in Dauphiné.

Your *Galium argenteum* [Vill.] [= *Galium pumilum* Murr. (1770)], is it not found growing here and there in grassy meadows? and your *Galium anisophyllon* [Vill.] in sterile sandy areas? *Galium pumilum* Murr., [**Prodromus stirpium**] **Gottingentium,** 44. *Galium album* [Vill.] [= *Galium album* Miller (1768)], *supinum multicaule, angustifolium, polyspermum.* Your *Galium gerardi?*[338] is quite clearly from Gérard, **Gallo-provincialis** 226, no. 2. Does it not call to mind *Galium lucidum* All.?

The chevalier de Lamanon, a skillful phytologist and physician from Salon [-de-

---

[336] Rodent damage agin; but it can be discerned that he also collected "on the beautiful mountain of Vars," and that he was hampered by considerable rainfall.

[337] This species was never published. In his list of plants for the Embrunais, Chaix gave only *Epilobium alpinum* L.

[338] Published as *Gallium gerardi* Vill., **Prosp. Pl. Dauph.** 19 (1779), but later as *Galium rigidum* Vill., **Hist. Pl. Dauph.** 2: 319 (1787). [= *Galium lucidum* All. (1773)].

Provence], paid me the honor of a visit while traveling in Dauphiné for a portion of the natural history of Provence. He gave me the list of plants on Mont Ventoux; I gave him the one for our Mont Aurouze. He talked to me about your description of Mont Vieux Chaillol published in the **Journal de physique** (can you make it available for me to read?).[339] When the winter snows drive him from the Briançonnais, he will go on to spend the winter in Turin in order to come back in the month of April.[340]

Monsieur Blanc, professor at Embrun, addressed me a certain number of sheets from a prospectus. I am sending one of them on to you. He has had some theses printed, defended by three of his students in his philosophy course, all very well conceived and preceded by a preface that must be admired by any man of morality and religion. But Blanc's note on Tournefort's system cannot be commended either for its novelty or for its correctness. Father Rossignol welcomed me very warmly in Embrun and has since written to me. Monsieur Gaude, with whom I recently made the trip to Durbon, charged me, as he always does when I see him, to extend his warmest regards to you.

Try to respond to me by post at the earliest (simply to my address) about what now concerns you, and to the items mentioned in mine; and I shall have your letter picked up at the post office. Also tell me if the printing of your work on the plants of this province will soon be underway. You had also spoken to me of a sketch concerning Haller's works, and I have heard nothing further about that. I wish for you enough health and leisure to fulfill public expectation. I shall find therein my own personal satisfaction, for no one is more closely devoted to you than I.

P.S. The very frequent rainfall is hardly favorable. May God have it succeeded by clear skies! Amen.

3 October 1783: God has heard clearly. *Serenum est. Serenum firmetur!*

64

Les Baux, 3 February 1784

After a silence on your part of four months since your departure from here in October, on Saturday, January 31st, I received your two shipments and two letters dated January 6th and 13th. Thank you for the purchase of the gold cross with heart for my niece and for the loupe that will be very useful to me. I will send you a exact reckoning for them. I have read your paper on our mountains with pleasure and profit. I should be delighted to learn about the outcome of your research on particular plants at different degrees of elevation. Mont Vieux Chaillol will become famous thanks to visits from so many illustrious naturalists. Permit me some remarks on your plants observed on that mountain in particular or on others in general.

1. *Pinus sylvestris* [L.], C. Bauhin, [**Pinax**] 491], *genevensis* J. Bauhin [I, p. 2, 253], grows at a lower elevation than the firs at Polligny, at Durbon, at Montmaur, and so on.

---

[339] Villars, "Observations de météréologique et de botanique sur quelques montagnes du Dauphiné." **Journal de physique** 22 (April 1783): 269-279.

[340] This is Robert de Paul, chevalier de Lamanon (1752-1787). See Villars, "Mémoire sur la prétendue découverte d'un volcan éteint dans le Dauphiné, annoncé par le chevalier de Lamanon." **Affiches du Dauphiné**, 7 November 1783. In October, the intendant had sent Villars, Ducros (Director of the Bibliothèque publique), and Dr. Prunelle de Lière to investigate Lamanon's claim that Vieux Chaillol in Champsaur was an extinct volcano.

In support of your observation, *Pinus montana* [Miller] or *genevensis* [= *Pinus mugo* Turra (1764)], is found around the village of Montgenèvre; and *Pinus pumilio* [Haenke] or *P. mugo*, is only found much higher towards alpine meadows. I could easily believe that these are only varieties, but that there is no necessity to merge them.

2. *Pinaster conis erectis* C. Bauhin; *pumilio* Clusius; *mugho* Matthiole; Haller, Hist., no. 1659; Scheuchzer, **It. alp.** 6, p. 460 [= *Pinus cembra* L. (1753)], grows above the firs at Le Noyer, Les Baux, Durbon, and Le Monêtier [-les-Bains] of Briançon, and so on.

3. *Salix lanata* L. Haller, Hist., no. 1651. I have never seen it in the marshes of the valleys. I formerly called it *Salix paludosa*. When you cite it in your work, keep in mind *Salix sericea* Chaix; Haller, Hist., no. 1643.[341]

4. I have not seen *Juncus campestris* [L.] on Mont Vieux Chaillol, which is found at Les Baux near the woods belonging to the Chapter of Gap, but rather *Juncus pilosus* L. var. E *alpinus*; Haller, Hist., no. 1326.[342]

5. You gave me *Carex leporina* auct. [= *Carex ovalis* Good. (1794)] from Allevard that I have not seen on Vieux Chaillot, but rather *Carex brizoides* [L.]; Haller, Hist., no. 1358.

Please have the kindness to tell me the value you set on these notes for my personal instruction. Monsieur de Lamarck will never put himself in a position to look for our *Artemisia insipida* [Vill.] near Grenoble.[343] I am not ambitious for the citation, but I am sorry about the claims of truth betrayed.

I am sending back the books by Durande and de Las with your paper and the loupe as you requested.[344] I have paid the shipping charges to Monsieur Pellissier as is proper.

If I have to yearn for news about you for as long as the recent interval, the moment of my satisfaction will be deferred until the beginning of June. Deprived of a pleasure that I no doubt do not deserve, I shall feed in secret the fire that I have dedicated to you, and I shall pray to God every day of my life for your health and your prosperity.

65
Les Baux, 17 February 1784

At the beginning of last October here, as you confirmed in person what you had proposed to me in the summer of 1782 concerning the herbarium now in question, I did not keep your [initial] letter, believing in good faith that it would not again be necessary to refer to its content.[345] I have retained enough of it in my memory, as well as of your verbal confirmation, to be able to attest to the matter in dispute. You proposed to me (and I accepted your offer) assembling an herbarium of about 500 plants for Monsieur Jourdan

∾

[341] *Salix sericea* Chaix (non Marshall), Pl. Vap. 69 (1785). Chaix in Vill., Hist. Pl. Dauph. 1: 373 (1786). *In borealibus alpibus*, at Orcières. Chaix in Vill., ibid. 1: 382 (1786). *In borealibus, juxta nives perennes*, at Monêtier. [= *Salix glaucosericea* B. Flod. (1943)].

[342] *Juncus alpinopilosus* Chaix, Pl. Vap. 14 (1785). Chaix in Vill., Hist. Pl. Dauph. 1: 318 (1786). Alps, Mont Vieux Chaillol. [= *Luzula alpinopilosa* (Chaix) Breistr. (1947)].

[343] *Artemisia insipida* Vill., Prosp. Pl. Dauph. 32 (1779). In Hist. Pl. Dauph. 3: 249 (1788), he says that the plant was found near Les Baux in the woods of Monsieur Mondet by Chaix and had not been found elsewhere.

[344] J. F. Durande, **Notions élémentaires de botanique** (1781), and de Las, **Phytographique universelle, ou nouveau systèm de botanique** (1783).

[345] This is the first indication that Chaix did not routinely preserve letters from Villars.

[premier secrétaire de l'Intendance], collected mostly in this province, upon being supplied with a bound folio, for an honorarium of 72 livres. I fulfilled my pledge and more; for I believe that I put more than 650 plants in his herbarium, and I added an alphabetical table to it that had not been specified in your letter. If anyone troubles to pay some attention to the collection, the drying, the arrangement, the pasting, the nomenclature, the selection of the plant species, that ought to convince anyone that the salary hardly exceeds the labor. Those knowledgeable, who examined the herbarium here, were surprised that I had accepted such [a modest] return. What is more, it would be extremely distressing if you were to become his dupe, [not compensated for] the paper supplied and bound, and for the shipping expenses, in order that I receive the full amount agreed upon. Such a liberal arrangement on your part would seem to me out-of-place.

The elder Monsieur Pellissier had the 48 livres that you entrusted to him regarding the herbarium delivered to me yesterday. In my preceding letter enclosed with your books and carried by the younger Pellissier, I acknowledged to you receipt of the gold cross for my niece. I recall that, in concluding it, I expressed to you my regret about the rarity of news from you; but your never-ending kindness to me has dissipated my fear. . . . I am very pleased to learn about the state of your beloved family and about the well-being of other notable people in Le Noyer [where Villars had visited on February 9th], your friends and, I dare to say, mine as well. You tell me that *you will come at Easter to cede your domain and assets to your brother from Villeneuve.* The term *to cede* seems ambiguous to me. Are you turning it over to him in perpetuity by establishing an annuity? or only as a lease with a fixed term? That is what I have not understood.[346]

Can one still write letters to you or have them addressed to you under the cover of the Monsignor Intendant, notwithstanding the departure of Monsieur de Marcheval?[347] I await a word from you in response to these matters and those about which I talked to you in my preceding letter if you have not already taken them up.

The abbé Pourret, out of great kindness to me, has offered me a benefice worth at least 1000 livres [a year] within the see of Narbonne, a living of which he himself may be the incumbent. While grateful for his concern, I told him in response that the mountains and the forests are my element, as in Lapland; and that, in all probability, I shall die there in my simplicity.[348] I have written to the chevalier de Lamarck about our *Artemisia insipida* etc. I do not know whether I will get a response from him.

The place I have for you in my heart will never permit me to lose sight of you. The household that you are going to establish in Grenoble will widen the distance between us, but it will take my heart only a moment to transport itself to where it knows you are.

                                           ~

[346] For reasons never directly explained, Villars had left his wife and children in Le Noyer when he went to assume his medical post in Grenoble late in 1781. In early 1784, he apparently concluded that he must move his family to Grenoble. His second younger brother, Pierre, who had married in 1774 and had left Le Noyer to reside in his wife's home in Villeneuve, hameau de Poligny, now returned to Le Noyer in 1784 to replace his elder brother, hence the arrangements questioned by Chaix.

Four of Villars' five children had survived: Pierre (1767), the second Dominique (1774), Marguerite (1777), and Marie-Anne (1780). Pierre was incompetent, and Villars established a pension for him with the proviso that he remain in Le Noyer (where he would die in 1795 as a laborer). In the margin of the manuscript of Villars' brief autobiographical sketch (1805), he would write the following sentence: I ask God's pardon, and that of my dear children, for having abandoned myself to my passion for botany, for study, and for having neglected their interests too much. See Manteyer, **Les Origines de Dominique Villars**, p. 204.

66
Les Baux, 19 March 1784

Montgolfier's device, Mesmer's secrets, and the politics of the Divan are of little interest to me even though I read accounts of them with great pleasure given the opportunity. Poor old botany better suits an inhabitant of the Alps of Upper Dauphiné. I do not know how to talk about anything else.

I am critical of you for giving a plant a trivial name already applied to another species. If one does not avoid this reef religiously, the result is chaos and confusion. Since you know the small sallow as *Salix lanata* L., why designat Haller's willow, no. 1643, with that name? and why not take, as Allioni and Scopoli always do, a name from the author's description, in this case, *Salix sericea* [Chaix]?

Carex leporina [auct.] [= *Carex ovalis* Good. (1794)] has adjacent spikelets (but not contiguous), alternating, forming a spike that is quite long and lax. In contrast, *Carex brizoides* [L.] has contiguous spikelets, gathered into a head. I have collected the latter at Chaillol, not the other one that you sent me from Grenoble. Do not conclude that I am confusing it with *Carex foetida* [All.].

After your examination of my galiums in the herbarium for Monsieur Jourdan, you conclude that your *Galium tenue* [Vill.] is Gérard's *Galium* no. 2, just as Monsieur Deleuze has always believed.[349]

"I have always feared," you say to me, "confusing your claimed *Ulex* with *Genista germanica* [L.], which resembles it; and I assure you that, if I had not found it in Adanson [-------][350], I would have believed it possible to make double use of this plant [-------]." I understand from what you say that you have not seen, or that you have not examined, the same shrub as I did. It resembles *Genista germanica* about as much as a billy-goat resembles a ram. *J. Bauhini pace dixerim.* I know this spiny broom very well (*G. germanica*). *Spartium scorpius* [= *Genista scorpius* (L.) DC. (1805)], which does not exeed a cubit according to Clusius, can resemble it in some respects. [--------] Clusius has seen it double a man's height and as thick as an arm between Bordeaux and Bayonne. Its branches and spines are so tufted that you cannot see through them. Its spines are robust, usually glabrous; those covered with down must be immature new spines. One sees leaves on them only in early spring, as they soon drop off. I saw it in that condition in Bellecombe [-Tarendol], not lower down but quite near the pass high up. Ray cites it for England, Flanders, France, and Germany: *In Gallo-provincia regione aestuosa ipsissimum hunc fructicem observavi.* Gérard also found it there, **Gallo-prov.**, no. 489, [*Provenit in gallopr. meridionalis montibus asperis. Frutex.*][351] Perhaps you are confused by the synonyms given by Tournefort who merged those of *Spartium scorpius* L. with those of *Ulex europaeus* L. As

⌇

347 Pajol de Marcheval, intendant for Dauphiné since 1761, was replaced in 1784 by Gaspard-Louis, baron Caze de la Bove, who would be the last royal intendant for the province.

348 The reference is to the admiration Linnaeus expressed for the simplicity of the Lapps' existence following his trip among them in 1732.

349 *Galium foliis linearibus, sulcatus, retrorsum scabris, pedicellis capillaribus.* Gérard, **Fl. Gallo-prov.**, p. 226.

350 Another section damaged by rodents.

351 Gérard had cited the Linnaean description for *Ulex europaeus* L.

I was traveling, I did not have the courage to collect any of it, because of its dreadful spines; but I do not despair of procuring it some day if God prolongs my life. If you can, send me *Spartium scorpius*.

You are going to accuse me of dabbling, but the desire I have to inform myself makes me devour everything. I am now quite ready to believe that the species sent to me by Monsieur Deleuze this year is *Bromus sterilis* L.: *glumis maximis, aristis subulatis, panicula erecta, maturescente secunda* (I am sending you its glume). And that our prior *Bromus sterilis* is really *Bromus arvensis* L. (*Festuca graminea, juba effusa* C. Bauhin).[352] As for the other somewhat smaller one that you called *Bromus arvensis* L., which perhaps Linnaeus passed over as a variety, it is certainly *Gramin phalaroides minus, erecta spicae, obliquis aristia* Barrelier, **Icones** 243. I call it *Bromus infractus*.

Here is the advice Monsieur Thouin asks me to pass on to you: "You really should convince Monsieur Villars," he says to me, "to publish at least the alphabetical table of his **Botanicum Delphinicum**, just as Vaillant did for the plants in the region around Paris. That would give botanists a foretaste of his great work and secure some subscriptions for him for sure." I quite support Monsieur Thouin's recommendation. Try to do it unless there are substantial reasons opposed to it. If I can do anything, let me know.

If you have any remaining seeds, let me share them. I received very few this year, and I am not expecting any more from anyone. In fact, are there any live plants you could remove from your garden without impoverishing it? Do you have *Helleborus niger* [L.] or *H. hyemalis* [L.] [= *Eranthis hyemalis* (L.) Salisb. (1807)]? I recall Monsieur Liottard telling me they are in Grenoble. I have asked for them for some time along with *Scheuchzeria palustria* [L.], but nothing ever escapes from that fine floral enclosure for my benefit. The famous line from Horace, *non cuivas homini contingit adire corinthum*, is confirmed by my case.

If your trip into your native country at Easter could bring you as far as my retreat, your presence would greatly enhance the joy for me that this holy time inspires.

67

Les Baux, 23 July 1784

Regarding the last of your letters, June 15th, which I received in due course, I am troubled on two counts. The first concerns the state of your health, altered, you tell me, by a fever brought on by the over-complexity of your work. I have still been unable to get assurance of your complete recovery, although I have sought information as frequently as possible through Monsieur Garnier. Please let me know at the earliest how you are. The second concerns Pollich's three volumes,[353] which you ought to have sent on June 18th by messenger, along with the copy relating to *Saxifraga repens* L., and which has not been been received here, contributing not a little to increasing my anxiety about the state of your health. If I have waited so long to respond to you, it is because I deluded myself into expecting more news of you from one courier or another in keeping with your promise.

The sudden death of our worthy confrere, Monsieur Arnaud, curé du Noyer and your

~

[352] Villars concurred, **Hist. Pl. Dauph.** 2: 116 (1787).
[353] J. A. Pollich, **Historia Plantarum in Palatinatu electorali sponte nascentium incepta** (1776-1777).

deserving pastor, has certainly much saddened me; and I sympathize with the part you have had to take in coping with this blow from providence. His good works were already known in heaven, and he will remain venerated here on earth by all those who knew him. *Thy righteousness shall go before thee; the glory of Jehovah shall be thy reward.* Isaiah 58. 8. Monsieur [Jean-Pierre] Tourniaire, who succeeds him, is very meritorious. The Lord, therefore, will always have this portion of His realm at heart.

We were at first told that your brother would become vicar in Ribiers, but I learned last week that he will be placed in that capacity in Tallard. His proximity makes me hope for the pleasure of seeing him a bit more often. He wrote me a word when passing through Gap on his return from Marseille. I have had no other news about him.

Please tell me whether the last edition of Linnaeus' **Genera**, **Species**, and even his **Philosophia botanica**, could be found in Grenoble or in Lyon, and what they might cost for the lot. Monsieur [Jean-Baptiste] Martin, curé du Saix, whom you know, would like to obtain these books. He has already a good knowledge of the **Elémentaire** that he had sent from Aix, and he needs some Linnaeus in order to progress in this science for which he has developed a great interest. He must accompany me into the Valgaudemar the first week of August, God willing. For the third time, I shall see this valley you have irrigated with your sweat, and which you have illustrated through your writings. On the basis of your information, Monsieur Martin will buy these botanical works.

I am sending you a small bundle from my botanizing this year, plus a few botanical notes on a separate sheet. Look it over at your leisure and have the kindness to give me your opinion about it.

A thousand compliments to your dear wife and to her two beloved pupils. A day does not pass that you and yours do not come into my mind several times. . . .

Botanical Observations for Monsieur Villars
Les Baux, 23 July 1784

1. Our geranium of the woods and mountains, which we formerly took to be *Geranium pratense* [L.], is, no doubt about it, *Geranium sylvaticum* L.: leaves very rugose, less deeply cut, lobes obtuse; corolla campanulate, violet or purplish, white on the claw, nearly marginate.

The one from Paris I am cultivating, sent under the label *Geranium sylvaticum* is, on the contrary, *Geranium pratense*: leaves deeply cleft, lobes narrower and sharper; corolla blue, white, or variegated (the claw not a separate color), quite entire.

According to me, the first is Haller's no. 932; Ray 1061. It appears that *Geranium malvaefolium* Scop. is only a variety of *Geranium pusillum* L.: *staminae 5, alterna minima, castrata fere nulla; ut bene* Murray, [Systema] 515. Common in villages and along roads.[354]

2. The plant that came up from seeds sent to me from Paris labeled *Crepis tectorum* [L.] set me off on new research. And I conclude from it that, without any doubt, the crepis you call *Crepis biennis* [L.] (and obliged me to call it that against my judgment) is *Crepis tectorum* L.; Haller, no. 31. The root does not descend deeply, the plant is gray-green, the terminal pinna of the cauline leaves is quite entire, narrow, lanceolate; and the affinity with

~

[354] Chaix was right about the geraniums. The *Geranium pratense* he was cultivating is not native to Dauphiné. Villars accepted the change but not the recommendation about *G. malvaefolium*.

Leontodon autumnale [L.] can readily be seen: the calyx hairs are sticky; the corolla is yellow below as above; the style free from the anthers and somewhat brown

The plant that you call *Crepis dioscoridis* [L.] is, on the contrary, *Crepis biennis* L.; Haller, no. 30. The root is long, connical, biennial; fistular plant, red below, glabrous or strewn with purplish prickles; leaves pinnate-lyrate, obtuse; terminal pinna large, angular, dentate; rays red below.[355]

The plant born here from the seeds from Paris is, in fact, *Crepis dioscoridis* [L.]: annual, nearly glabrous; leaves narrow, glabrous, smooth, dentate, rays red below.

I have not given much importance to their opening and closing. That varies according to whether the plant is exposed earlier or later to the rays of the sun.

3. I am quite tempted to agree with Monsieur Danthoine that our mountain *Matricaria chamomilla* [L.] [= *Chamomilla recutita* (L.) Rauchert (1974)] is *Chrysanthemum inodorum* L. [= *Matricaria perforata* Mérat. (1812)]. The plant grows here rather close to the mountain houses, notably around La Grangette and Chaudun. It has no odor; the scarious involucral bracts reveal an edge that is withered, wasted, brown in color, the *margine obsoleto* of Linnaeus. It is perhaps Haller's plant no. 101.[356] I must admit, however, that the seeds have appeared to me to be bare, excrescens. What leads me toward Monsieur Danthoine's opinion is that the plant he sent me for *Matricaria chamomilla* appears to me to be different from ours. It is the true camomile according to the authors (without speaking about *Anthemis nobilis* [L.]); it has an agreeable odor; and Monsieur Gouan locates it near the sea, but my claimed *Chrysanthemum inodorum* (alias *Matricaria inodora*) at the edges of the wheat-grasses. I am sending you a bit of the specimen from Danthoine. Compare it with yours and tell me your opinion of it. It will be necessary to obtain seed from Paris or Provence in order to sow it.

4. In addition to some of the specimens in the small bundle I am sending you, my trip to visit Monsieur Deleuze in Valernes in the spring produced the following for me:

Cochlearia draba [L.] [= *Cardaria draba* (L.) Desv. (1814)];
Astragalus incanus [L.];
Cytisus argenteus [L.] [= *Argyrolobium zannonii* (Turra) P. W. Bell (1968)];
Geranium molle [L.], although I already had it from Paris;
Anthemis cotula [L.];
Hyoscyamus albus [L.];
Linum campanulatum [L.];
Alsine mucronata [auct.] [= *Minuartia mutabilis* Schinz & Thell. ex Beckerer (1938)];
Psoralea bituminosa [L.];
Caucalis platycarpos [L.];
Antirrhinum arvense [L.];
Lathyrus angulatus [L.]; and the euphorbias in the bundle.

My trip to Sigottier, near Serres, produced the following plants:

Pistacia terebinthus [L.];
Rhamnus alaternus [L.];
Saxifraga hypnoides [L.], *foliis omnibus trifides, petalis candidis*, distinguishing it from
    Saxifraga caespitosa [L.];

∽

---

[355] Villars accepted Chaix's revision.
[356] Haller's no. 101 was *Matricaris chamomilla* L.

Carex drymeia [Ehrh.] in L. fil. **Suppl.** or *Carex sylvatica* [Hudson].

At a pond called Les Egaux at St.-Marcellin-lès-Veyne, I collected *Utricularia vulgaris* [L.]; *Myriophyllum verticillatum* [L.]; *Potamogeton crispus* [L.]; and *Alopecurus geniculatus* [L.].

I found *Carex leptostachys* [Ehrh.] in L. fil. **Suppl.**; Haller, no. 1390, according to my limited judgment.[357]   Do you have it from your botanical fieldwork?   *Hieracium auratiacum* L., *alpinum* C. Bauhin, which I had wanted very much, also revealed itself to me at La Grangette in the little alpine meadow called the Clot des Tiniers. I greeted it with great pleasure. Have you also met it?

As a favor, have a specimen of *Spartium scoparium* L., which you have around Grenoble, sent to me. Our broom in the Gapançais and Upper Provence, as you have described it very well, is not Linnaeus's. It is surely within the Linnaean genus *Genista*, but clearly not *Genista linifolia* L. All its leaves are simple, and never are they ternate. I quite believe that [Pierre] Garidel took it to be *Cytiso-genesta scoparia vulgaris flore luteo* Tournefort. That is to say, for *Spartium scoparium* L., just as Messieurs Danthoine and Deleuze understand it today. But we must be guided by the rules, not the examples.[358]

You must have *Pedicularis palustris* L. in Grenoble. Try to get it for me. We only have *Pedicularis sylvatica* L. in our marshes here.

Have you collected *Sonchus plumieri* [L.] [= *Cicerbita plumieri* (L.) Kirschleger (1852)] in the Grande Chartreuse? Monsieur Pourret had sent it to me from the Pyrenees. One cannot confuse it with *Sonchus alpinus* [L.] [= *Cicerbita alpina* (L.) Wallr. (1822)], which is abundant at Chaudun, in the valley of the Pleynet, along with *Arctium personata* [L.] [= *Carduus personata* (L.) Jacq. (1776)] that Haller appropriately made a *Carduus*, and with *Aconitum cammarum* [Jacq.] [= *Aconitum variegatum* L.].

What do you think of using *Saponaria officinalis* [L.] as a remedy for the syphilis virus as published by Monsieur Seguy, physician to the king, in the **Journal de Paris**, no. 34, last February 3rd? If experience verifies it, Europe will no longer covet *Lobelia syphilitica* [L.] from Virginia. . . .

[Following Chaix's signature, Villars appended the list of plants sent with the above letter, indicating they reached him on July 30th and adding highly abbreviated comments.]

Poa rigida: good

Triticum gerardi aristatis: perhaps new

Triticum gerardi muticum: perhaps *unilaterale*?[359]

Aegilops aristata: good

Carex arenaria L.: Is Haller, no. 1360

Juncus pilosus L.: name *lazulinus*?

Carex drymeia L. fil.: syn. our *silvatica, patula* Scop.

Alopecurus geniculatus: good

Carex phyllophora L. fil.: *C. pulicaris*?

~

[357] The following year Chaix called the plant *Carex ferruginea* Scop., Hist. Pl. Dauph. 1: 312, repeating Haller, Hist., no. 1390, which Villars did not recognize.

[358] This became *Genista scoparia* Chaix, non Lam., Pl. Vap.   39 (1785). Chaix in Vill., Hist. Pl. Dauph. 1: 343 (1786).   [= *Spartium cinereum* Vill. (1779) = *Genista cinerea* (Vill.) DC. in Lam. & DC. (1805)].

[359] *Triticum gracile* Chaix, Pl. Vap. 10 (1785). Chaix in Vill., Hist. Pl. Dauph. 1: 314 (1786).   [= *Triticum unilateralis* L. = *Vulpia unilateralis* (L.) Stace (1978)].

Carex leptostachis L. fil.: not new

Carex caespitosa Hall.? good. Affinity with La Roche *Carex montana*?

Avena spicata Valerna: *pratensis* L.? *montana* N.[360]

Bromus nemoralis Ch.: *giganeus* N.

Scabiosa integrifolia L.: dubious? *fol. lyrata.*

Centaurea calcitrapoides L.?: too many fixed opinions it seems. Thanks.

Trifolium striatum L.: good, thanks.

Euphorbia pilosa L.: thanks. *palustris*

E. esula L.

E. rubra danthonii[361]

E. platiphylla L.: is too small

Lapsana stellata L.: from Bellecombe etc.

Ranunculus pallidor var. *R. bulbosus?*[362]

Galium jussiei Danthoine: perhaps *tenue* N. perhaps *pyrenaeum* Gouan?

Saxifraga hypnoides? var. *caespitosa?*

Crepis dioscorides: good

C. biennis L.: perhaps good

C. tectorum? *biennis* preferably

Gentiana nivalis: good

Hedysarum alpinum: good. *stipula longiores* than elsewhere.

Hieracium aurantiacum: unsure between *Hieracium aurantiacum* and *H. cymosum* L.

Teucrium folium: good

T. capitulum: thanks

Hypecoum procumbens: thanks

Geranium pratense and *silvaticum*: thanks

Cytisus supinus: good

Crucianella latifolia L.: thanks

Myriophyllum verticillatum: good

Matricaria cammonmilla Danthoine: perhaps a sweet-flowered *Matricaria* N. and L.

Chrysanthemum inodorum: perhaps *Matricaria chamomilla* L. I was quite inclined to
accept 2 plants: *M. suaveolens* and *M. chamomilla*. . . . But when I have seen 3
plants in **Flora Suecica**, no. 764, 765 , and 766, I do not know what to say. We
must wait.

Serapias longifolia: good

～

*Villars was in his fortieth year in 1784. On two earlier occasions, when he had been gravely
ill and not expected to survive, he had each time drawn up a will and testament. On 19
September 1784, although apparently recovered from illness noted in Chaix's letter of 23 July*

～

[360] *Avena calycina* Chaix & Vill. in Chaix, Pl. Vap. 11 (1785). Chaix & Vill. in Vill., **Hist. Pl. Dauph.** 1: 315 (1786): invalidly published. [= *Helictotrichon sedenense* (DC.) J. Holub. (1970)].

[361] *Euphorbia rubens* Chaix & Vill. in Chaix, Pl. Vap. 44 (1785). Chaix & Vill. in Vill., **Hist. Pl. Dauph.** 1: 348 (1786). Villars omitted this species in the 3rd volume of his flora, evidently believing it to be a synonym of *Euphorbia characias* L.

[362] *Ranunculus pallidor* Chaix, Pl. Vap. 31 (1785). Chaix in Vill., **Hist. Pl. Dauph.** 1: 335 (1786). [= *Ranunculus sardous* Crantz (1763)].

1784 and in the letter below, Villars returned to Le Noyer to complete his third will and testament, a suggestion that he lacked confidence in his longevity.[363]

It would appear that this legal action occurred at the end of a circular field trip Villars took in 1784 when he guided the new intendant, Caze de La Bove, through the Oisans, the Briançonnais to Mont Dauphin, to Embrun, where they visited the area of Boscodon and finally the Gapençais. Villars asserted, Hist. Pl. Dauph. 3: xiii-xiv (1788), that he had been glad for the opportunity for the trip. Having suffered several months of daily or intermittent fevers, and having learned from 25 years of living in Champsaur that the pure, dry, cold air of the Alps prevented such attacks, he had hoped for a cure. He claimed that the trip did rehabilitate him.

68

Les Baux, 14 September 1784

You never stop making me feel the effects of your friendship. The work by Pollich gave me great pleasure. I use it a great deal, especially for the mosses where I am still very backward. I was waiting for you to let me know its cost, since you were willing to let me have it; but you have made it a gift to me. At the front of the book I found the mark of your benevolence, and the imprint of it on my heart will be indelible. Another sign of your friendship is the ingenuousness with which you ask for my advice on your manuscript on the trees of Dauphiné.[364] A poor curé, relegated to the foot of a mountain, is hardly in any condition to judge the turn of phrases, the correctness of style, or the value of a work. Nevertheless, I am going to speak my sentiments to you with all the frankness I owe to the best of my friends.

By not flattering you, Monsieur de Jussieu treats you as a true friend. Your manuscript, good as it is, is not ready for printing. It does not meet what the public expects from someone of your reputation. If this is to be only a simple listing, then you would need nothing more than the citation of authors and the native locations. But that would be insignificant, and the work would lack merit. If you want the reader to sense that it comes from the hand of a botanist of distinction, then his particular style must be evident in every species of plant treated, as it is in your sea-buckthorn and your willows. You have omitted quite a number of plants that I have substituted, as you will see, for some that you have. Your copy contains a great number of spelling errors (I have corrected some of them) that would annoy the printer when it comes to the correction of galley-proof.

In sum, I should prefer, adjacent to the Latin name and on the same line, the reader to see the French name; after which a generic character when there are several species; or, when there is only one species, a short but sufficient [generic character] should appear in its description.

Your explanations in reply to Monsieur de Jussieu have no merit. Are you only writing for about fifty wealthy seigneurs? Are you citing your plants only for those who already know them? Do you expect that, on the basis of your citations, your readers will

⌒

---

[363] Manteyer, "Les Origines de Dominique Villars," p. 140.
[364] This would become "Liste et observations sur les arbres de la province du Dauphiné," not published until 1787.

have recourse to Linnaeus, to Haller, to Duhamel, to Reichard, books very difficult to procure according to your letter of 3 August 1784 from Grenoble? Can you presume that they own all these authors? You are putting them off until [the publication] of your general history. Why, then, put a work into your readers' hands that is inadequate for them? Moreover, take care: you are praising both your general history and your riches of Dauphiné too highly in advance. If that is necessary at all, you must insinuate it, but not be the panegyrist. Unless you mean, like Haller, to invent new species of *Polyporus* and *Amonita*, your flora will hardly have more than two thousand native species.

The title of your work does not appear to be quite precise. Look at the one I should like to see substituted for it. The text from Virgil is well chosen. Your preface is prodigiously too long, and its style lacks polish. And so on.

You have treated the willows in a learned, critical manner for which I applauded you while making some corrections that I have noted. If the descriptions and observation were to be retouched, it would be better to publish the willows alone.

I sense that your delicate health and your duties will not permit you to recast the bulk of this work. If you approve of my manner of writing it and my treatments, which you will see in the enclosed notebook, I would very gladly undertake to rework it in its entirety. In that case, you could send the whole thing back to me, and I believe I could complete [the revision] before All Saints' Day. Following which you could make your own corrections and have it published.

But if you should decide to abandon it entirely, please send your manuscript back to me. I would turn it into a work in my own words, which I would enjoy doing, and which I could subsequently publish in my own name.

You will see from this what freedom of speech and action I take when it comes to you. If anything herein seems to you to be unwarranted, consider the source of it, remembering that I have no other aim than to prove to you the affection I owe you.

Your brother, vicar of Tallard, came to see me August 22nd. From here, he went through Chaudun in order to reach his relatives in Le Noyer. I saw him again in Gap on the 26th. He appears to be happy with his position. I believe he has notified his patron about his degrees.

Try to restore your health. I never stop asking the Lord for this favor. Monsieur Gaude, who was here last week, sends you a thousand compliments and prays to God for you every day.

I am deferring to another time our dispute over the two *Crepis*. I have not yet been able to convert myself to your position on the matter. Give me your news at the earliest.

P.S. . . . If you see the abbé Jullien, tell him, after having offered him my respects, that I received his small bundle, and that I shall respond to him by sending him something in kind. But have him collect for me a branch, either in flower or in fruit, of the medlar tree that you locate at Seyssins, his country. . . .[365]

[365] Abbé Jullien was curé de St.-Georges-de-Commiers, a village on the Drac southwest of Vizille.

69
Les Baux, 21 September 1784

You asked for my opinion about your work. I could not, without betraying the friendship that unites us, disguise it from you. But my judgment is of too little value to make an impression upon you. You are, therefore, free to put together and to publish your work when and how you please. You will receive with this [letter] your manuscript left in care of [Dr.] D'Heralde. Have the kindness to return to me the poor mess of notes I had assembled.

I am quite convinced, and I have always been, that I do not know how to write, much less to speak, in French; and from the time that I have had the honor to converse with you and to write you letters, you surely ought to have become convinced of it, too. But I believe that I understand French a bit. You shall see whether I have understood the sense of your letter from Gap dated 16 September 1784. In it, you say to me: "You sometimes put the adjective before the subject and the verb." I know that one says, for example, *des hommes illustres*. But I did not know that one must say *des hommes grands, un arbre beau, le mort de mon père bon, la fête de Jean saint,* etc. I doubt whether I have used an adjective before a verb. I have never seen the agreement of the adjective with the verb.

You say: "You truncate some sentences when wanting to shorten them." As I only studied nearly all the little botany I know in Linnaeus, it is not surprising that, when writing about plants, I feel the effects of his precision and his technical terms. [You say:] "You cannot reform my title in this manner: **Arbres, arbustes** etc., a faulty construction since this is a catalogue, a list, a book, and not trees." What was Monsieur Séguier thinking about when entitling his book **Plantae Veronenses**? since he was presenting a book and not plants! What did Monsieur Gilibert have in mind when he entitled the book he is composing, according to you, **Linnaeus Europae**? as Linnaeus is gone and as Gilibert only wants to publish those plants of Europe reported by the Swedish botanist.[366] Authors mislead us when they write as book or chapter headings *the earth, fire, birds, trees* [-------] *plants* etc. [367] (In the woods, [-------] in the region of [-------], at Le Buis: The native place of the plant is sufficiently indicated by such phrases that I believe it would abuse the reader's patience if I were to confront him with *it comes from, it grows, it is found,* and so on for every species.)

The efforts you made on the willows are appreciable, perhaps to be preferred over everything written about them before you; but anyone who might conclude that you have left nothing further to be researched on the subject could be mistaken.

(Continuing with what you tell me: "It is necessary [-------] to put into the title, on the title page, *by Monsieur Chaix prior-curé etc. and Villars physician etc.*; or *by Monsieur Villars, reviewed and augmented by Monsieur Chaix,* whichever will please you." Do you mean by this to have the frog burst by forcing it to inflate to a point of equaling the ox in size?) "It seems as if you have not read my preface correctly. It is the province whose riches I mean to praise, not the work I am announcing." I have understood that, but I feared that your announcement could have in the public mind, so often unjust, a reference to the fable: *The mountains are in labor; a ridiculous mouse will be born.* Horace, **Ars Poetica**, 139.

~

[366] J. E. Gilibert, **Caroli Linnaei botanicorum principis Systema plantarum Europae** (1785-1787).
[367] Rodent damage in this letter.

Everything I said and did was done with the best intentions. It appears from the analysis I have just made of your letter that you did not take it for that. I am in despair for having offended the best and the oldest of my friends. Burn my notes! Forget my imprudence! And let the friendship, that you have lavished upon me to date, continue for the remainder of my life. As for me, nothing will change the feelings of respect and affection that I have pledge to you until my last breath.

~

In **Notice historique sur Dominique Chaix** *(p. 304), Villars recalled this brief flare-up as the only moment of bad temper on Chaix's part during the 33 years of their friendship. As Villars remembered the incident, he had completed a catalogue of rare plants and shrubs of Dauphiné. He hastened to send it to Chaix for review, who noted a number of mistakes in Latinity. Given Villars' natural reluctance to revise his own work and the pressures upon his own time, he asked Chaix to undertake the corrections in spare moments during the winter, committing himself to publishing the work under Chaix's name or under both names, whichever Chaix preferred. Chaix was greatly displeased, and Villars was both surprised and angered that his proposal had provoked an unjust interpretation. He replied with a severe letter. Chaix backed away in his following letter, and the friendship survived the unfortunate moment.*

70

Les Baux, 8 October 1784

Nothing angered me in your letter written from Gap, and nothing in anything that will come to me from you will ever be able to anger me. I shall always take pains to distinguish the deeds of the best and the most sincere of my friends. Even so, I shall not conceal from you that, as you yourself asked for my opinion about your manuscript, and having given you exactly what I thought, I was not expecting the criticism of my critique. But you, having concluded that I had strayed off the path of truth in my judgment, chose to judge my judgment yourself, in order to get me to admit my error, rather than appealing to a third party who would have been unknown to me, as a consequence of which I never would have been enlightened. Such is the effect of close friendship. How right the wise man was to say: *A faithful friend is a strong defense: and he that hath found such an one hath found a treasure.* **Ecclesiastes** 6. 14. It was probably quite imprudent of me to show my head, but a charitable rap from your discipline has put me back into my shell until the dust of my grave. I prefer to grope along the traces of the amiable botany that unfolds before me than to declare myself its standard-bearer openly. I shall have nothing to fear if walking step by step in accordance with the feebleness of my powers. An incommensurate effort could head me for a mortal fall. I shall pray in the shadow of my solitude for those who are obliged to make their way in the full light of day for the benefit of the public.

Since, in your opinion, the French language cannot accommodate the technical and characteristic terms used by Linnaeus, the botanist who wants to write in French must substitute others for them that are either better or compensate for them. Otherwise, he ought not become involved in writing in his own language. . . .

Linnaeus calls the algaes *vernaculi*, the slaves; and the mushrooms *nomades*, the dregs for the country people. But this learned observer of nature had admired greatly the treasures from God's power and magnificence whether in the *Mucor* or in the *Adansonia*. I know that you are already extremely busy. I am sending you a small bundle of them, and, when you have the leisure, please take a glance at the names with which I have labeled them. I do not doubt that I have often taken one for another, having only some inadequate diagnoses or descriptions, without drawings and lacking instruction from any master. You will give me great pleasure if you will tell me, for example, if your *Lichen opuntioides* is really *Lichen* [--------] etc.

I am no more skillful with mosses, but I am withholding them to show them to you another time, if you will have the time to glance at them for my personal advancement.

For the first time, I have found three agarics (*Boletus* L.) on an old larch that have produced a half-pound for me. If you want to make use of it in medicine, I shall send it all to you.[368] If you have any need for *Lichen islandicus* [L.], I could also pick up a supply for you. The resin on the larch, *resina larigna*, which is abundant on old trunks, seems to me to have such an agreeable odor that I am going to try to use it in the censings during holy services in place of the alleged incense sold to us in the shops. I have no doubt that it could be used in different ways in medicine.

. . . I send greetings to your dear wife and all your family with all my heart and soul. I only regret to know that you are not entirely recovered. Give me your news as often as possible. In anticipation, I remain, with the sincerity of all the affirmations of friendship and affection that I can have made to you when concluding the prodigious number of letters I have had the honor to write you dating from our acquaintance, your very humble and very obedient servant.

71

Les Baux, 29 October 1784

The corrections that you were good enough to make in my *Lichen* have given me the greatest pleasure. I am applauding myself for being mistaken on only about one-third of them. I have put all the cryptogams I have been able to find in good order so as to make a separate herbarium of them. I have hardly more than one species on each quarter of a sheet of paper in order to represent the detail better, the mosses in particular. Everything you can send me about this class will give me the greatest pleasure inasmuch as I am more advanced in the other parts of botany. All the shoddy specimens that I sent to you were meant to be yours. I was merely asking for the right names when they were lacking. But it was my fault for not explaining myself clearly.

If anything among them would give you pleasure, and you would like to have them back, I will have them sent to you; for I have everything in duplicate or even quadruplet. I am enclosing a new bundle here in which you will perhaps find something of interest. When you have had time to examine the entirety in your moments of recreation, please write down your opinions of it for me: a short list on paper. And when I know that you will be able to look over my mosses, I shall also have them sent to you.

One could not be more delighted than I have been to get the news of the title Monsieur Allioni has conferred upon you.[369] Despite your modesty, your merit has cut through the veil you use to conceal it, and your reputation extends more amd more. I bless

---

[368] The *polypore du mélèze*, or white agaric of the pharmacists found on larch, has been known variously as *Polyporus officinalis* Fries, *Boletus officinalis* Balsch, *Boletus laricis* Jacquin, and *Agaricus laricis* Lam.
[369] Correspondent of the Reale Academia delle Science.

the author of all gifts for it. When the [**Flora Pedemontana**] by this celebrated academician, and our neighbor, goes on sale, I shall do my utmost to obtain a copy of it for myself, along with the **Linnaei** [**Systema Plantarum Europae**] by Monsieur Gilibert. Works of this sort provide the only pleasure I have in the desert of my wretched solitude after I have acquitted the duties that my station requires.

I am sending you the description of all the roses I know. It could be of use to you in your work on the trees and shrubs of the province. Send me your news as often as you can.

72

Les Baux, 30 November 1784

As your letter of October 12th, sent unofficially, reached me only a month after its date, I was unable to respond to it in mine accompanying a small bundle and a list of my roses that you should have received via the intendance. Therefore, you will find herein a little of the *Boletus laricis* for use in medicine.[370] I still retain about a third of it. If you need it, I shall send it upon your request. I have added about fifteen small packets of seeds to the package. If any of them merit being sown in your garden, I shall be happy. Although newly resident of Dauphiné, *Saxifraga sarmentosa* L. fil. *semina vel plantam vivam*; *Helleborus hyemalis* [L.]; *Helleborus niger* [L.]; *Spartium scoparium* [L.], etc. are plants I have desired.

This summer in Valgaudemar, I crushed *Osmunda crispa* [L.] underfoot without knowing about its medicinal use, about which you speak in your letter, only bringing back herbarium specimens.[371] I am quite angry not to be of use to you regarding this plant. A propos of *Osmunda*, I am aroused against Linnaeus for having put *Osmunda lunaria* and *Osmunda crispa*, as well as *Osmunda spicant*, into a single genus, when their fructification is so different.[372] Haller did not do the same with them. What is your view?

Again, I am sending you some specimens that, if imperfect, are recognizable. When you have the time, please have the kindness to tell me what you think of them.

Festuca bromoides [L.] [= *Vulpia bromoides* (L.) S.F. Gray (1821)]: You had given me an excellent example of this, which relates very strongly to the description and to the note in the **Species Plantarum**. Why, then, does Linnaeus, in **Mantissa 325**, say that it is close to *F. ovina* [L.], to *F. amethystina* [L.], and to *F. rubra* [L.]? I am tempted to believe that *bromoides* was written by mistake while looking at *F. duriuscula* [L.]. Murray, his ape, transcribed without discernment all that he found in his master. What does it seem like to you?

Festuca duriuscula [L.] [= *Festuca lemanii* Bast. (1809)]: For this plant you have given me a small specimen collected near La Mure, if I remember correctly, which is the same as *Festuca pumila* Chaix collected in Chaudun.[373] If I adequately understand Pollich, *Festuca duriuscula* is common in our dry pastures in Les Baux, different from *F. rubra* [L.] because of its narrower leaves, its green spike, rarely violet.

∽

[370] See letter no. 70.
[371] Villars would say, **Hist. Pl. Dauph.** 3: 839 (1789), that *Osmunda crispa* L. [= *Cryptogramma crispa* (L.) R. Br. ex Hook. (1842)] was a proven antitussive to be used to combat colds.
[372] *Osmunda lunaria* L. = *Botrichium lunaria* (L.) Swartz in Schrader (1802). *Osmunda spicant* L. = *Blechnum spicant* (L.) Roth (1774).
[373] *Festuca pumila* Chaix, **Pl. Vap.** 12 (1785). Chaix in Vill., **Hist. Pl. Dauph.** 1: 316 (1786). [= *Festuca quadriflora* Honckeney (1782)].

Agrostis verticillata Vill. I believe to be *Agrostis alba* L. ex Poll.[374]

My *Agrostis villosa* [Chaix], which we collected in a meadow near the town of Vallouise, although close to *Agrostis arundinacea* [L.], differs because of abundant silky awns at the base of the follicle. *A. arundinacea* only has them on one side.[375]

Oenanthe peucedanifolia Poll., [Hist. Pl.] Palat., grows in La Saulce. I also saw *Poa salina* Poll., **Palat.** no. 92, near the saline spring there. But I had the misfortune of losing my specimens while collecting other plants in the rock.[376]

I have not been able to distinguish your *Festuca phoenix* [Vill.] from *Festuca elatior* [L.].[377] I do not know if it grows here. Also, Pollich does not mention it. I am surprised that they do not have our *Festuca arundinacea* [Schreber] either in Germany or in Sweden, or our *Bromus pratensis* [Ehrh.] [= *Bromus commutatus* Schreber (1806)], very common here.

I have put all the grasses, rushes, sedges, and allies that I have to date in good order in my herbarium. I can find them in an instant and consult them. This winter I shall put together a table of all the plants I have in my herbarium, indicating our natives with a sign.

Catalogues in botany, whose usefulness you defend in your letter to me, are only reference tables in my opinion. So scanty are they for amateurs that such works remain new for a long time in their libraries. Because Gérard added new plants and descriptions to his work, it is useful; whereas Gouan's is virtually neglected, because he only copied descriptions from Linnaeus, Tournefort, and Sauvages. I could have eaten the berries of belladonna a hundred times instead of whortleberry in our forest of Loubet if I had only possessed a catalogue of the plants of that place, or if someone had not taught me about the different qualities of the two plants. A fleabag sometimes has a sign as attractive as that of an excellent inn. One needs information to obtain decent lodging. Therein lies the difference between the simple botanical catalogue and any other work in this genre that describes plants. Remember that you yourself raise your voice against the spinners of floras or catalogues in your preface on the trees of this province, citing quite appropriately the judgment the celebrated Monsieur Adanson rendered about them.

The abbé Jullien sent me a nameless *Carex* within a small bundle, and I cannot determine it unless it is *Carex baldensis* [L.]: A bisexual spike is seated in the axil of a bract about six inches long. That is to say, female below and male above. About one inch higher up, there are two or three more sessile spikes, similarly bisexual, each one subtended by a bract, very close one to another, nearly clustered; the scales yellowish, slightly beaked, bifid. Seen from a distance, one might take it for *Carex flava* [L.]. I do not know whether you can follow me from this poor description, but I am inclined to believe that it is *Carex*

---

[374] *Agrostis verticillata* Vill. has been called *Agrostis alba* L. var. *verticillata* (Vill.) Fiori and, more recently, *Polypogon viridis* (Gouan) Breist. (1966).

[375] *Agrostis villosa* Chaix, Pl. Vap. 74 (1785). Chaix in Vill., Hist. Pl. Dauph. 1: 378 (1786). [= *Secale villosum* L. = *Dasypyrum villosum* (L.) P. Candargy (1901)].

[376] Villars evidently assumed Chaix had seen *Poa aquatica* L. [= *Puccinellia distans* (L.) Parl. (1850)]. *Aira miliacea* Vill. is another synonym, as is the invalidly published *Aira brigantiaca* Chaix.

[377] Both are regarded as synonyms of *Festuca arundinacea* Schreber (1771).

baldensis as I recall that you cited it from Allevard in your manuscript.

Please inform me about the state of your health, and if, finally, it has been restored.

P.S. I pointed out to you in my notes on your trees that it seemed to me that you had slipped some mistakes into your sentences on *Prunus cerasus* L. and *Prunus avium* L. No doubt you will have paid some attention to the matter.

Prunus cerasus L. In Paris, and in Sweden as a consequence, its fruit is called *cerise* [cherry]. In Dauphiné and in Provence, it is called *griotte* or *aigriotte* [morello cherry]. In the Gapençais, we have the variety *austera* L. growing wild, **Species plantarum** [p. 474], to which we graft the variety *caproniana* L., ibid. We call *Cerosa hispanica* Lob. the *aigrotte* of Catalonia. I do not know of any others here.

Prunus avium L. In Paris it is called *guigne* [heart cherry], but here *cerise*. To this species, Linnaeus, **Mantissa** 397, has properly related the [variety] *bigarella* [L.], the *bigarreau* [white-heart cherry]; and *duracina* [L.], the *duroa*. I think he would not have done badly to relate it also to *dulcis, cerasa alba duleia* J. Bauhin that I believe to be our *aigrattioux*. The red and black wild cherries in our woods are the principal varieties of *Prunus avium*. Please give me your opinion on these observations.

As the package is somewhat large and unshapely on account of the agaric, I have not dared send it to you as in the past under the intendant's cover.

73

Les Baux, 17 December 1784

The wind from the east (*Eurus*), which is called the Lombard here, having dominated us for five days, the 8th to the 13th of this month, brought us such a great quantity of snow in passing over the Alps that we have three feet of it here. And the cold that followed in its wake has so hardened it that, from all appearances, the earth will not shed this blanket for some months. My feet take no other walk than the short span that separates my cottage from the church. But my mind pays you a visit in Grenoble a hundred times a day. In a moment, it can descend from Le Lauraret to Sisteron, from Montgenèvre to Le Buis; and, in taking me pleasantly the length of its route, it allows me to collect all sorts of pretty plants. Thus, my time, filled with the duties of my station and with this pleasant occupation, does not seem too long to me. I now conclude this prelude that is useless to you.

Your charming letter of the 6th instant reached me on the 12th. Everything in it is instructive and interesting. But the news you give me concerning the coming publication of your work on the plants of our province has been a complete joy for me. I am the more sensible to it for having awaited it for so long. Ah! How delighted I should be if I could contribute in some way to its completion! Have no doubt that I would undertake any difficulty or any study for that purpose. I accept very gladly the task you ask of me: to draw up a list of the plants I have collected in the Gapençais, by which I mean from the Valgaudemar to Le Poët and to Serres. I would be determined to arrange them according to Linnaeus's natural orders (just as I want to do for the catalogue of my herbarium). I shall add the notes and remarks I have to make immediately after each species. I shall use Latin to be more precise and in order not to offend the French language with uncommon expressions. You may adopt from it whatever appears to you to be well founded, but I shall not risk putting anything in without serious reflection. Please have the kindness to indicate whether you approve this procedure. If you have another one in mind, I shall conform to it when you give me your opinion. I do not see that naturalized plants ought to be omitted, even though not native, such as the wheat-grasses, the vines, the nuts, etc., as those are the plants of more general interest. Would it not be to disparage them by saying nary a word about them in such a work?

You tell me that I am not in agreement with Linnaeus, who refers the *guignes*, the *bigarreaux*, and our *agrattioux* to *Prunus cerasus*. You would not make this criticism to me if, when writing, you had had my letter either in your memory or in front of you. In this letter, I shall excuse myself from repeating what I wrote to you in the previous one, but which attributes all *griottes* and *cerises* (I mean as those names apply in Dauphiné) to a single species. Linnaeus did not fail to correct his error in his observations. In a word, apart from the different varieties of *griottiers*, every *cerisier*, cultivated or wild, is *Prunus avium* [L.]. Nevertheless, I do not know what Monsieur [Henri-Louis] Duhamel [du Monceau] has written on the subject. Consult him and give me his opinion.

I sent you, it is true, *Centaurea calcitrapoides* L. and my [*Centaurea*] *hybrida* [= Chaix, non All.];[378] but I did not send you my *trispinosa*, which is quite another plant, having received only a defective specimen of it from Monsieur Deleuze.

I have no doubt that *Chrysanthemum inodorum* L. is the plant I designate by that name.[379] All the characters reported by the authors fit it. It may very well be that the plant in Grenoble you know under the name *Matricaria suaveolens* [L.] is *Matricaria chamomilla* [L.]. They differ from each other only in the direction of their rays. It remains for you to find out whether yours has a reflexed ray. Please send me some seed of it. Yours from here has no odor. It was never the camomile sold in shops. It is, therefore, really *Chrysanthemum inodorum.*

It may be that your *Galium anisophyllum* [Vill.] is *G. sylvestre* Poll. [= *Galium pumilum* Murray (1770)]. But if I were to rely on the report and the specimens of it sent by Monsieur Danthoine, I am surprised that you define this galium of Pollich and *Galium jussiei* [Vill.] as very close.[380] The former is weak and slender, often over a foot tall. The latter, in Monsieur Danthoine's report, and in the appearance of all his specimens, is rigid, tufted, not exceeding an inch and a half in height. Compare this with the drawing of [*G.*] *jussiei* that I have never seen. I am enclosing two specimens of it here. Moreover, are you convinced that the number of leaves, used by Linnaeus, Haller, and by you, is a [truly] distinctive character except in the species with quaternate leaves? By using that [character] with the descriptions alone, one cannot settle on either your species or those of Linnaeus. A different character must be found.[381]

Who has assured you that your single specimen of *Carex montana* [L.] from Le Sapey

~

[378] *Centaurea hybrida* Chaix, non All., Pl. Vap. 61 (1785). Chaix in Vill., Hist. Pl. Dauph. 1: 365 (1786). Chaix called it an authentic hybrid of *C. calcitrapa* and *C. aspera*, collected by Deleuze on level ground at Ventavon. [= *Centaurea* x *chaixiana* Rouy = *C. calitrapa* L. x *C. aspera* L. ssp. *aspera*].
[379] See letter no. 67.
[380] *Galium jussiei* Vill., Prosp. Pl. Dauph. 20 (1779), may be a ssp. of *Galium pumilum* Murry or a good species.
[381] Villars would segregate his *Galium* into three groups: 1) Species with smooth seeds; 2) those with erect, cylindrical stems; 3) those with hairy seeds.

is the true *montana*, to the exclusion of the one I was naming [*C.] verna*? Pollich clearly reduces the latter to the former, but I am not rejecting it.[382]

Bromus montanus Poll., **Palat.** p. 116, no. 116; Haller, **Hist.**, no. 1506, is *Bromus ramosus* Murray, **Syst.**, p. 102. . . . (Is it not an inconvenience to use the same trivial name for two diverse plants as Murray does in this case and in that of *Geranium pyrenaicum*?) I have sent *Bromus montanus* Poll. to you.[383]

Bromus versicolor Poll., **Palat.** p. 109, no. 109; Haller, **Hist.**, no. 1503. I believe this is the species you want to call *Bromus arvensis* [L.]. But the true *Bromus arvensis*, Haller, no. 1509; Pollich p. 113, no. 113, produces the awn from the very top of the lemma, as the *Festuca*, according to Pollich; whereas your alleged *arvensis* produces its awn very visibly below the top of the lemma. Monsieur Danthoine also disagrees with your opinion.[384] I have reported the citation of these numbers carefully here, because I see that your citation does not correspond.

You ought to ask the abbé Jullien to send you his plants from Dauphiné. He may well have found some that you have not. Let me know what has become of him. I made a shipment to him about which I have as yet heard nothing. I wish you a happy and holy Christmas.

74

Les Baux, 7 January 1785

The two copies of the brochure announcing your history of Dauphiné were promptly sent to me.[385] I have had the greatest pleasure reading it, and I have been most keenly aware of the very civil testimony you wanted to make in it in my regard. I have retained one copy of it as a monument to your friendship; and, according to your desire, I have sent the other one to the comte de La Roche, who will pass it on to the comte de Revilliasc [later Revigliasc], seigneur de Veyne, to the abbé d'Aspres, etc. The quantity of snow has not yet permitted me to go down [to La Roche] to see him; but he sent word that he will gladly become one of your subscribers, and that he will enlist his friends to do likewise. In your accompanying letter, you describe the conditions that you have had to accept for the

⁓

[382] *Carex verna* Chaix, non Lam., **Pl. Vap.** 8 (1785). Chaix in Vill., **Hist. Pl. Dauph.** 1: 312 (1786). Haller, no. 1381. [= *Carex caryophylla* Latourr. (1785)].

[383] See *Bromus giganeus* Vill., **Hist. Pl. Dauph.** 2: 118 (1787). Villars was able to exploit this Linnaean name because of his prior, and successful, transfer of *Festuca gigantea* (L.) Vill., ibid. 2: 110 (1787).

[384] As a consequence, Villars would change his equivalent of Haller's no. 1503 to *Bromus nemorosus* Vill., **Hist. Pl. Dauph.** 2: 117 (1787), citing as synonyms *B. nemoralis* Huds. and *B. asper* Murray. [= *Bromus ramosus* Huds. (1762)].

[385] The 8-page announcement for the **Histoire naturelle des plantes du Dauphiné** in 3 volumes was unfortunately called a prospectus. The work was offered by subscription at 24 livres for all 3 volumes including the plates. The first volume was projected to go to press in May of 1785, the printing to be completed in December in order to appear on 1 January 1786. Subscribers were asked to respond by 1 May 1785 so that the number of copies to be printed could be determined before going to press. The announcement was worded in such a way as to suggest that subscribers send their orders directly to the printer, J.-L.-A. Giroud, rather than to Villars, a source of later confusion. A copy of the announcement is preserved by the Bibliothèque Municipale de Gap. The importance of the impending flora had already been recognized officially by the subdelegate Delafont in his report to baron de Caze de La Bove as he passed through Gap on 15 September 1784. See Delafont, **Mémoire sur l'état de la subdélégation de Gap en 1784.** AH-A, 8 324, pp. 15-16.

publication of your work. Your dedication to the public weal and your disinterestedness are always the two incentives that guide your actions. God will make up for that which is lacking in the justice of mankind.

On the 5th of this month, the same day that your letter of December 28th reached me, I completed my list of the plants that I know to be native or commonly naturalized in the Gapençais, from the Valgaudemar to Serres and Le Poët. I have found about 1550 species within about 460 genera, but some genera of cryptogams are incomplete. I am sending it to you in its present form. As the brochure simply announces the list of plants from the Gapençais and from the Briançonnais, to be inserted in the first volume, the few notes I have added to it will really belong only in [the later volumes devoted to] species, along with any correction or modification you will deem necessary. I think that all observations, all analysis, should be reserved for *that* portion of the work, even the indication of the native locale of a plant. It appears to me, therefore, that you will be obliged to repeat the locale of your plants in the region of Grenoble in the later volumes if, as is stated in the announcement, the first volume contains the list of your botanical field trips in the region around that city. Even so, in making this observation to you, I do not mean to intrude upon your work.

I have indicated the native area of the plant when I believed it to be rare or little known. I have even written descriptions when I have not found any in the authors or believed them to be inadequate for the plant. You may make use of them as you judge to be appropriate in your later volumes on the species; and it will not be difficult to translate my observations into French whenever you value them to be judicious and necessary. If, however, you should want my list of plants in another form, indicate that to me when returning the notebook to me in the package for Monsieur Delafont. Two weeks of work will settle the matter. I have added one other short list of rare plants from the Briançonnais dating from our trip in 1779, and mine in particular into the Embrunnais in 1783, with the native locations, that could be of some use to you.

Trivial names are the touchstone in botany. Thus, do not create one when the plant already has one in some author. Monsieur de Lamarck has only sown confusion by wanting to give new names to plants that already had them. I shall never forget the maxim of the illustrious Linnaeus: May the same trivial name never be given to two plants--a difficulty not always avoided by Monsieur Murray. If I have altered some trivial name of yours [on my own list], it is because I thought I could do it before the printing of your work, substituting one more suitable for it. For example: How do you find the form of the narcissus flower in our beautiful mountain garlic? I prefer to call it *grandiflorum*.[386] Several *Bromus* are perennials. Why not, instead of your *B. perennis*, use *pratensis* taken from C. Bauhin's synonym? etc.[387]

I am delighted that you have finally come to understand me about the *Prunus cerasus* and *avium* L. What you tell me about *cerosa dulcia alba* C. Bauhin (that I believe to be our *agrattious*) is what I told you in my penultimate letter. I have not given up hope that you

~

[386] *Allium grandiflorum* Chaix, non Lam., Pl. Vap. 16 (1785). Chaix in Vill., Hist. Pl. Dauph. 1: 320 (1786). [= *Allium narcissiflorum* Vill., Prosp. Pl. Dauph. 18 (1779)].
[387] *Bromus pratensis* Chaix, non Lam., Pl. Vap. 12 (1785). Chaix in Vill., Hist. Pl. Dauph. 1: 316 (1786). [= *Bromus perennis* Vill., ibid. 2: 122 (1787) = *Bromus erectus* Huds. (1762)].

will come around similarly in the future to my view of *Crepis tectorum* and *biennis* to which I am still holding despite your contrary arguments. It seems to me that Gouan says there is an imperceptible transition from *Crepis dioscoridis* [L.] to *Crepis tectorum* [L.], which accords easily with my view. But *Crepis dioscoridis* will not assimilate with *your Crepis tectorum*, which is *Crepis biennis* in my opinion.

As for *Bromus arvensis* [L.], I should like to believe you are right; but I clearly see our small *Bromus* with narrow spikes, hardly a line and a half in length, nearly purplish, in Pollich, no. 109 (*versicolor* that this author places immediately following *B. secalinus* [L.] and before *B. mollis* [L.], quite appropriately as it has the greatest affinity with *B. secalinus*; and it does not have oval-oblong spikes.) *Bromus arvensis* L. must have affinity with *B. sterilis* [L.] as it follows it immediately, and as Rudbeck has called it *festuca arenaria sterilis elatior circa upsaliam*. The one that Pollich describes by that name, and gives its native location, is totally unknown to me. It may be presumed that Pollich, being separated from Switzerland only by Alsace, consulted Haller. As for *Bromus montanus* Poll., which has some affinity with *B. sterilis* [L.], it is truly perennial.[388]

The *Phyteuma* that Monsieur Liottard got from me, *foliis radicalibus cordato-oblongis, obtuse dentatis*, I have always believed to be a variety of *Phyteuma orbiculare* [L.]. Those leaves led me to call it var. *betonicaefolia*. It grows at Rabou and Chaudun *in alpinus*. The more common *P. orbiculare* merely has oblong leaves, including the radicle. *Phyteuma charmellii* [Vill.] has radicle leaves that are cordato-rotunda, like the March violet.[389]

Here is something more serious, Monsieur. The wife of my nephew in La Grangette is afflicted by a kind of dropsy in her stomach and legs, brought on (I think) by the weakness of her stomach that only a little food overloads, and by fatigue initially, having difficulty with her respiration. Her condition is possibly similar to that of your wife last year. Please have the kindness to respond as soon as possible about what we can do to avoid dangerous consequences, a favor I beg of you very earnestly. I wish you and your wife a happy new year.

P.S. If it is possible for you to send me a bit of your *Bromus arvensis* L., which you say is quite different from my *B. versicolor* Poll., even if it should only be a spike, you would give me great pleasure.

75

Les Baux, ? February 1785

[*Only a fragment of this letter remains.*]

You have rendered the greatest service to me and my relatives by prescribing and sending me the remedies for dropsy for my niece in La Grangette as quickly as possible. Despite the remedies prescribed for her by local alleged experts, the swelling in her legs, thighs, and stomach remained extraordinary, to a point that she could urinate very little and only with great difficulty. The three doses of *ipecacuanha*[390] did not have a noticeable effect upon her evacuation; but whether through their action, or through the use of the

---

[388] See letter 73.

[389] *Phyteuma betonicifolium* Vill., **Hist. Pl. Dauph.** 2: 518 (1787). *P. charmellii* Vill., ibid. 2: 516 (1787).

[390] The Brazilian name for the root of *Cephaëlis ipecacuanha* (Stokes) Baill., used medically in general as an emetic.

elderberry infusion with solium nitrate [saltpeter], the swelling has considerably diminished, and the urine has resumed a freer and more abundant flow. It appears that her stomach is functioning more normally, for the patient has a satisfactory appetite and is getting enough sleep. I can no longer find any *arnica*, and I sought it in vain in Gap, to help her sleep according to your prescription. Therefore, will you please send me a small packet of it by return messenger with instructions to have it delivered immediately to Monsieur Garnier, who will pay the shipping charge and from whom I will get it. Please have the kindness to include any instructions that you deem necessary. The patient is still in serious condition, and we must neglect nothing to restore this poor mother of a family--this relative who is very dear. She is not pregnant, and the usual flows have not appeared for some time. My nephew will try to express his gratitude to you, and this mark of your friendship for me augments the number of benefits you have never stopped piling on me.

I think, Monsieur, that you should not fail to put the epithet *runcinatus* in your glossary of botanical terms. Linnaeus often used it, and he did not explain it. Those who came after him (at least all those who have fallen into my hands) have not said a word about it. In my judgment, understanding this expression is not so easy.

My catalogue of plants from the Gapençais, already sent to you, is not ready for printing as such. As you are expecting to include it in your first volume, please read it with your usual attention, make your criticism of it forthrightly, and send it back to me: so that I can revise, take advantage of your observations, add to it any species that have escaped me and that you will take the trouble to note, above all regarding the final families in the Cryptogama. Between now and Easter, I will have it sent back to you. The expression *errant parisienses* is intolerable. It left my pen for your eyes alone, not for others. I shall indeed be able to modify it, as well as numerous others.

As for the two *Crepis*, the more I read the **Species** by Linnaeus and my extract from Haller's **Historia**, as well as Pollich, the more I hold to my opinion. You have told me a hundred times in your letters that the one you call [*Crepis*] *biennis* "has a flower yellow below, but larger than the other one." The second part of your description is incorrect: the flower is smaller by a half. You have my *Crepis tectorum* L., with a flower like *Lapsana*. . . . You still criticize me for making my *Crepis biennis* (corolla red below) with a larger flower than the other one. I only *say* that, but it is Nature that has *made* it larger. Take the trouble to compare the specimens with the figures in J. Bauhin. . . .

76

Les Baux, 12 March 1785

My poor niece from La Grangette died February 11th from her dropsy. I expected this end for some time, having seen in what condition she was; but I could not neglect anything in order to prolong her days. That is why, after having seen the uselessness of the remedies that were brought to her from Gap and from Veyne, I turned to you. Your last shipment of *ipecacuanha* and *arnica*, reaching me as quickly as possible, was only delivered to me on the day of her burial. I am obligated to you for your care and your expense. The loss for her family, which is mine, is very great. Her first child, a girl, was stillborn. She then had seven boys in succession, of whom only four remain, the last being only eighteen months old. May God in His pity grant her eternal rest and his consoling help to her relatives: *amen.*

In accord with your directions, I shall make two columns on each page of my catalogue; and I shall put my observation, which I shall write out to the extent possible for me, at the bottom of the page. I cannot avoid adding a few more of them to those in the first draft of the manuscript. A bare catalogue rather discredits its author and is nearly useless for the reader. Even so, I believe that it will barely exceed 35 pages in the format of the **Prospectus**.

Your definition of the adjective *runcinatus*: "pinnatifid leaves whose segments form a triangle, with a straight line and transversal at their back part, oblique, and facing toward their point on their foreside," does not encompass all the runcinate leaves in Linnaeus. How many of them are only sinuate-dentate? I do not know what Monsieur de Jussieu will say to you about the matter. I shall venture to give another, and more general, definition of the term and insert it into my observations.[391]

I am embarrassed by the trouble you are taking to enlighten me about our two *Crepis*, *dioscoridis* [L.] and my *tectorum* [L.]. But it is not surprising as these two plants have affinity. Yet, my *Crepis biennis* [L.] is not close to them. Everything you transcribed from Gmelin for me on *Crepis tectorum*, [**Flora sibirica**] 2: 28, no. 25, the stem, leaves, seeds, all fit perfectly with *Crepis tectorum* such as I know it; and we have no dispute about *Crepis dioscoridis*. Perhaps we are in agreement and have not been understanding each other.

As a favor, take the trouble to examine the two specimens that I sent you in my last letter as well as the names that I wrote under them. My *Crepis biennis* is truly biennial. I have observed it: its leaves are pinnatifid or pinnate-lyrate; its conical seeds terminate in a very thin point. The two plants are so well defined and described in the **Species** of Linnaeus that I cannot make a mistake there. I leave to the scholars the task of making a concordance of synonyms, which is beyond my powers, lacking both books and discernment. I fear, indeed, that my labels in the two herbaria I sent you do not reflect my present opinion.

You have inserted *Arenaria saxatilis* [L.] in my catalogue. If this is Linnaeus's *calicibus ovatis*; Guettard's *obtusis*; or Haller's no. 867, *petalis calice longioribus*, please tell me where it grows. I have never seen it. The only thing I know that approaches it is [*Arenaria*] *larcifolia* [L.]. As for [*Arenaria*] *tenuifolia* L., abundant on our slopes, you will note in Murray its annual character. Mine, which *is* Linnaeus's, is very much a perennial, and I could never make it into *A. saxatilis*.[392] As for the synonym in J. Bauhin and in Séguier, I refer it to *Alsine mucronata* [L., **Mantissa**] [= *Minuartia mutabilis* (Gouan) Schinz & Thell. (1938)]; as Séguier's synonym referred by Linnaeus to *Alsine mucronata* belongs to *Arenaria fasciculata* [L.].

I know too well, dear and affectionate friend, the state of your sentiments in my regard not to be assured of the pleasure I would bring you were I able to go to visit you during the fine season; and I have been confirmed more and more in that belief by your reiterated

⁓

[391] Villars did investigate the matter further and would publish a revised definition in Hist. Pl. Dauph. 1: 40 (1786): "leaves pinnatifid, the segments triangular, straight or perpendicular on their petiole, oblique in back, and facing the point at the front, as in dandelion or *Sisymbrium irio* [L.]. The character is not explained in **Philosophia botanica** [1751], but it is in the first volume of **Amoenitates academicae** [1749]."

[392] *Arenaria tenuifolia* L. = *Minuartia tenuifolia* (L.) Hiern. = *Minuartia hybrida* (Vill.) Schischkin (1936). See Vill., **Prosp. Pl. Dauph.** 48 (1779). While *Arenaria saxatilis* L. would appear in Chaix's catalogue, he cited Villars as justification for its inclusion.

solicitations. But, as the years have gone by, I have become increasingly ponderous;[393] and, beyond my bearing, my speech and my accent in your city would reveal me to be simply a bear pulled from the mountains and the woods. But I shall see what I might do.

So, Messieurs de Bruno still remember me! It is more than thirty years since I left that house, but my sentiments have never separated from it. Perhaps not a day passes that my heart does not go back to that amiable family in Grenoble on to their country place in St. Joine....Ah, if only I had had a little taste and talent for botany when I was really there, the plants of that pleasant country up to the Duiers River and to Pont-de-Beauvoisin, the marshes of Chirens, of Massieu, and of the Lac de Paladru, would not have escaped me. Are Monsieur and Madame de Bruno, the father and mother, still alive? Their elder son, my pupil, whom we called Monsieur de St. Severon, died a few years after my departure. The second son, Joseph, must have become an ecclesiastic. The third son named Claude, the father's name, should now be head of the house. I did not know the others born later. Please inform me about them and present my very humble respects to those who remember me.

You ask my advice about the manner of numbering your plants. I prefer the one you are proposing to Haller's way. For the purposes of citation, who can learn by heart numbers carried into the several thousands? Thus, follow the practice of Linnaeus as you suggest to me.

Has Monsieur Gilibert published his **Linnaeus** yet?[394] What is your opinion of it? Through our friend the abbé Blanc, I shall procure Monsieur Allioni's latest work if the price is not beyond my slender means. In addition to the subscription of the comte de La Roche, you will have one from Monsieur [Jean-Baptiste] Martin, curé du Saix, and possibly one from the Abbé d'Aspres [Jean-Joseph Corréard]. I shall also approach Monsieur Bertrand of La Freissinouse.

Since December 8th, we have neither seen nor walked on the ground; and the carpet of snow that conceals it from us is still not pierced at any point. It is a good 20 inches deep around my village. The weather, even though less cold this month, makes us very desirous of fine days and a robust economy. When it shall please the sovereign moderator of Nature to give His orders, the fogs will dissipate, the ices melt, and the waters will be seen flowing on all sides. *He sendeth out his word . . . and the waters flow.* **Psalms** 147. 18-19.

77

Les Baux, 9 April 1785

Everything you sent to me by registered package reached me safely. I received initially your three live plants: *Spartium scoparium* [L.] [= *Genista scoparius* (L.) DC. in Lam. & DC. (1805)], *Cicuta virosa* [L.], and *Saxifraga sarmentosa* [L.] [= *S. stolonifera* Meerb.[395]]. I send you my thanks for them, and I shall send you what you ask for when the snow finally

∾

393 Chaix had reached the age of 55.
394 J. E. Gilibert, then at Lyon, published seven volumes of selections from Linnaeus between 1785 and 1787 entitled **Caroli Linnaei botanicorum principes Systema plantarum Europae**. Villars' Latin translation of his **Prospectus**, entitled **Flora delphinalis**, was issued separately by Gilibert about June of 1786, but with a publication date of 1785. Gilibert contributed a short preface. The work was organized according to the Linnaean sexual system.
395 Introduced from East Asia and occasionally naturalized in southern Europe.

disappears. At the moment, everything around us is still under snow, as for more than the past four months, with the exception of a few worthless slopes with southern exposure.[396] Subsequently I received an envelope from Paris, originating from Monsieur Pourret, enclosing a printed instruction for the pursuit of butterflies. As my age does not allow me this exercise, I shall send it on to our friend the abbé Blanc in Embrun, who, by employing his pupils, can obtain whatever this distinguished amateur may want from that realm. I am led to hope for further shipments from [Pourret] and seeds from Monsieur Thouin. When they reach you, please give them your usual care.

You will find my catalogue enclosed. I have revised and abridged it as much as possible for me; but I have not been able to avoid adding some observations to it that I have believed to be interesting. The entirety will perhaps occupy a few more pages than you intended for it; but I hope that you will be persuaded, despite your expectation, to oblige me. It is the first, and it will be the last, published essay to appear under my name. In the future, it could be of some value to me, at the very least for the sale of my herbarium after my death. It is the sole fruit that I expect from more than 25 years of work in botany. It could be tightened up by using small letters, in particular for my notes. I do not believe I have offended any other author. I have only felt it necessary to criticize Monsieur Murray a bit on the matter of the *Ranunculus gramineus* [L.] and *pyrenaeus* [L.] in order to defend Monsieur Gérard and in the interest of truth. Where I have disagreed with your opinions, for example, in *Galium, Crepis,* and *Plantago,* I have not named you. I could not do otherwise without betraying my own thought. I have introduced my work with a short preface that I hope you do not find extravagant.

Please respond to me about the 8 or 9 specimens of cryptogams also enclosed, whose names you may insert in my catalogue if you do not find them already there. You had promised me a small bundle of your mosses. You will give me the greatest pleasure if you will steal an hour from your serious occupations to select them and then present them to me. It is nearly the only part of botany where, groping along, I falter virtually every step of the way.

Yesterday, April 8th, we buried the comte de La Roche, dead of a dropsy of the chest that ultimately became generalized at about the age of 63. Monsieur D'héralde, who treated him, did not leave him until after his death. I administered the sacraments to him. We have all been edified by his piety, by his patience in suffering his illness, and by his submission to the Sovereign Providence. May the merits of Jesus Christ have been applied to him! He is generally mourned, and his loss has been so keen for me that my heart will be griefstricken for a long time. Everything preaches to us a detachment from this wretched life. . . .

If you do not have the *Arenaria saxatilis* described by Linnaeus, do not use the trivial name *saxatilis* for the plant you have in view, even though the synonyms lead you there. You will only sow confusion in people's minds; because one understands, by the Linnaean trivial names, the species indicated by his descriptions. It follows from this principle that I shall always apply *Arenaria tenuifolia* L. to the plant his description was given, even though I am quite convinced he took his trivial name, *tenuifolia,* from Jean Bauhin who was speaking of quite another plant. If you do not want to conform to Linnaeus, then

~

---

[396] A reference to hillsides denuded by deforestation.

invent a trivial name that suits you. That is my way of thinking. As an illustration of my practice, I could never have called our common ornithogalum *Ornithogalum umbellatum* L. if Murray had not corrected Linnaeus's description. The same reason prevented me from revising *Rumex acutus* L.[397] But I am giving a lesson to my teacher. Excuse my boldness!

78
Les Baux, 9 May 1785

My troubles with the *Crepis* appear to me to be resolved thanks to the following solution, where you will find your own calculation in large part:

1. The *Crepis* with leaves glabrous, deeply runcinate, and even lyrate; stem always red at base; involucre farinose; corolla red below; seeds sharp-pointed; pappus nearly stipitate; in the meadows at Les Baux; which I have stubbornly called *C. biennis*, is certainly *C. dioscoridis* L. *Hieracium majus erectum latifolium*, rather than *angustifolium*, C. Bauhin, Pin. 127. Gérard, **Gallo-Prov.** [170], no. 3. During our debates, I was always of the opinion that you wanted it to be *C. tectorum*, which I could not digest. If I was wrong to believe that, you ought to have led me to *C. dioscoridis*, which you never did.

2. *Crepis tectorum* L. with leaves narrow, glabrous, sinuate, dentate; flowers small, corolla red below; seeds oblong, striate, very small; does not grow around Les Baux. I cultivated it last year from seeds received from Paris under that name. I have collected it in the field in Valgaudemar, but I erroneously labeled it *C. dioscoridies*. I do not know whether you have it around Grenoble.

3. Your *Crepis biennis* L. is indeed now my *Crepis biennis*. The two others being placed, I no longer have any doubt about this one; and I find myself in agreement with everyone: with Linnaeus, Haller (who has no *C. dioscoridis*), Gmelin, Vaillant, Gouan, Chabrey, etc.

Therefore, please change my trivial names at once in my catalogue of the Gapençais. That is to say, write *dioscoridis* in place of *biennis*; *tectorum* in place of *dioscoridis*; and *biennis* in place of *tectorum*. As for *C. dioscoridis*, add *calyces farinosi*; and, after *semina subulata*, add *pappus subotipileatus*. If, in my descriptions, I made any comparisons of one species to another, please make the analogies correspond to the changes I have just made.

Why are you endeavoring to correct Guettard's synonym of *Arenaria saxatilis* L.? Linnaeus cites a synonym from Guettard to his *A. saxatilis*, and he has cited another one from the same Guettard for his *A. tenuifolia*. I want to believe with you that *Alsine saxatilis et multiflora, capillaceo folio* Vaill. belongs to *A. tenuifolia* L.[398] Monsieur [Antoine] de Jussieu, on Barrelier, does not mention the latter species. Monsieur Guettard has to have known two different species as he characterized both of them separately. Moreover, Haller, Hist., no. 867 [= *Arenaria saxatilis* L.], says that the petals are longer than the calyx. *Arenaria tenuifolia* L., on the contrary, has petals a little shorter than the calyx. So, I leave you to extract yourself from this difficulty. It suffices for me to see my own plant in Linnaeus's *tenuifolia*, and I am settled about the matter.

⤙

[397] A possible hybrid of *Rumex crispus* L. x *Rumex obtusifolius* L., as in the case of *Rumex pratensis* Mert. & Koch.
[398] Linnaeus gave it as a synonym for his *Arenaria saxatilis*.

The bundle of mosses from Monsieur Jullien, which you have verified, has given me great pleasure. You may send me yours when you have the leisure, and please send on to him what I am enclosing here for him. They are seeds for his garden.

The shipment from Monsieur Thouin finally reached me one month and a half after its date. I do not know whether it passed through your hands in Grenoble, . . . but I received it with pleasure. Monsieur Serre, apprentice surgeon, ought to have delivered the live plants to you that you requested from me. After making the above corrections on my *Crepis*, will you also remove the citations from Haller that I could have written in the descriptions of the three *Crepis*. In addition, please erase completely *Ulex europaeus* [L.] as an item in the catalogue. From the specimens of it sent by Monsieur Jullien, I see that my spiny shrub from Bellecombe is not *Ulex*, but *Spartium scorpius* [L.] [= *Genista scorpius* (L.) DC. in Lam. & DC. (1805)] as you have argued. Whatever it may be, I ought not to draw upon it for my catalogue, it being beyond the territory that I am embracing.

I enclose herein four subscriptions for your work, namely, from the comtesse de La Roche and the curés Gaude, Martin, and [Jean-François] Roüy [de Châtillon-le-Désert]. I have grounds to expect two or three more of them. Dom prieur de Durbon, whom I have seen, gave his subscription directly to Monsieur Giroud.[399] He would have readily subscribed to you directly had he known that to be your intention. The matter being between Monsieur Giroud and you, as you did me the honor to explain, why did you tolerate the arrangement for subscriptions printed at the end of the announcement – having them taken only in Monsieur Giroud's name? I feel that not taking them in your own name will cost you subscriptions.

Monsieur Martin, curé du Saix, who has already made progress in botany, begs you order for him – from Lyon or elsewhere – **Philosophia botanical** Linnaei, **Genera Plantarum** L., and **Species Plantarum** L. He wanted to give me money to this effect to send to you. If that is necessary in order to make the purchase, let me know the approximate sum. I will have it sent on to you. You will find his commitment herein along with his subscription.

You gave me great pleasure by sending me news of Madame de Bruno and her sons. Please pay them my respects when you have the occasion. Apparently Monsieur de Bruno, the father, is no longer living, as you have never said anything about him. . . .

P.S. *Crepis tectorum*, which I have just worked out in this letter, and such as I have grown it from seeds sent from Paris with that name, has the greatest affinity to *Crepis virens* [L. nom illegit.] [= *Crespis capillaris* (L.) Wallr. (1841)] because of its leaves, the smallness of its involucres, and especially because of its very small seeds. Consequently, I would very much like that the one should follow the other immediately in my catalogue. I believe that Monsieur Gouan has spoken of this affinity.[400]

~

[399] Alexandre Giroud, the printer in Grenoble who was also known for his knowledge of metalurgy.
[400] The actual order in the catalogue would be *tectorum, dioscoridis, virens*, with a statement that the latter two had affinity. Chaix in Vill., **Hist. Pl. Dauph.** 1: 367 (1786).

79

Les Baux, 27 May 1785

I learn from your letter of May 13th about the difficulties to be overcome in order to become an author: much work, criticism from some censors, and uncertainty about public response.[401] One must have lofty purposes, such as those that have inspired your work, in order to consume all those obstacles. I hope that God, who has inspired your purposes, will give you the strength to carry them through. I realize that your said letter was already en route by the time you ought to have received my last letter. . . .

Have you *Lycopsis pulla* Clusius in your plants of Dauphiné? Monsieur Danthoine sent me a poor specimen of it. He says it is different from the *L. pulla* L. that I have from Monsieur Gouan. The abbé Blanc has found some stalks of a borage at Baratier near Embrun. He sent me a good specimen that I believe to be the said *Lycopsis pulla* of Clusius and Danthoine. He has transplanted it in his garden, and we will have seed from it. The discovery is new for me.[402]

Please have the kindness to tell me the names of the enclosed mosses, by the numbers that I have written for them, and add them to my catalogue.

Dominique Bonnardel, to whom you will please have the letter addressed to him delivered, is the son of Antoine Bonnardel of Les Baux, my closest neighbor. He formerly served Monsieur Delafont as a domestic. Subsequently he served in Lyon for several years. But as the air in Lyon was unhealthful for him, and as his wife was a native of Grenoble, he has removed to the latter city with her; and he seeks to find a new master to serve. He will have the honor to present himself to you and explain his situation to you. I beg you to take an interest in getting him placed in some house where you know that a domestic is wanted. I hope that his masters will be pleased with his service.

I am delighted by the good reports you give me about Monsieur Serre, and even more so about the proposals you expect to make to the intendant to have him admitted to the hospital as an apprentice surgeon. His parents are very grateful for your kindness to their son.

Your reiterated solicitations to get me to agree to come see you are most seemly, the most unmistakable expressions of your feelings about me. I am as aware of them as it is possible to be. But I lack the courage to undertake the trip. I am no longer such as you knew me. I have become quite slow-moving; the slightest effort puts me into a sweat; my sight has weakened; my teeth do not serve me well. Thus, even though I enjoy good enough health, thanks to God, I feel myself slipping into old age. Nevertheless, if God conserves for us, each of us, more life time, I have not given up the idea of going to spend a few days with you in the future, perhaps in the fine season next year. If I were of some use to you relative to your work, the intendance provides us an easy way to communicate, and you may want to employ me to the extent of my limited powers.

I hope you have been happy with my revisions in the *Crepis*. . . .

∾

[401] The first volume of Villars' work was sent to the Société Royale de Médecine in Paris for its approval and authorization to publish. As was customary, a committee of three was appointed to review the work: Geoffroy, Jussieu, and Tessier. Their criticism, of which more later, along with a recommendation to publish under the "privilege" of the Society, would be printed as part of the introduction to the first volume.

[402] Neither Chaix nor Villars listed any other species in this genus than *Lycopsis arvensis* L. Both *L. pulla* L. and *L. vesicaria* L. were specifically excluded from the flora of France by Grenier and Godron in 1852.

~

On the back of the second sheet of the above letter, notes in Villars' hand apparently reflect his immediate reaction to the cryptogams sent by Chaix with the letter.

1) *Jungermannia dilatata* [L.] ?

2) *Jungermannia lanceolata* [L.] is *Hypnum complanatum* L., Haller, no. 1771.

3) *Bryum hypnoides* [L.]. Uncertain given the absence of fruit. But its leaves do closely resemble it; and, if it really is [*hypnoides*], it must be the species and not a variety, as the others have pointed leaves.

4) *Fontinalis squamosa* [L.] ? This may rather be *Fontinalis minor* [L.]. It is true that its anthers are at the extremity of the branches instead of in the leaf axils as I have seen in your specimen; but the accompanying follicles are pointed, a character that only fits *Fontinalis minor*, Haller [no. 1795]. . . . If the operculum is elongate-pointed, it would be *Fontinalis minor*, for it is conical, or rather mamillate, in *Fontinalis squamosa*.

5) *Bryum carneum* [L.]. It resembles it very much, but it could also well be a variety of *Bryum caespiticium* [L.].

80

Les Baux, 21 June 1785

The abbé Villar, returning from Le Noyer, turned aside [to visit me] at home. I was delighted to see him. My joy was compounded by the reception of your letter and by its happy news about your entire family, since confirmed by your second letter, June 4th, to which I am now responding.

[We are now in agreement on the *Crepis*], and the matter is closed. Please have the kindness to put the trivial names where they belong in my catalogue according to our common opinion. It took a perseverence as firm and as learned as yours to extract me from my error, into which I had been led by the confusion in Haller's descriptions.

All the respect I owe you aside, I cannot relate the species I believe to be *Jungermannia tamarisci* [L.] to *J. dilatata* as you do. The latter, having affinity to *J. complanata* [L.], is firmly attached to the bark of trees; whereas the former is a large moss on rocks or on the base of trees in the shade.

I recently found a violet that I am sending you with the name I have given it. It appears to me to be a new species. If it appears to be so to you, it would be appropriate to record it next to *Viola nummulariifolia* Vill. [1779].[403] Tell me your opinion of it. It is not in the group of violets with pinnatifid stipules; it cannot be referred to *V. montana* [L.], which rises a foot and a half in my garden, with large, oblong-cordate leaves. If you can spare me a specimen of *V. palustris* [L.], you would please me. I have never seen it. Please insert my find into my catalogue.[404] Also add *Cardamine pratensis* [L.] that I found at the same time near my said *Viola pumila* in the swamp called Chastelas at Corréo de La Roche where *Ranunculus gramineus* [L.] and *Avena calycina* [Vill.] [= *Danthonia alpina* Vest (1821)] are also found.

If you would be willing to send me the copy of Linnaeus for the curé du Saix that you

~

[403] A species that Villars, Hist. Pl. Dauph. 2: 663 (1787), would say he owed to Chaix's collection.

[404] *Viola pumila* Chaix, Pl. Vap. 35 (1785). Chaix in Vill., Hist. Pl. Dauph. 1: 339 (1786), from a swamp called Chastelas near La Roche.

tell me you have already acquired, even if you have not yet procured a copy of the **Philosophia**, I will carry it to him immediately so that he can have the use of it before his plants wither. I will have your money sent to you. And please give me news of poor Bonnardel.

How sad the season is in this country this year! The winter ruined a great part of our winter wheats. Our spring wheats or *transailles*[405] are completely dying from the drought, there having been virtually no rain in the months of May and June. The meadows are nearly without hay, and there is no grass in the countryside: a season hardly favorable for the botanist and miserable for the poor people. Added to that, some detestable harpies have gotten hold of what wheat there is in the region in order to exercise a dreadful monopoly, suffocating the indigent through excessive prices for this important commodity. *He that withholdeth grain, the people shall curse him.* **Proverbs** 11. 26. May God defeat the designs of the impious and extend his blessings upon those who invoke them!

I put my catalogue entirely at your disposition. Make use of it in whatever way you judge to be appropriate. But have they not begun printing the work? It appears to me that time is running out to give it to the public at the beginning of 1786 in accordance with the advertisement in the published announcement. As for my herbarium, I am increasing it every day. I correct it, and I shall draw up a catalogue of it this coming winter. You yourself will be its appraiser and adjudicator, but I shall still keep it for some years if God prolongs my life.

Please let the directors of your local seminary know about my objection to the language in our breviaries, and pass on to me any explanation for it that they may be willing to give me. I respect that language, but I do not understand it, taken in a literal sense. I have already spoken about the matter in Gap and in Embrun. Excuse the many burdens that I give you.

P.S. I have been obliged to reopen my package, sealed under the intendant's cover, in order to make known the following to you:

1) Having gone into the woods before dinner to gather vulnerary plants for a sick Carthusian of Durbon, I found *Viola mirabilis* [L.] in Monsieur Mondet's brushwood, where it is not unusual. I have had it for several years in my garden without remembering from where I had collected it. Therefore, will you please put it into my catalogue along with its native location? If you do not have it in Grenoble, it will be easy for me to get it for you. I am sending you a specimen of it. *Viola nummulariifolia* Vill., *Viola pumila* Ch., *Viola mirabilis* L., *filiae agri Bauxiensis.*

2) *Phyteuma betonicifolium* Vill. is also found mixed with *Phyteuma orbiculare* [L.], *foliis lanceolatis*, in said brushwood. I suspect that one is only a variety of the other. I am sending you specimens of them.[406]

3) *Orchis incarnata* L. is found at Corréo in the marsh called the Grand Sagne: *bulbi bifoli; flores carnei; nectarium integrum servatum, puniceo rubrus pictum.* Add it to my catalogue.

4) *Orchis latifolia* [L.], from the same place: *non maculata. Porro macula ta passim in paludosis.*

~

---

[405] The name used in the Midi for the sowing that is done after the spring harvest and before the autumn sowing.
[406] Villars kept them segregated as they still are today.

81

Les Baux, 1 July 1785

Just last Saturday I received the three volumes by Linnaeus along with your letter of June 9th. The items therein I shall respond to below. The following day, Sunday, in the afternoon after concluding my parish offices, I carried said volumes to Oze where Monsieur Martin had come [from Le Saix] to get them. He received them with great pleasure and charged me to express his gratitude to you for them, not having had the time to write you as he means to do. He hopes you will continue to seek a copy of **Philosophia botanica** for him along with a small loupe such as you once obtained for me. He will soon be in condition to become a botanical correspondent. He already has a good collection of plants that he has had me examine. He paid me 30 livres, to wit, 25 livres for the **Species** and **Genera**, and 5 livres for the promised **Philosophia**, in accordance with your letter. On my return, I met Monsieur Pellissier on the road, who told me that he would be leaving for Grenoble on Thursday, June 30th; so I entrusted the said 30 livres to him to carry to you. Please acknowledge reception of the money when responding to my present or my preceding letter if you have not already done so.

You will have without doubt inserted all the items in my catalogue about which I wrote you recently. Be good enough to insert the additional ones I enclose herein: two *Galia* and this siliquous plant that I do not know; plus *Phleum pratense* [L.] *a Oze prope Moletrinum*; plus *Lepidium latifolium* [L.] *circa Aspres.* that I have seen in Monsieur Martin's collection. I also noticed therein your *Phyteuma charmelii* [Vill.] *in rupibus circa Le Saix et Clausonne*, as well as *Doronicum pardalianches* [L.] *in monte de Laric.* I believe that these last two plants have been reported in my catalogue.

Given the description that you make of it, I have no doubt that the *Carduus* about which you consult me is my *Carduus aurosicus* [Ch.]. Could you have had the seeds that produce it for you from anywhere other than from me? I am delighted that it retains the characters that I observed at the foot of Mont Aurouze when grown from seeds in the garden. Those same seeds have produced, for me, only the second year after having been put in the ground. That is to say, last year. What plants remain do not promise me a stalk this year. Its glabrous leaves permit no comparison with *Carduus carlinoides* [Gouan], *foliis lanatio* Gou., **Ill.**, *tomentosus* Tourn., not to speak of the scales on the calyx that are not open in his figure, as you say.

You tell me that you obtained a variety of *Carduus medius* Gouan from me that is slightly pubescent. The *Carduus medius*, which I have formerly cultivated from seeds sent by Monsieur Pourret, and which is now in flower in my garden, has a stem and leaves that are completely glabrous, and is only slightly downy at the top of the peduncle. The one that came to you from Lyon is perhaps the pubescent plant that you tell me about, and which may well be a species different from my *aurosicus* and from *medius* Gou.

Please send me the description for *Delphinium azureum* Lamarck, as the seeds from Monsieur Thouin with that label have produced *Delphinium elatum* L. along with another very hairy plant with leaves like *Aconitum cammarum* [Jacq.], whose flowers are not yet sufficiently developed to determine whether it is a delphinium or an aconite. Its beauty has aroused my curiosity. Also indicate the synonyms Lamarck assigned to his *Delphinium azureum.*[407]

Hyssopus lophanthus [L.] is abundant this year around here.[408] Please define the Greek word *Lophos* for me. *Digitalis purpurea* [L.]: the beautiful flower!

Cneorum tricoccon [L.] gives me flowers tetrapetalous and tetrandrous. This genus, therefore, established by Linnaeus, can mislead a beginner.[409]

82

Les Baux, 15 July 1785

Since the theologian who read my epigraphs had no theological explanation to give, he might get out of the situation by inserting a literal French explanation that even the weakest schoolboy in the fourth form would be able to understand.

Androsace villosa [L.]. I am joining this variety with a corolla nearly flesh-colored to the species I already had from you with a white corolla. But neither one of them has any pleats between the segments, and their lobes are simply obtuse and not emarginate. Where, therefore, is Haller's no. 620?[410]

Arnica scorpioides [sensu Jacq.] [= *Doronicum grandiflorum* Lam. (1786)]. *Semina cum centralia, tum marginalia pilose papposa, obovata, stricta, pubescentia.* Mss.

Doronicum pardalianches [L.]. *Mihi ut Hallers semiflosculorum semina papposa non sunt; flosculorum vero pappo coronatur.* Mss.

Orchis sambucina [L.] [= *Dactylorhiza sambucina* [L.] Soó (1962)]. In marshes with [*Orchis*] *latifolia* [L.]. *Petala carnea; nectarium integrum, serratum; bracteae floribus longiores.* Mss. *Hinc alteram ab altera distinguo.*

Pollich has evidently reversed the synonyms from Haller by relating Hall., Hist., no. 53, to *Hieracium* [x] *dubium* [L.], which I believe should be *Hieracium auricula* [L.] [= *Hieracium x floribundum* Wimmer & Grab. (1829)]; and by relating the latter to Haller, no. 52, which is really *Hieracium dubium.* What could have brought on his error would have been the epithets *pauciflorus* and *multiflorus* in Haller. For here in Les Baux, *Hieracium dubium* is only 2-3-4-flowered, which is *pauciflorus;* and Pollich himself says that, too. I have found your *Hieracium piloselloides* [Vill.] in the meadows around Montmaur, really appearing as a multifloral umbel, as Haller describes for his no. 52, not throwing out any stolons; leaves narrow with nothing but a few hairs; stalks glabrous; calices small, but that one could not call glabrous (the latter character preventing referring this species to Haller, no. 54). . . .[411]

Hieracium auricula differs from *dubium* because of its stalks, *pilis eminentibus scabris;* because of its leaves, *hirsutus oribus, majoribus, magio ad lancelatum,* by its *calices hirsutissimis* [rodent damage]. I cultivate it. It is quite singular and rare, the stem usually simply bifid or twice bifid, the flowers solitary. I have readily found it around Les Baux in

---

[407] This was probably not a European native, but rather *Delphinium azureum* Michx., **Fl. Bor. Am.** 1: 314 (1803), as the Michaux material that survived had been brought to Paris. [= *Delphinium carolinianum* Walt.]

[408] This could have been one of several subspecies of *Hyssopus officinalis* L.

[409] The genus, and this species, are 3-4-merous. Linnaeus had put the genus within Triandria Monogynia.

[410] Villars would give Haller, Hist., no. 620, as a synonym for *Androsace villosa* L.

[411] Vill., **Hist. Pl. Dauph.** 3: 100 (1788), does give Hall., Hist., no. 54, as a synonym for *Hieracium piloselloides* Vill.

the Mondet woods and the Devez groves. Add this minor note to [the species] in my manuscript. I am sending it to you herein.[412]

After closer consideration, I believe that *Galium sylvestre* Poll. [= *Galium pumilum* Murray (1779)], Haller, **Hist.**, no. 715; Barrelier, **Icones** 57, is the plant you want to call *Galium montanum* [Vill.], according to what you have written me. And I suspect that your *Galium anysophyllon* [Vill.], blooming very early and common in our mountain pastures and even around Les Baux in the meadows, has a great affinity with your *Galium tenue* [Vill.] and is a variety of said no. [715] of Haller, *Galium album supinum multicaule angustifolium polyspermum* Rupp. p. 4. Therefore, please leave the first [*Galium*] in my catalogue under the name *G. sylvestre* Poll. *in pascuis montanis*, because I cannot put your name there, *Galium montanum minor* (which you have not yet published, and which you relate properly to *Mollugo montana*), *Galio albo fimilis*, Ray, **Hist.**, 482. For the second one, if I called it *Galium sylvestre* Poll. in my last letter, I retract that; and if you cannot call it your *Galium tenue*, please write *Galium ruppii* **Fl. jen.** 4, *in siccis glareosis apricis, floret julio*.[413]

I no longer have much doubt that the *Crepis dioscoridis* and *virens* are the same species. The seeds of the latter, which I brought back from the Valgaudemar last year, give me exactly the same plant as the seeds of the former, which had been sent to me from Paris labeled *Crepis tectorum*. Notwithstanding the respect due to the great Jussieu, our true *Crepis tectorum* will never degenerate into either one or the other of those two species. Speaking of *Crepis*, see if the faithful Murray copies again the typographical error, *pappus plumosus*, for the genus in his last edition of the **Systema**, which leaps out from the page. We owe our masters respect, but we are not slaves.

[The following paragraph is meaningless because of substantial rodent damage at the top of the page.]

It is the great Jussieu who has given the name *Delphinium azureum* at the garden in Paris and not the singular Lamarck. I should like to know the difference he has established to distinguish it from *Delphinium elatum* L. and the synonymy that he relates to it. The species I collected on the Col de Vars is not as hirsute as the variety given me by the seeds from Paris labeled *Delphinium azureum* nor as glabrous as the other variety coming from the same seeds. Beyond which I still see nothing sufficiently striking to justify making a species distinct from *Delphinium elatum* L.[414]

My *Rosa montana*[415] *foliis glabris ovatis obtusis, pedunculis germinibusque glanduloso-hispidis* [damaged section] *frutex Rosa caninae magnitudine; rami purpurescentes aculesti.* Differs from *Rosa canina* [L.] *foliis ovatis obtusis, non acuminatis, graminibus hispidis, non glabris*; from *Rosa rubignosa* [L.] *foliis non rubiginosis*; and from *Rosa sylvestris* Poll. or

∾

[412] Villars followed these instructions faithfully; but later, in Vill., **Hist. Pl. Dauph.** 3: 100 (1788), he remarked that "Monsieur Chaix has been tempted to give this species the name *Hieracium hybridum* Ch., believing it to be a cross of *H. cymosum* L. and *H. pilosella* L."

[413] Instead, Villars, **Hist. Pl. Dauph.** 1: 361 (1786), inserted *Galium obliquum* Vill., *in apricis sterilib.*, giving as a reference Haller, **Hist.** no. 715, var. g., which reads *Galium album supinum, multicaule, angustifolium, polyspermum* Rupp. p. 4.

[414] *Delphinium elatum* L. can be glabrous or hirsute. But there remains the possibility, see letter 81, that there had been some confusion with the North American plant, *Delphinium azureum* Michx.

[415] Chaix, **Pl. Vap.** 42 (1785). Chaix in Vill., **Hist. Pl. Dauph.** 1: 346 (1786). From Rabou and La Grangette.

Haller, **Hist.**, no. 1102, *stylis sessilibus, petalis incarnatis.* In thickets in Les Baux, Rabou, and La Grangette. Please remove *Rosa sylvestris* Poll. from my catalogue and substitute my *Rosa montana* [Ch.] for it; or insert it under whatever name you have already given it if you have it in your notebooks with the above distinctive characters.

When you have taken the trouble to make the above corrections in my catalogue, as well as the other I have begged you to make in prior letters, take a firm hand to *Festuca duriuscula* (auct.) and to *Frageria sterilis* [L.] [= *Potentilla sterilis* (L.) Garcke (1856), which I have not yet met in this region.

For the last three or four days, I have suffered from a violent pain because of a gumboil on the last lower molar, which, for some time, has been weakened. Nothing has given me much relief except for putting my feet in hot water. How sweet is the relief after such suffering!

All the good things you tell me about young Serre give me great pleasure and increase the regard I have for him. You will not grow weary of sheltering him with the protections he needs to attain his goals. To those reasons you have for recommending him to the intendant, add the fact that there is now no surgeon between Gap and Serres. Pateas's son has moved to Spain, and the father can no longer do anything. The public weal necessarily requires the residence of a surgeon in this district.

83
Les Baux, 30 July 1785

The information from you telling me the intendant's time of arrival and the approximate days for the examination of the surgical students has given Monsieur Serre real pleasure. He is now reassured about the timing, and I think he will leave next Monday, August 1st, to be present. I continue to add my prayers to the good intentions you have to assist him; and I reiterate my request that you explain to the intendant the current need for a surgeon in the canton of Veyne, none being established between Gap and the town of Serres. The public should realize a great benefit in the person of Monsieur Serre, who would settle either at his home in La Roche or in Veyne.

Speaking about botany, if Monsieur de Jussieu *regards method as the science itself,* according to the expressions in your letter of July 7th, he is wrong. Method is the key that opens the path to knowledge; but the knowledge itself is the attainment of that end to which one aspires. If you yourself say that method *is only the basis for the names* (again your expression in said letter), you are equally wrong.

I can envision a young man in Le Noyer, fascinated by plants and gifted with high intelligence, who will succeed, by himself, in creating a method, referring all umbels to one category, the legumes to another, the pentandrous to still another, and so on. And who will observe some particular characters for each species. Thus, he will be a methodist and will acquire a knowledge of plants that consists of knowing how to distinguish between them and how to characterize them, yet be ignorant of the names given to them by the authors. I bear down on this matter, because it has appeared to me that the abbé Blanc, to whom you may have communicated your opinion on this subject, has been unsettled by it; for he begged me to tell him how I see this disagreement between Monsieur de Jussieu and you. I have given you my view quite directly.

I hope that, upon your return from the trip to Savoy, you will be willing to make your discoveries known to me, and to come back to my last letter to which you had no time to respond, pressed by your imminent departure. In it, I sent you a specimen of *Hieracium auricula* L., Haller, no. 53 according to me, and which is the true *Hieracium auricula* described by Pollich. Herein, I am sending you (1) *Hieracium dubium* L. with stolons. See if it could be Haller, no. 52, *floribus umbellatis*. It is not umbellate in our region, being only 2-3-4-flowered. I am also sending you (2) the *Hieracium* that we are calling *piloselloides* [Vill.]. I do not know whether this is the one that Haller describes in his no. 52, for it is clearly umbellate; but I have never seen it with stolons.[416] I have just found it in the Montmaur preserve, the flowers very frequently distorted, no doubt from the bite of some insect, which is rare in the semi-flosculars. Give me your opinion of them.[417]

84
Les Baux, 2 September 1785

The fine things you tell me about your trip make me envious, not of you or your illustrious companions, but to have been a member of the party. I have to console myself in a different arrangement made by Providence and through the communication of your rich discoveries that you have the kindness to make. What you tell me about the tomb of St.-François de Sales, of St.-Bernard, founder of the magnificent hospice in the highest of the Alps, about the pious personages who live there, and about your perspective from the top of that mountain, delights me. And, after having blessed God for your safe return, I turn next to plants, one of His most beautiful productions.

I should not like the *Sisymbrium* from the Lac du Bourget to be designated by the epithet *asarifolium* (as the resemblance is not exact), but rather by the name *heterophyllum*.[418]

In which of his works did Haller have your beautiful willow *oleaefolia* engraved?[419] I did not see it in the **Nomenclator**. I am sorry that you did not observe *Salix amygdalina* [L.], Haller, no. 1636, at the edge of Lac Léman. *Salix hastata* [L.], Haller, no. 1654, has very wide leaves and very large stipules. How can you relate it to the one from Chantaussel, which I call [*Salix*] *spadicea* [Chaix], and which I would relate to Haller, no. 1655?[420]

I ought to have inserted my opinion about *Pedicularis tuberosa* [L.] in my catalogue. It is the only one I know to flower whitish with the *foliosa*. It is abundant on our mountains

~

[416] *Hieracium piloselloides* Vill., **Prosp. Pl. Dauph.** 34 (1779). Two subspecies are now recognized, one more frequently having stolons than the other.

[417] Villars, **Hist. Pl. Dauph.** 3: xiv (1788), indicates he botanized in the valleys of the Faucigny and around Chamonix and Mont Blanc in 1785 before entering Switzerland through the Col de Balme and the valley of the Trient. From there he went to the Grand St.-Bernard, staying in the hospice that was open to all travelers without exception. On the return trip, he passed through Martigny and St.-Maurice; and then through Geneva on his way to Grenoble.

[418] Neither epithet seems to have been used by Villars for a *Sisymbrium*.

[419] *Salix oleifolia* Vill., non Sm., **Hist. Pl. Dauph.** 3: 784 (1788). A single specimen, without flowers or fruit, survived in the Chaix Herbarium and was examined and described by Timbal-Lagrave, "Observations critiques et synonymiques sur l'Herbier Chaix," **Mém. Acad. Sci. Toulouse**, ser. 4, 6 (1856): 149. Although the species was not found again in Dauphiné, Grenier and Godron, **Flore de France** 3: 133, who examined not only that Chaix specimen, but one he had sent to Pourret, considered the species valid. Collected by Chaix near Gap.

[420] Chaix, **Pl. Vap.** 69 (1785). Chaix in Vill., **Hist. Pl. Dauph.** 1: 373 (1786). Villars suspected this to be a hybrid, probably *S. lanata* L. x *S. hastata* L. Chaix's description suggests *S. caprea* L.; Haller, no. 1653.

in Les Baux and Rabou. Its spike, at the beginning of flowering, is short and obtuse; but it subsequently lengthens even to five or six inches. It is Haller, no. 323. *Pedicularis folius bipinnatis, calic non cristato, floribus ochroleucis in spicam nudam congestis.* Allioni, [**Specimen stirpium**] **pedemontii** [tab.] 11, f.l. . . .[421] I do not know the *Pedicularis comosa* in Allioni with yellowish flowers that approaches your *P. gyroflexa* [Vill.]; Haller, no. 324. . . .[422]

As for roses, you tell me in your letter of 19 July 1785 that "I am not ready to segregate them." If you reunited this group as only so many varieties of one species, something I would be unable to approve, why do you segregate so liberally within the *Galium*, the *Salix*, the *Hieracium*, the *Plantago*, and so on? Too lenient for these, do you want to be too severe on the former?

In your prospectus, why did you include *Veronica allionii* [Vill., **Prosp. Pl. Dauph.** 20 (1779)] with the tetrandrous when it is surely diandrous? You could be criticized for this mistake.

Nothing is more common in this area than *Scabiosa gramuntia* [Vill., non L.] *corollulis quinquefidis*, as Murray and Gérard have noted very well.[423] Linnaeus [*S. gramuntia* L.] made the corolla four-lobed, guided no doubt by the judicious Gouan; who, himself, made it trilobed. Please be good enough to add it to my catalogue.[424]

Whenever I cited the small pond between Veyne and Aspres, for instance for *Alopecurus geniculatus* [L.], *Utricularia vulgaris* [L.], *Potamogeton serratum* [L.], *Myryophyllum verticillatum* [L.] etc., I wrote *les Egaux*. Kindly spell it correctly, *les Aiguaux*, as well as say *ab aquio*. Note: the topographers of Dauphiné, when speaking of the pass between Oisans and the Briançonnais, write *Lautaret*, which indicates nothing. I think it should be written *l'Hôteret: ab hospitiolo in summo monte constructo*, which gives a meaning.

I have often seen *Hypericum perforatum* [L.], as well as *H. hircinum* [L.], to be tetragynous when grown in my garden.

Seeds labeled as *Arabis thaliana* Crantz have produced plants from which I send you a specimen. Should it be your *Arabis serphyllifolia* [Vill.]?[425]

Have you seen in the garden of Paris or Grenoble a *Physalis* that the abbé Blanc sent me to determine, which I do not find in Linnaeus, and which I characterize as follows: *Physalis nora; foliis lanceolatis, dentatis denticulatis; floribus solitariis; annua; folia alterna, glabra; corolla magna, coerulea, ad umbilicum albida; anthera pallidae.* It is not *Physalis angulata* [L.]. I cannot say anything more about it.[426]

⁓

[421] This was Allioni's description for *Pedicularis comosa* L., which is correctly listed in Chaix's catalogue.
[422] *Pedicularis gyroflexa* Vill., **Hist. Pl. Dauph.** 1: 283 (1786), appears as a substitute for *P. tuberosa* L. in Chaix's catalogue, further indication that Chaix was confused when he wrote this passage. But his description above certainly hints at *P. tuberosa* L. One might deduce that Villars arbitrarily altered Chaix's list, but see letter 85.
[423] Gérard, **Fl. Gallo-Prov.**, p. 220, no. 6, did in fact write *Scabiosa corollulis quinquefidis, foliis duplicato-pinnatis, foliolis setaceis.* Villars, **Hist. Pl. Dauph.** 2: 294 (1787), citing this as a synonym, miswrote *trifidis* for *quinquefidis;* yet described this plant as *quinquefidis.*
[424] The species does not appear in the catalogue, at least not under that name. It may well have been *Scabiosa columbaria* L., common around Les Baux.
[425] Villars recognized *Arabis thaliana* Crantz to be a synonym of *Arabis thaliana* L. [= *Sisymbrium thalianum* (L.) Gay = *Aridopsis thaliana* (L.) Heynh. in Hall & Heynh. (1842)].
[426] This may have been *Atropa physalodes* L. [= *Nicandra physalodes* (L.) Gaertner (1791), a cultivar from Peru that occasionally escaped to become naturalized in Europe.

Please buy two books and send them by [the messenger] Pellissier: **Voltaire parmi les ombres** and **Institution divine des curés**.[427] I have read both of them, and I should like very much to own them. I think they should be found in Grenoble. If you have not read the first one, read it before sending it to me. You will be pleased by it. Let me know the price. I think that Monsieur Martin will be more than delighted to keep the copy of **Philosophia botanica** that you are willing to give him.

Chloris Lugdunensis[428] is a very minor work, only a simple catalogue. I should not like to have the adjectival or trivial names in your book begin, as they do in the above title, with a capital letter, thus distinguishing them from the [generic] nouns.

I am working on an herbarium of about 1200 plants . . . on a commission Monsieur Blanc provided me. I do not know for whom I am working, but I have been promised Monsieur Allioni's three volumes in exchange. I ask your pardon for the many tasks I have put upon you, being able to do nearly nothing from my side to serve you. You must content yourself with my good will and most affectionate attachment that will bind me to you until my death.

85

Les Baux, 18 September 1785

The joy that your letters give me is based both upon the renewed pleasure of reading your thoughts and upon the new insights they never cease to bring me. My heart is gratified by them and my mind instructed. I have your last letter of the 13th instant before me, and I am responding to a few items.

1) Haller led me astray from my Linnaeus on the yellowish lousewort of our mountains as on the *Crepis*. It is indeed *Pedicularis comosa* L.; *calcibus quinquedentalis* All., tab. 11, fig. 1. I also believe it is *Crista galli montana floribus pallidis in spicam congestis* Ray 770, no. 7, which that author found in the high mountains of the Grande Chartreuse. Please take the trouble to write *comosa* in place of *tuberosa* in my catalogue. Its radicles, however, are truly tuberous; the white spine, about which Gmelin speaks, is very apparent.

The yellow color of the corolla, the glabrous calices with simple segments, Haller, no. 323; and the analagous figures in J. Bauhin and Barrelier for *Pedicularis tuberosa*, do not cover your *P. gyroflexa*, whose corolla is always red; the calyx hirsute with the segments crenate or crested; the radicles less tuberous than in *P. comosa*. Barrelier, however, has said: *flores (tuberosae) modo purpurei, modo albi*. If Haller had engraved the calyx segments either dentate or crenate in *P. tuberosa*, it would approach your *P. gyroflexa*, separating it from the figures in J. Bauhin and Barrelier. I have always believed that our *P. gyroflexa* was Haller, no. 324, found by [A.] Gagnebin near Mont Dauphin, the flower red, calyx crested but glabrous according to him. The crenate calices of *P. tuberosa* L. (and crenate in Murray) strongly suggest our *P. gyroflexa*, to which I would refer Allioni tab. 11, fig. 2, because of its leaves, found near Annecy in Savoy on the mountain called La Tourrette. But its calyx without crest and its yellowish flowers prevent that. Have you see it in this country? From all this I conclude, with you, that *P. tuberosa* is only a variety of *P. comosa* on the basis of

∼

---

[427] The latter work was by Gabriel-Nicolas Maultrot (1714-1803), an attorney attached to the Parlement de Paris, who took an active part in the Jansenist movement.
[428] Published by Marc-Antoine-Louis Claret de Fleurieu de La Tourrette of Lyon in 1785.

Haller's description and the figures in J. Bauhin and Barrelier; and a variety of our *P. gyroflexa* in accordance with the description in Linnaeus and Haller's figure. You will discuss all this in your work.[429] As for me, it will suffice to give *P. comosa* in my catalogue in replacing *P. tuberosa*, and to use your *P. gyroflexa* in raising doubt about *P. tuberosa*.

2) In your penultimate letter of August 26th, you tell me, "I saw *Carum bunias* L. [= *Ptychotis saxifraga* (L.) Loret & Barrandon (1876)], *Athamantha oreoselinum* [L.] [= *Peucedanum oreoselinum* (L.) Moench (1794)], and *Pimpinella hircina* L.[430] along Lac Léman. Only I conclude that this first plant is really *Seseli saxifragum* L., as Gérard and others have presumed, unless I am mistaken." Let me point out to you that Gérard has not mentioned *Seseli saxifragum* in his flora, and that the Linnaean discription, *foliis duplicato-ternatus linearibus* (*Sesali saxifragum*) does not fit *Carum bunias*. Did you mean to tell me *Pimpinella peregrina* L., Gérard, p. 256? The Linnaean description, *foliis radicalibus pinnatis crenatis* (*Pimpinella peregrina*) quite fits our *bunius*, but then what becomes of the synonymy particular to each plant? Could you not have seen *Seseli saxifragum* along the lake at Geneva that Linnaeus indicates as it locale?[431]

3) Your *Sisymbrium asarifolium* could very well be only a variety of *Sisymbrium nasturtium* [L.] [= *Nasturtium officinale* R. Br.]. But understanding you to say *asarifolium* suggests, does it not, a species with leaves entirely simple? Nevertheless, the first leaves on your plant are like that, something I have also seen on *Sisymbrium nasturtium*.[432]

If the pass between Briançon and Oisans takes it etymology from the adjective *altare*, high in French, it would be necessary to write the *Hauteret*; if from the noun *altare*, altar, it would be necessary to put an apostrophe between the article and the noun to write *l'Autaret*; and if from it small hospice, *l'Hôteret*.[433]

Did Haller mean by *Vallis ursaria* the Vallée oursine, which is in the Faussigni [Le Faucigny] in Savoy; or *urserine* from the Urserin parish, *Andermatt* in German, in the canton of Uri below the Pont du Diable on the Reuse by which you reach the hospice of St.-Gothard? I believe that Scheuchzer, speaking of the alpine trips into this latter valley, calls it *Vallis ursaria*. You have these works and will be able to verify the location. I bring the matter up because Haller, in his **Enumeratia [stirpium . . . in Helvetia]**, does not go beyond Swiss territory.

5) Can you not provide me with a specimen of *Salix amygdalina* [L.] [= *Salix triandra* L.]? Do not forget to speak to me about *Salix lapponum* [L.] when you have received the leaves promised to you.

I do not yet possess the copy of Allioni in exchange for the herbarium I put together. Therefore, do not be in a hurry to part with your extra one. I would take advantage of your

∾

[429] See Vill., Hist. Pl. Dauph. 2: 426-430 (1787), one of the longest treatments of a species in the entire flora!

[430] If he meant *Pimpinella hircina* Mill., the plant might have been either *P. saxifraga* L. or *P. magna* L.

[431] There are several confusing aspects to Villars' statement: 1) He had already published *Seseli bunius* (giving *Carum bunias* L. as a synonym) and *Seseli carvifolium* in Prosp. Pl. Dauph. 24 (1779); and 2) He would note in Hist. Pl. Dauph. 2: 587n that *Seseli annuum* L. was the *Seseli* he had found along the lake at Geneva.

[432] To his description of *Sisymbrium nasturtium* L., Villars added, Hist. Pl. Dauph. 3: 340 (1788), that he had frequently found a small variety of it, or a creeping species, with leaves simple or ternate, the siliques shorter, the seeds biseriate, along the Lac du Bourget in Savoy. Although he knew Crantz's work on crucifers, he does not seem to have consulted him on *Sisymbrium*. This may be *Nasturtium austriacum* Crantz [= *Rorippa austriaca* (Crantz) Besser (1822)]. Also see letter 84.

[433] The usual spelling, *Lautaret*, appears in Chaix's list in 1786. *L'Autaret* has also been regarded to be correct.

kind offer if they go back on their word. The modesty of my means drove me to accept this arrangement in order to obtain a work I want very much.[434]

Please have the goodness to send me my catalogue as you have promised so that I can make improvements before they print any of it, and so that I can make use of the notes that you will have taken the trouble to add to it. I discover some mistake every day, that is, beyond those I have already mentioned to you. *Acrostichum marantae* [L.] ought to be *A. ilvense* [L.][435] Perhaps one ought to write Oursiere, and not Orciere, as I have written it [Orcières on the Drac Noir].

As I have resolved to go to Gap tomorrow for the festival of St.-Arnoul, patron saint of the city and the diocese, I am hurrying to do this letter today in order to carry it to the post myself. Take care of your health.

86
Les Baux, 28 October 1785

I am sending my manuscript back to you, having made some minor excisions and a few small additions, requiring some scratching out and alterations; but I hope they will be able to read all the words and to understand their references. If Monsieur de Jussieu is not happy with the order I have followed, he cannot criticize it: the authority of the illustrious Linnaeus protects it. If he criticizes my notes, I will be under obligation to him, as I shall benefit from his censure. Moreover, as I am presenting little more than a simple catalogue, using natural families (even though less perfect than those more recently established, and which I do not sufficiently know) that have appeared to be to be preferable to using artificial methods to carry out my intent.[436]

It would be improper to present my short preface in French, the catalogue being entirely in Latin: it would be a laughable mixture. Consequently, I beg you to insert it following your own catalogue just as it is, as well as the content following the preface. You will do me much honor and give me great pleasure. (Excepting any corrections, of course, that you should deem to be necessary.) As this is a small work, set off from yours and in another's hand, its different style will not mar your book.[437]

Monsieur Allioni's **Flora pedemontana** in 3 volumes was sent to me on the very day that I sent the herbarium in exchange, into which I had put 1150 plants within two volumes: herbs, trees, and a few mosses. It took me a good deal of time and gave me much trouble. I shall not undertake such a task again for under four louis. The work gave birth

~

[434] Chaix already owned a copy of Allioni's **Specimen pedemontana** (1755), the basis for his prior references to Allioni's plants; and he was on the verge of acquiring the coveted new **Flora pedemontana** in 3 volumes (1785).

[435] Chaix, but not Villars, recognized the presence of *Acrostichum ilvense* L. in Dauphiné, *in alpibus frigidissimis Valgaudemar*, Hist. Pl. **Dauph.** 1: 377 (1786) [= *Woodsia ilvense* (L.) R. Br. (1815)].

[436] He means that he has followed the natural sequence of families recommended by Linnaeus, the incomplete order he called "Fragmenta methodi naturalis," published in **Philosophia botanica** (1764), which Chaix owned. The passage further reveals that Chaix knew Jussieu was perfecting natural classification but had insufficient knowledge of the refinements. Jussieu had exposed the system in a paper dated 13 April 1774, but full publication of his **Genera Plantarum secundum Ordines Naturales** would be delayed until 1789.

[437] It seems evident that Villars, whose flora was to be in French, had raised a question about the propriety of an inconsistent preface in Latin. These paragraphs underscore, what is apparent in earlier letters, that Chaix's catalogue, if somewhat different in format, had from the start been written for inclusion in Villars' first volume. Hence he expected to be susceptible to Jussieu's censure.

to a plan to make one of them for myself next year (God willing) in four volumes of three quires of paper each, in which I will only put one specimen of each plant species, and that without any injury to my large herbarium of unbound sheets. I do not know for whom I was working. The abbé Blanc made all the arrangements. I hope the person will be pleased with it. As for me, I am quite happy to have earned the Allioni work that I had greatly wanted. I thank you for the offer you made me of your copy. I have still not gone through very much of it, but enough to have observed a number of mistakes in the assignment of synonyms. He has not always included the names in your **Prospectus**. That should not surprise, those names having been presented bare [without descriptions]. The flora comprises 2813 numbers, including those in the appendix. Although the territory covered is greatly variable: the plains of Italy, the seacoast in the County of Nice, and the chains of the Maritime, Cottian, Graian, and even the Pennine Alps, I am not sure about the reality of all his plant species. The engravings are good, but they appear to me to exceed the natural habit in height. Please tell me what the two umbels, tab. 43 and 63, are in Linnaean nomenclature.

Whether you esteem Monsieur [P.-J.] Buchoz as much as Monsieur [Joseph] Dombey, if you do not already have the latter's *Villaria*, you have had the former's *Villaria* for the past few years. I read about it in the public announcements. . . .[438]

The cold from Aquilon this night has obliged Flora to lower her flags in this country. This lovable goddess bids us adieu until next year. Take care of your health and give me your news from time to time.

87

Les Baux, 4 November 1785

I approve very strongly the course you are inclined to take in the matter of numbering your plants. It is certainly sufficient, and much easier for the reader, to have to count only the genera from beginning to end, rather than having to count the species within the genera as [other authors] have done. One will be grateful to you for having preferred that form.[439]

I have corrected a few spelling errors. I am satisfied with your classes, and the sections therein, except for what I am going to tell you here: If I were doing it, I would prefer to sacrifice the natural series a bit, which you have meant to preserve, to avoid complicating your Class III.[440] Therefore, I would combine the triandrous *Crocus* with *Iris* to reduce the range of sections. I would have the valerians and so on follow them; but I would transfer the rushes and the lilies to Class VI [Hexandria], as well as *Alisma* and *Veratrum*; and *Butomus* to Class IX [Enneandria]. As for Class X [Decandria], Section 2 [the Caryophyllaceae], Subdivision d. [genera either apetalous or with petals entire that have fewer than 10 stamens], the difference in the number of stamens is quite acceptable because of the small number of species in each genus, and because of a nearly natural exterior affinity. I would say the same thing for joining *Sempervivum* to *Sedum* in Section 3. But

⌒

[438] Neither author's *Villaria* is recognized in J. C. Willis, **Dictionary of the Flowering Plants and Ferns**, 8th ed. (1973).

[439] For volumes 2 and 3, the main body of the flora, Villars would number the genera, rather than the species, consecutively from beginning to end. Allioni, to give only one example, had numbered his species consecutively but gave his genera no numbers at all.

what became of *Alsine?* Have you cut it out or forgotten it?[441] Let me add, to reinforce my opinion about the lilies, they have no more affinity with the grasses (to require being placed after them) than do the Caryophyllaceae.[442]

Although only knowing your method from what is in your **Prospectus**, I could never agree with you about making my catalogue conform [to your method]. Perhaps I would have had some reluctance to adopt it to its full extent; besides which, readers would perceive too close a collaboration between us. The natural orders of Linnaeus are known. While some of them are badly arranged, the greatest number are quite well done. As I had only a fairly simple catalogue to put together, I believed, and still believe, that these Linnaean orders are sufficient.

The only native *Chenopodium* I know are those in my catalogue. I do not believe you have seen *Chenopodium urbicum* [L.] in Dauphiné, which has leaves 5-6 inches long and 4 inches wide.[443] I have expressed the difference I see between *Chenopodium rubrum* [L.] and *C. murale* [L.]. I am now calling *C. album* [L.] false orache, all too common in our gardens, and which is widely used in soup during the springtime when the true orache [*C. bonus-henricus* L.] is unavailable. I do not yet know whether *C. viride* [auct.] differs from it. I have never seen *C. glaucum* [L.]. I am nearly sure that the *Chenopodium* in Allioni, **Flora Pedemontana**, no. 2007, and *C. deltoideum* Lam., are only *C. murale* L., for *C. urbicum* has branches that are filiform, fructiferous, very long, without leaves, and perhaps grows only in northern countries. Haller did not find it in Switzerland.

I am enclosing herein seven small slips of paper, each on indicating a plant that I omitted in my catalogue. Please take the trouble to attach them in the designated place with a little paste, also providing a reference mark so that the reader cannot miss them. I apologize for all the trouble I am giving you.

The following is a short list of plants drawn from Monsieur Allioni that I am transcribing for you along with my opinions. You will judge where I am right or wrong:

Carex myosuroides Vill. is *Carex bellardi* All., no. 2293.[444]
Carex gynobasis Vill. is *Carex alpestris* All., no. 2329.[445]
Viola pumila Chaix is not *Viola valderia* All., [no. 1644].[446]
Carduus autareticus Chaix is *Cirsium ochroleucum* All., no. 546.[447]
Asperula saxosa Chaix is *Asperula hexaphylla* All., no. 48, on my slip of paper.[448]

∾

[440] Plants with 3 stamens or 6 stamens inserted on the petals. Consequently, Villars did not call this class Triandria.

[441] *Alsine* would only appear as a synonym in various genera in Caryophyllaceae.

[442] Villars, Hist. Pl. Dauph. 2: xix-xx (1787), did not make any of these recommended changes.

[443] According to Villars, Hist. Pl. Dauph. 2: 564 (1787), Chaix found this *Chenopodium* near Le Poët. What follows in this paragraph seems inconsistent with that statement.

[444] He is correct. Both are now synonyms for *Kobresia myosuroides* (Vill.) Fiori in Fiori & Paol. (1896).

[445] He is correct: both species = *Carex hallerana* Asso (1779).

[446] He is correct: both remain good species.

[447] He published this as *Cnicus autareticus* Chaix, Pl. Vap. 77 (1785). Chaix in Vill., Hist. Pl. Dauph. 1: 381 (1786), found at Lauteret. Villars later republished it, without citing Chaix, as *Carduus autareticus*, Hist. Pl. Dauph. 3: 12 (1788). This letter makes it clear that Villars failed to alter the genus in Chaix's catalogue. [= *Cirsium* x *purpureum* All. (1785); probably *C. heterphyllum* All. x *C spinosissimum* (L.) Scop.].

[448] *Asperula saxosa* Ch., Pl. Vap. 57 (1785). Chaix in Vill., Hist. Pl. Dauph. 1: 361 (1786). [= *Galium saxosum* (Ch.) Breistr. (1948)].

Mentha austriaca [Jacq.], All., no. 73, on my slip of paper.[449]

Veronica tenella All., no. 272 [= *Veronica bellidioides* L. ssp. *lilacina* (Townsend) Nyman] is a variety of *Veronica serpyllifolia* [L.]. I have observed it for a long time.

Primula viscosa Vill. is not *Primula viscosa* All., no. 338, but rather *Primula hirsuta* All., no. 337.

Phyteuma scorzonerifolium Vill. is *Phyteuma michelii* All., no. 427.[450]

Scabiosa pyrenaica All., f. 12 [= *Scabiosa holosericea* Bertol. (1810)] may be *Scabiosa lucida* Vill.

Centaurea seusana Chaix, on my slip of paper, may be *Centaurea triumfetti* All., no. 579.[451]

Inula britannica, or is it rather the one you found near Lac Léman?[452]

Hieracium auricula Haller, Hist., no. 53 [recognized by Allioni 1: 213 as a synonym for *Hieracium dubium* L.].[453]

Iberis garrexiana All., no. 920, is my *Iberis linifolia* discovered [at Laric] by Abbé Martin, very distinct from *Iberis linifolia* L.[454]

Myagrum erucaefolium Ch., [**Pl. Vap.** 46 (1785); Chaix in Vill., **Hist. Pl. Dauph.** 1: 350 (1786)], is *Crambe corvini* All., no. 937.[455]

Brassica cheiranthus Vill. [= *Rhynchosinapis cheiranthus* (Vill.) Dandy (1957)] is *Sinapsis recurvata* All., no. 963.

Sisymbrium erucastrum Poll. [**Poll.**, no. 628] is *Sisymbrium dentatum* All., no. 1001.[456]

Trifolium thymifolium Vill., [**Prosp. Pl. Dauph.** 43 (1779)] is *Trifolium saxatile* All., no. 1108 [1774].

Saxifraga biflora All., no. 1530, is known in the highest mountains near Argentière.

Dianthus vaginatus Chaix, [**Pl. Vap.** 26 (1785). Chaix in Vill., **Hist. Pl Dauph.** 1:330 (1786)], is *Dianthus atrorubens* All. [no. 1545].[457]

At your leisure, please be good enough to tell me your opinion about all this; and whether you know the umbel on Allioni, Tab. 63, and the other umbel on Tab. 43. The latter cannot be *Ligusticum austriacum* L. [as Allioni gives it] for which Linnaeus specifies

~

[449] This was probably a ssp. of *Mentha arvensis* L.

[450] Both *Phyteuma* are now in the *michelii* group.

[451] He is correct: *Centaurea triumfetti* All. ssp. *t.* (1773).

[452] Villars, in Hist. **Pl. Dauph.** 3: 214 (1778), would recognize *Inula britannica* La Tourrett, not *Inula b.* L. One cannot tell whether Chaix was questioning the choice, or whether he had *I. hirta* L. in mind.

[453] Villars, **Hist. Pl. Dauph.** 3: 99 (1788), gave Haller, Hist., no. 52, as a synonym for *H. dubium* L.; and Haller, Hist., no. 53, as a synonym for *H. piloselloides* Vill., Pros. **Pl. Dauph.** 34 (1779).

[454] *Iberis garrexiana* All. [= *Iberis sempervirens* L.].

[455] *Myagrum erucaefolium* Chaix [= *Erucastrum nasturtiifolium* (Poiret) O. E. Schulz (1916)].

[456] *Sisymbrium erucastrum* Poll. [= *Erucastrum gallicum* (Willd.) O. E. Schulz (1916)].

[457] Both are synonyms of *Dianthus carthusianorum* L., a highly variable species much subdivided.

the leaf segments as entire. Should it be your *Ligusticum gmelini?*[458]   As for me, I cannot distinguish your *Ligusticum lobelii* [Vill.] from *Ligusticum austriacum* L.[459]

88

Les Baux, 22 November 1785

I am going to have the honor to make my observations to you concerning your last letter of November 12th with all the freedom that the close friendship, which has united us for so long, entitles me; after which I shall try to respond to what you ask me.

[You write] "In separating the lilies from the grasses, I had the sedges, the rushes, and the aphyllanthes on one side, which approach [the lilies] in habit; and the orchids, the iris, and the crocus on the other side, which approach [the lilies] in stamen number." I have not understood the meaning of your sentence. What definition is to be expected from the natural orders of Linnaeus? The entirety of the definition lies in understanding the title for each section so that any botanist with a minimum of Latin can comprehend easily. Any complaint that could be voiced would only revolve around an inadequate affinity between certain genera, or on the inadequacy of the character assigned in the heading.

"Why," you write me in reference to my catalogue, "have you not written a preface on your travel for this treatise?" Because, I say to you in response, I owe as many plants to my sedentary observations as to my botanical trips. . . . It would be seen as absurd to claim so many exertions for having, in fact, been only four steps from my cottage.[460]

Pes anserinus Dalech. I, **Tabernaemont.** [p. 427], and *Atriplex dicta pes anserinus* I, J. Bauhin [II: 975], are surely the same plant, which is *Chenopodium rubrum* L. But Linnaeus was wrong to refer J. Bauhin's synonym [also] to his *Chenopodium murale* [L.]; for the latter must have *pes anserinus* II, **Tabern.** [p. 428], as its synonym, *Atriplex dicta pes anserinus alter, sive ramosier,* J. Bauhin, fig. 2. Ray and Ruppius, who established these two species, as well as Linnaeus, used Bauhin's synonyms quite as I have just indicated. I know that Haller did not segregate these two species, perhaps with good reason. . . . I am sending you branches of both of them such as I have them.[461] As for what you say that, according to your notes, *Chenopodium rubrum* has obtuse leaves, perhaps you made those notes in Paris about *Chenopodium purpureum* Juss. I have it in my herbarium, grown from seeds sent by Monsieur Thouin, which appears to me to differ a bit from our species, with leaves with more obtuse angles.[462]

I see nothing in *Cirsium ochroleucum* All. to separate it from our *Carduus autareticus,*

⮁

[458] Both *Ligusticum austriacum* L. and *Ligusticum gmelini* Vill. are synonyms of *Pleurospermum austriacum* (L.) Hoffm. (1814).

[459] In Hist. Pl. Dauph. 2: 612 (1787), Villars replaced his *Ligusticum lobelii* Vill., Prosp. Pl. Dauph. 24 (1779), with his new *Ligusticum cicutaefolium* Vill. [= *Cnidium silaifolium* (Jacq.) Simonkai]. He remarked that he owed to Allioni the recognition of *Ligusticum lobelii* to be *Danaa aquilegifolia* All., no. 1392, Tab. 63 [= *Physospermum aquilegifolium* Koch = *P. cornbiense* (L.) DC. (1830)].

[460] Villars, no doubt, raised the question as he was devoting 30 pages of the preface to his first volume to his botanical itineraries. Whereas Chaix was reflecting the importance of his garden.

[461] Linnaeus had recognized and named three related species of *Chenopodium* that remain good today: *rubrum, murale,* and *hybridum.* Chaix recognized all of them in the pays de Gap. Haller recognized only *rubrum* and *hybridum,* his numbers 1583 and 1581 respectively. Villars recognized only *murale* and *hybridum.*

[462] No *Chenopodium* survived in the remnants of the Chaix herbarium.

but I have nothing to be compared to his *Cirsium purpureum* [All.].[463]

Up to now, I cannot make out the difference between my *Asperula saxosa* [Chaix] and [*Asperula*] *hexaphylla* All. Its flowers are nearly sessile between the last leaves; and its seed, convex on one side, is flattened on the other side. I have just brought in some heads of it from Mont Aurouze for my garden.[464]

Could *Galium glaucum* [L.] not be an *Asperula ab corollam infundibuliforme?* I believe that I saw *Mentha arvensis* L. at St.-Sauveur near Vizille, a good specimen of which either Monsieur Liottard or you sent to me. But my *Mentha austriaca* [Chaix] is something else.[465] According to what you tell me about *Scabiosa pyrenaica* All., it should be the variety of *Scabiosa arvensis* [L.] around Les Baux, which I have also see at Valernes, and which I call *Scabiosa purpurea*, Haller, no. 207.[466]

When Linnaeus attributes very entire leaves to *Centaurea amara* [L.], can he have seen its different varieties? I have one of them whose lower leaves are notably dentate, as in Gouan's Ill. 73. It differs principally from *Centaurea jacea* [L.] by its decumbent habit. Otherwise, they have affinity as Gérard remarks.[467]

In addition to my *Centaurea seusana* [Chaix] [= *Centaurea triumfetti* All. ssp. *t.*], I have still another one that I cannot relate to any in Linnaeus, and which I am calling *Centaurea menteyerica* [Chaix]. Here is a description of it: *perennis, pedalis, subtomentosa, multiflora, folia lanceolata, omnia petiolata, hirsuta, denticulis glandulosis ad margines instructa, squamae calycinae fuscae, ciliis argenteis circumseptae.* At Menteyer [Manteyer] along the road that leads to the Combe Noire woods. If you will take the trouble to add it to my catalogue, my description of it on a separate slip is enclosed.[468]

Should not the *Centaurea* you have from the hills along the Rhone near Lyon be the narrow-leaf variety of *Centaurea montana* [L.], with thinner stalk, less downy, readily branching, the scales edged with black cilia, as in *Centaurea montana?*[469] Monsieur Liottard had sent it to me, and I still have it in the garden.

Will you leave *Centaurea rhapontica* [L.] among the centaureas? I have never seen sterile florets on it, nor did Haller. All of them are bisexual.[470] Do you know *Jacea* in

∼

[463] See letter 87.

[464] *Asperula saxosa* Chaix, Pl. Vap. 57 (1785); Chaix in Vill., Hist. Pl. Dauph. 1: 361 (1786): *foliis semi-linearibus, floribus terminalibus subsessilibus. Tota glabra, debilis ramosa spithamea, folia summa confertiora, cor. 4 fida carnea: sem. oblonga majuscula in radicibus.* From Mont Aurouze. [= *Galium saxosum* (Chaix) Breistr. (1948)]. *Asperula hexaphylla* All., F. Pedem. 1: 12 (1785): *foliis senis linearibus, floribus umbellatis terminalibus subsessilibus. . . . Semina sequum oblonga compressa; tota planta glabra est.*

[465] Chaix, non Jacq., Pl. Vap. 52 (1785); Chaix non Jacq., in Vill., Hist. Pl. Dauph. 1: 356 (1786), is one of many infraspecific taxa, separated usually on vegetative characteristics, which have been lumped as *Mentha arvensis*.

[466] Haller, Hist., no. 207, was *Scabiosa pyrenaica* All., Fl. Pedem. 1: 140, which Villars did not treat. Also see letter 87.

[467] *Centaurea amara* L. is now held to be a synonym of *Centaurea jacea* L., a species with considerable morphological variation.

[468] *Centaurea menteyerica* Chaix, Pl. Vap. 61 (1785); Chaix in Vill., Hist. Pl. Dauph. 1: 365 (1786). [= *Centaurea scabiosa* L.]. The species is said to be variable in branching and leaf-shape.

[469] Villars commented on the variety and his reasons for not segregating it as a new species in Hist. Pl. Dauph. 3: 51 (1788). [= *Centaurea lugdunensis* Jord. = *C. triumfetti* All. ssp. *lugdunensis* (Jord.) Dostál (1976)].

[470] This was later recognized as *Rhaponticum heleniifolium* Godr. & Gren. = *Leuzea rhapontica* (L.) J. Holub (1973). The florets on *Centaurea* are rarely all bisexual and fertile; but Villars, despite asserting that the florets were all fertile, maintained it as a *Centaurea*: Hist. Pl. Dauph. 3: 44 (1788).

Haller, **Hist.**, no. 184?[471] Do you know *Centaurium majus, folio non dissecto*[472] C. Bauhin 117; Chabrey 345; Daléchamp 2: 560? The latter is surely not *Centaurea rhapontica* L. Although the editor uses J. Bauhin's synonym, he provides a different figure and description. I quite believe that *Centaurium majus enulae folio subtus hirsuo, et incano* J. Bauhin [Hist. 3:4l] is our *Centaurea rhapontica* L.; but the leaves of our plant are *ovato-obloga denticulata* as in L., **Spec. Pl.**, rather than *subcordata oblonga* as in L., **Mantissa** 478. If I did not put said *Centaurea rhapontica* in my catalogue, have the kindness to put it in: from La Grangette and Rabou on subalpine hills, *ob plantae folia helenii affinia*.

The error Gouan made by using the name *Sisymbrium erucastrum* for what is really *Sisymbrium murale* L. should not prevent you from giving that name to *Brassica erucastrum* L., which is truly a sisymbrium. But I beg you to give it another trivial name to avoid getting into collusion with my *Sisymbrium erucastrum*, which is Pollich's no. 628.[473] Could you not call it *Sisymbrium brassicastrum*, thereby recalling the former Linnaean name?

Gentlemen botanists, when will you stop playing tennis with our dear umbels? I am really sorry for them, seeing them tossed back and forth between you.. Each one of you picks up the ball on a whim and feels obliged to place it more appropriately. But I come back to the species in question between us. In speaking about *Ligusticum austriacum* L., you tell me: "I must ask you where you have seen Linnaeus give it very entire leaflet segments? On the contrary he calls them confluent." I answer you that I have seen, and am at the moment reading, in **Species Plantarum** L. [p. 250] *foliolis confluentibus incisis integerrimus*: a description the author did not alter in his **Mantissa**, and which is unchanged in Murray's **Systema** with a synonym from Lobel after Séguier's citation. Whether the synonym is well applied I cannot be sure. But my plant, whose specimen I am sending you, indeed appears to me to be *Ligusticum austriacum* L. according to his description. Your description of *Ligusticum lobelii* [Vill.] does not accord with [Linnaeus's] description except for *caule pauciflora*: for my plant often has several umbels, quite full, fertile throughout the disk, often subtended by a leaflet, with involucres linear-lanceolate, caducous. Compare it to *Danaa* All. where you will certainly be rebutted.[474]

I know *Ligusticum peloponnesiacum* [L.] [= *Molopospermum peloponnesiacum* (L.) Koch (1824)], which I grow in my garden from seeds from Paris. I have never seen *Ligusticum austriacum* All., tab. 43, [no. 1323], nor your *Ligusticum gmelini* [Vill.]. You will have to verify them. . . .

For that which is *Ligusticum seguierii* [Vill.], Séguier described it very well as *Ligusticum alpinum perenne, ferulae folio, floribus albis*, Seg. [Ver.], 2: 41, tab. 13. But the leaves in his figure are too large and poorly depicted, as Gouan has said. It is certainly both *Ligusticum pyrenaicum* All. and *Ligusticum ferulaceum* [All.].[475] Out of respect owed to this respectable man [Allioni], we must consult him by sending him the leaf of our plant that

~

[471] *Jacea foliis radicalibus semipinnatis, caulinis ovato lanceolatis.* Hall., Hist., no. 184.
[472] It appears that he has combined or confused two synonyms of *Centaurea rhapontica* L.: 1) *Rhaponticum folio Helenii incano* C. Bauh., Pin. 117; 2) *Centaurium majus folio Helenii* Tournefort, Inst. 449.
[473] See letter 87. Villars accepted *Sisymbrium erucastrum* Poll.
[474] For the beginning of this dispute, and its resolution, see letter 87.
[475] *Ligusticum seguierii* Vill. = *L. lucidum* Mill. (1768). *L. pyrenaicum* All. = *L. ferulaceum* All. These species are very affinate.

I enclose here. Séguier says that the herbalists of Verona call his ligusticum false meum. (Those who know it here call it the same thing; but, of course, that is not *Meum adulterinum* Lob.) He makes no mention of Tilli.[476] He only adds that he still must check whether *Ligusticum ferula folia* in Tourn., Inst., no. 324, is the same as his plant.

Pollich, **Palat.** 1: 289, no. 292: *Oenanthe peucedanifolia* [Poll.], *foliis omnibus linearibus, radicalibus bipinnatis, caulinis pinnatis, involucro universali nullo, radicum napulis ovatis sessilibus.* . . .[477]

My *Viola pumila* [Chaix][478] has lightly crenulate leaves, very small flowers as in our variety of *Viola palustris* [L.]; whereas *Viola valderia* All. has very entire leaves, bent but widespread, and flowers as large as those of *Viola calcarata* [L.].

Do not fear that my *Carduus aurosicus* [Chaix][479] is only a variety of *Carduus defloratus* L. Last week I went to dig up several roots of it in its native area and have transplanted them to my meadow. I had a good opportunity to see the difference between them. Beyond what I have said about it, here again are some characters: *Carduus aurosicus radix perennis, simplex, albida, tota obrosa, sue diffusa*, quite cleft or ragged all around. *Carduus defloratus radix perennis, composita, laevis, cortice fusca.* If you wish to add something from that to my manuscript, fine. If not, you may use it in your history.

I had resolved to go next Monday to Oze and to Le Saix to ask Monsieur Martin for a specimen of *Iberis linifolia* [L.] [= *Iberis stricta* Jord. (1864)] for you, which I named for him in the determination of his plants. But it has rained so much these past few days that the swollen streams would be an obstacle for me. I shall carry out my project when it seems to be possible, and I shall not forget your request. I enclose a small specimen of *Carduus aurosicus* here, along with a branch of *Carduus nigrescens* Vill. that you asked for. If it is insufficient for you, I will bring some radicle leaves from near Gap that I will send you with said *Iberis*.

Have the kindness to add *Polypodium rhaeticum* [L.] to my catalogue for which you will find a slip enclosed.[480]

Take care to have your genera and species correctly spelled before going into print. What a ridiculous thing it would be, for example, to write *Gysophylla* in place of *Gysophila!* The first is a contradiction; the second makes good sense.

When speaking of *Fagus sylvatica* [L.], you could well say that its leaves, falling dried in the autumn, provide a very good bed (*culcitra*) during the year for the country people of Upper Dauphiné.

The springs of live water in our high mountains, in the opinion of all those who have gone near them, are very cold in summer, but temperate in winter, to a point of melting the snow at the place where the springs come forth. I have believed, as do most natural philosophers, that these springs retain the same degree of coolness in all seasons, and that

<hr/>

[476] Michaelis-Angeli Tilli, **Catalogus plantarum horti Pisani** (1723).

[477] Villars did not recognize this species as distinct from *Oenanthe fistulosa* L., whereas Chaix located it at La Saulce.

[478] *Viola pumila* Chaix, Pl. **Vap.** 35 (1785); Chaix in Vill., **Hist. Pl. Dauph.** 1: 339 (1786). At Corréo [La Roche], in the marais des Chastelas.

[479] *Carduus aurosicus* Chaix, Pl. **Vap.** 60 (1785); Chaix in Vill., **Hist. Pl. Dauph.** 1: 364 (1786). Mont Aurouze above Matacharre; La Cluse-en-Dévoluy.

[480] The species was not added. Villars, **Prosp. Pl. Dauph.** 51 (1779) held it to be var. *a* of *Polypodium polymorphum* Vill. See *Athyrium* Roth.

the different sensation derives from the different condition of our organs, those affected by cold and heat, just as our cellars seem cold to us in summer and warm in winter. But if that be true, when one reaches these springs in winter, when drenched in sweat and quite overheated, and then puts a foot or hand into the water, they ought to feel as cold as they do in summer when the body is in similar condition. Experience, however, proves the contrary to be the case. This has driven me to seek another physical cause, and here it is: I say that these springs are cold all summer, because they provide water from the winter snows that are impregnated with a large amount of nitrous particles; and that they are tempered all winter, because their reservoir of nitrous water, having been nearly entirely exhausted by the flow during the fine season, can only give during the winter waters provided by the very abundant autumnal rains,before the melting of the snows begins again. I do not know whether my physical explanation is new or whether any other naturalists hold this view. Please give me a word about it, and take the opportunity to mention it in your history.

It is time to finish. . . .

℃

Letters 89 to 125 cover nearly three years: beginning in December of 1785, when Chaix received separate copies of his catalogue; and ending in late 1788, by which time the first part of Villars' third and final volume had been published. The second part of the third volume would be delayed until the autumn of 1789. By then, the beginning of the French Revolution had intervened to complicate the lives of our two botanists most drastically. [1785-1788]

89

Les Baux, 13 December 1785

I went to spend the first week of this month at Oze and Le Saix. The two worthy pastors of those parishes, my true friends, are extremely devoted to you; but especially Monsieur Gaude [of Oze] who never forgets you, above all in his prayers and in the holy sacrifice of the mass. He continually presses me to express his sentiments to you. Monsieur Martin [of Le Saix], to whom I extended your invitation to come to see you (that you made to me), has promised to accompany me during next May, God permitting. He is a truly learned priest. He makes progress in botany, but he knows a number of other disciplines quite well: mathematics, astronomy, geography, history, and both canon and civil law, working on them continually. I took the specimen of *Iberis linifolia* from his plant collection, collected around the mountainous woods of Laric, which I am sending to you. If it is not *Iberis linifolia* L., it differs little from it. I would relate it to *Iberis garrexiana* All. [= *Iberis sempervirens* L.].[481]

He found a fleshy plant that, already dried, at first appeared to me to be close to *Sedum album* [L.]; but he assures me it must be a *Sempervivum* because of its 6 petals, 6 styles, and 12 stamens. Should this be *Sempervivum sediforme* Jacq.? See Allioni, **Fl. Ped.**, no. 1940 [= *Sedum sediforme* (Jacq.) Pau (1909)]. If you presume it is, honor him for it in your history.[482]

Aster acris [L.] [= *Aster sedifolius* L.], which I may have placed at Laric by mistake, grows abundantly at Sigottier among the vines. *Mespilus germanica* [L.], which I have not yet seen in the region around Gap, also grows at Sigottier. I have eaten its fruit and preserved its seed.[483]

If my *Scabiosa gramuntia* L., **Syst.** l: 280 [= *Scabiosa triandra* L.] from Les Baux is your *Scabiosa lucida* [Vill.], why should it have been called *lucida* and have glabrous leaves? Because of a short pubescence, mine is an ash-green; the cilia on the seeds barely surpass the coronula; and it is perennial. I do not see any affinity in it with *Scabiosa columbaria* [L.]. You point out to me that its cauline leaves are tripinnate in L., **Sp. Pl.** But those in the **Systema** are only bipinnate, having been corrected. Mine is the same as a specimen that Monsieur Pourret had sent me from Narbonne; the same as came up for me from seeds received from Paris under that name. And it is also the same plant you described for me, collected at St.-Paul-Trois-Châteaux.[484]

[481] This may be *Iberis linifolia* L. ssp. *villarsii* Jordan. See F. Lenoble, **Catalogue raisonné des plantes vasculaires du Département de la Drôme** (1936), p. 185.

[482] Villars would make no mention of Allioni's no. 1940 in his flora.

[483] A cultivated and naturalized shrub from southeastern Europe in the Rosaceae.

[484] Villars was convinced. He cited a collection of *Scabiosa gramuntia* L., **Hist. Pl. Dauph.** 2: 294 (1787), from St.-Paul-Trois-Châteaux.

I am sending you my only specimen of *Centaurea menteyerica* [Chaix], as I know where it grows and can collect it again in the fine season. This trivial name does not please me any more than it does you. Give it what seems to you a better name.[485] After seeing it, you will be a long way from referring it to *Centaurea triumfetti* All., which is *Centaurea seusana* [Chaix], close to *Centaurea montana* [L.].[486]

Since you call to my attention that *pes anserinus alter seu ramosior* J. Bauhin is *Chenopodium hybridum* L., I must agree with you, following Haller, that the synonyms referred by Linnaeus to his *Chenopodium rubrum* and *C. murale* are simply the same plant. Consequently, the specimen that I sent you labeled *Chenopodium murale*, which you say resembles the figure in Buxbaum [**Catalogus plantarum circa Hallam saxonium sponte nascentium (1721)**], is *Chenopodium urbicum* L. Monsieur Deleuze collected it at Le Poët, and the specimen I received from him was at least four feet high with triangular leaves. The lower leaves are three inches in diameter from one angle to the other, notably dentate and farinose; the spike is one foot high, erect, narrow, leafless, and composed of spikelets parallel to the stalk. I have compared it to two specimens grown from seeds sent to me and see no difference. The description of Pollich's *Chenopodium urbicum*, no. 246, corresponds completely to what I have just said. The two species, *Chenopodium rubrum* and *murale*, established by Linnaeus, are what led me astray. If there is a way of adding an end-leaf to the printed copies of my catalogue, with the alleged corrections and additions herein, that would be desirable. But it that cannot be, please take care to dicuss these plants in the appropriate descriptions in [the later volumes] of your history.[487]

As for my *Ligusticum austriacum* L., your *lobelii*, I do not know what led Scopoli to merge *Ligusticum austriacum* and *L. peleponnesiacum* L. But I do not see as much difference between the *Ligusticum peleponnesiacum*, which I have raised from seed, and my above said *austriacum*, as I do see between the former with figure 43 in Allioni, and between the latter with figure 63 of the same Allioni.[488] I persist in saying, however, that my *Ligusticum austriacum* [L.] is *Seseli montanum cicutaefolio, glabrum* C. Bauhin [**Pin. 161**]. Note that, what Allioni calls *Danaa* today, he had called *Coriandrum aquilegifolium* in his **Auctarium** [1762].

What prevented me from relating my *Carduus aurosicus* [Chaix] to *Carduus carlinoides* Gou. is that the pinnules of the latter are palmate-quadrifid, covered with wool; the stem terminates in a corymb bearing clusters of flowers. Whereas mine has decurrent leaves to be sure, but glabrous, pinnatifid; its floral heads in fact are borne on short, woolly peduncles; but they are not in clusters. At most, one should be able to test against any species of Gouan. You will find here the radicle leaves of your *Carduus negrescens* [Vill.].

Your care to have some copies of my catalogue run off for me is most courteous, and I am all the more moved as I was not anticipating it. I shall offer a copy of it to our Monsignor Bishop, which could bring me some credit with him; and another copy to

[485] *Centaurea menteyerica* Chaix, **Pl. Vap.** 61 (1785); Chaix in Vill., **Hist. Pl. Dauph.** 1: 365 (1786). At Manteyer near Combe noir. [= *Centaurea scabiosa* L.].

[486] See letter 88.

[487] By then, it was too late to make this alteration in Chaix's catalogue.

[488] *Ligusticum austriacum*, no 1323; and *Danaa aquilegifolia*, no. 1392, respectively. See letters no. 87-88 for this controversy.

Monsieur Delafont [the subdelegate in Gap] to whom I owe many obligations. I shall be pleased to give other copies to my dearest friends. I embrace you with all the affection in my heart.[489]

90

Les Baux, 23 December 1785

I am delighted by the manner in which they have printed my catalogue of plants. The whole thing has been done quite beyond the merit of the work.[490] I only regret not having been able to read one copy in proof before the final printing. I would have eliminated many errors that have slipped in. To repair them and to perfect this work, would you have the kindness to have a sheet provided at the end for the *supplenda* and the *errata* that I am sending you herein? The reader whom you want to attract [to the flora] will convince you of this necessity. I am going to correct the copies that you have had the kindness to send me in my own hand before offering them to those to whom they are destined. You have given me a great pleasure. Accept all the thanks in my power to bestow.[491]

It is unworthy conduct by Monsieur Gilibert to cast ridicule on your new species after having paid tribute to you. The action enhances neither his judgment nor his honesty. But was it not being decent of you beyond prudence to furnish others with the material for the construction of an edifice that you, yourself, had in mind? I can only see in that an excessive kindness on your part.[492]

~

[489] On pp. 15-16 of Delafont's report to the baron de La Bove in 1784, he remarked that the discovery of medicinal productions in the region had largely been the responsibility of Monsieur Villars, a physician of Grenoble. But that "in the diocese we also possess a very learned and profound botanist, hard-working, tireless, and of a rare modesty. He is Monsieur Chaix, prieur-curé des Baux, a man much esteemed." Published by Joseph-Hippolyte Roman, **Bull. Soc. Etudes Hautes-Alpes** 18, 2nd ser. (1899): 73-93, 167-186, 247-264. Ibid. 19, 2nd ser. (1900): 19-53.

[490] Not the least of the merits of the catalogue was Chaix's decision to employ a natural order rather than Villars' "mixed method." The Linnaean natural order that Chaix acknowledged was not the **Methodi naturalis Fragmenta** Linnaeus had published in 1751, but rather the revised natural order he published in 1764 as a supplement to the 6th edition of **General Plantarum**. See Chaix in Vill., **Hist. Pl. Dauph.** 1: 310-311 (1786).

[491] Villars' kindness in having a few separate copies of Chaix's catalogue run off for him has created a quite unanticipated and regrettable problem for botanists. As is evident from the prior correspondence, the catalogue was compiled at Villars' invitation for inclusion in the first volume of **Hist. Pl. Dauph.**, where it can be found, properly credited to Chaix. This volume appeared early in 1786, sometime before February 13th. The separate copies, meanwhile, had been given a title page with an abbreviated title and new pagination; and both Chaix's correspondence and the dates in the presentation copies prove that they had been run off in early December, the precise date again uncertain, but 1785: **Plantae Vapincenses sive Enumeratio in Agro; Vapincensi Observatarum Stirpium**. This alternate publication has long been known [see Adolphe Rochas, **Biographie du Dauphiné** (Paris: Charavay, 1856-1860), l: 194]. But no botanist saw any significance in the publication of these separates until P. Perret and H. M. Burdet, "Les 'Plantae Vapincenses' de Dominique Chaix et les travaux floristiques de Dominique Villars en Dauphiné." **Candollea** 38 (1981): 400-408. Given the arbitrary international rules, they recognized that the existence of the separate publication altered the traditional assumption of priority. That is why two citations have always been given here for each of Chaix's new species. But it needs to be remarked that no independent publication was intended by either Chaix or Villars, nor were the separates meant for scholarly distribution. For the actual distribution, see letter 91.

[492] Jean-Emmanuel Gilibert (1741-1814) had undertaken the publication of **Caroli Linnaei botanicorum principis. Systema plantarum Europae**, published 1785-1787 in 7 volumes. He had also been given permission to edit Villars' Latin translation of **Prosp. Pl. Dauph.** (1779), which Gilibert issued separately as **Flora delphinalis** in June of 1786, but dated 1785. It is evident from this letter that Chaix thought Villars ought to have published the Latin edition himself.

Do not believe that we have the *Danaa* All. in Dauphiné such as that botanist has described and drawn it. I do not doubt that the plant is indeed the same as Lobel's, as J. Bauhin said that Lobel saw it in Piedmont near the Po.

I would like to believe that the true *Ligusticum austriacum* corresponds to Allioni's figure, but have you seen it conform to that figure in Dauphiné? In that case, my *Ligusticum austriacum* is a new plant that would have to be called *Ligusticum cicutaefolium* [Vill., **Prosp. Pl. Dauph.** 24 (1779)], even though Séguier gives assurances that Lobel's figure portrays the plant better than the figure in Clusius. But again, my plant is not the *Danaa* in Allioni, but rather the one in Séguier.

As for our *Ligusticum seguierii* [Vill.], it is unquestionably the *Ligusticum pyrenaeum* Gou. and All. [= *Ligusticum lucidum* Mill. (1768)], but I am inclined to believe that it is not Séguier's plant. He had leaves engraved that are more fennell-like, more enlarged, more spread out, and with longer leaflets than ours, much as Allioni has in the figure for his *Ligusticum ferulaceum* [All., 1774]. Beyond which, Monsieur Allioni seems to settle it quite clearly by referring Lamarck's *Ligusticum ferulaceum* to *Ligusticum pyrenaeum*; and he makes it quite understood that we do not have his *Ligusticum ferulaceum* in France. It should be necessary, therefore, for you to retain the trivial name *pyrenaeum*.

Monsieur Allioni refers *Saxifraga retusa* Gou. to his own [*S.*] *purpurea*. Is that your feeling? In that case, I am in error; for I took a plant found with you in the Alps of Valgaudemar to be *Saxifraga retusa* Gou., which is closer to [*S.*] *oppositifolia* L. than to [*S.*] *purpurea* All., and which I would more readily believe to be [*S.*] *biflora* All. Please give me your opinion of it.[493]

Monsieur Serre's progress gives me increasing satisfaction. Yesterday, I read what you had indicated to me to Madame de La Roche and to the relatives of our young surgeon. You talked to me some time ago about some allowance for him from the intendant. I do not doubt that you have done everything possible about it. Has that expectation been completely defeated?

I shall see Monsieur Gilibert's work that you promise me with much pleasure. I will send this work back to you as soon as I have taken any notes from it that interest me. I am quite aware that my letter cannot reach you soon enough to wish you a good Christmas celebration; but for the New Year I make for you the same wishes I desire for myself: *Sanctam, tranquillam vitam, felicitate aeterna coronandum.*

P.S. Do you approve that your name in print is *Villars*? Up to now, your signature has been *Villar*. In the first instance, your name in Latin would be rendered *Villarsus* instead of *Villarus*. If it is not your intention, you must warn the printer.

∾

Chaix to Delafont,
Subdelegate of the City of Gap
Les Baux, 26 December 1785[494]

I have provided Monsieur Villars the catalogue of the plants of the Gapençais for inclusion in his general history of the plants of Dauphiné. He has had the kindness to have some copies of it run off separately for me. I am taking the liberty of presenting you one of them as testimony of my deepest gratitude and of my firm dedication.

P.S. I have thought it to be wise to have the *Geranium* promised to you spend the winter with me. I shall send it to you when the chills of winter are no longer to be feared.

91

Les Baux, 31 December 1785

How I hope you will be willing to provide a separate sheet for my *supplenda* and *emendanda*, to be placed at the end of my compilation, to give it the degree of completion and correctness that depends upon us. I beg you to add the material written separately on a list to the *supplenda*. It would be quite annoying to me if four fine plants, *Cucubalus italicus* [L.] [= *Silene italica* (L.) Pers.], *Cucubalus otites* [L.] [= *Silene otites* (L.) Wibel], *Lychnis flos-cuculi* [L.], and *Lychnis viscaria* [L.], omitted by an inadvertence that astonishes me, were not to be reported in my compilation, being, what is more, quite common. Since you have done so much to bring out my small work, you will no doubt make this additional effort.[495]

It is up to you to dispose of the remaining copies of my plants that you tell me you still have. I am already under enough of an obligation to you for having sent me six of them. I offered one of them to the Monsignor Bishop; one to Monsieur Delafont; one to the abbé Blanc; one to Monsieur Gaude; one to Monsieur [Jean-Pierre] Reynaud, curé de La Roche, my dear neighbor; and I am keeping the other one for myself. If I had had a seventh, or if I could have it, I would send it to Monsieur Deleuze along with some plants that are meant for him.[496]

I am flooding you with letters. But, in shortening this one, I hurry to finish it by rendering thanks to God for the benefits He has granted us during this year that ends today and praying Him to favor us during the new year that begins tomorrow. . . .

92

27 January 1786

Your thoughts on the two louseworts, [*Pedicularis*] *comosa* [L.] and *tuberosa* [L.], are very judicious, very well founded. I gladly approve them; and I doubt if you have another critique in your entire work that can give greater pleasure to learned botanists that this one,

⁓

[495] Villars, in **Hist. Pl. Dauph.** 1: 467 (1786), did add a sheet entitled "Fautes d'impression et additions" at the end of his first volume. A substantial percentage of the additions were to Chaix's catalogue, including the above four species.

[496] Delafont's copy survives, along with the above letter, in the departmental archives of Hautes-Alpes. The copy inscribed to the Curé Gaude is also known to survive. See Perret & Burdet, "Les 'Plantae Vapincenses' de Dominique Chaix," p. 400.

if you should publish it.[497]   In truth, Haller must not have omitted Allioni's *Pedicularis comosa* [L.], which would be [Haller, **Hist.**] no. 323, tab. 10. This would also be *Pedicularis bulbosa* J. Bauhin [3: 438], representing its roots very well in the figure; and described by Ray, [**Hist.**] p. 771, no. 10.[498] Even though the figure in Barrelier, **Icones**, 469, does not show the tuberous roots, the very long spike with entire bracts is sufficient explanation; and when this author observes on p. 22, no. 210, that the flowers are sometimes purplish and sometimes white, he has probably confused the two species. I am criticial of Monsieur Allioni for not saying a single word about the roots of his specimen where he has described it; and his figure does not show them sufficiently tuberous, as they surely are!

Our *Pedicularis gyroflexa* [Vill.], which is indeed the *tuberosa* of Linnaeus's description, even though in truth it does not have clearly tuberous roots, is certainly Haller's [**Hist.**] no. 324, tab. 11, also in Ray, **Hist.**, p. 770, no. 7: *Cristagalli montana, floribus pallidis in spicam congestis* that he claims to have collected on the mountain tops of the Grande Chartreuse: roots simple, fiberous; flowers pale yellow, each one having leaves similar to those below; different from *P. bulbosa* in J. Bauhin. Do you remember having seen it with this color on those mountains? If so, you could make use of this synonym from Ray.

You have had the kindness to have my *supplenda* and *emendanda* printed. I thank you for it, and I am pleased with it, although I perceive that I still omitted *Arabis turrita* [L.] *in sylvis caduis circa Vap., biennis.* I did not dare to have *Aretia helvetica* [L.] [= *Androsace helvetica* (L.) All. (1785)] entered into my catalogue, because, when botanizing with you around the crag of Bure, you gave me to understand that it was only an aged and dry variety of *Aretia alpina* [L.] [= *Androsace alpina* (L.) Lam. (1778)]. On rereading my notes, comparing them with descriptions in Haller and Allioni, and seeing in your flora of Dauphiné that you locate it on Bure, I am sure that it is one of our local denizens. The allegation you will make about both plants consoles me about my omission.[499]

Reading Allioni, no. 289 [*Veronica romana* L.], I perceived that the veronica I have called *verna* [L.] is really *romana: folia opposita, oblonga, 2-3-5-dentata; floralia integra, flos evidentur pedunculatus*; whereas *Veronica verna*, which I have cultivated from seeds sent from Narbonne, has leaves *omnia digitato-partita, etiam, floralis in quorum axillis haeret flos sessilis, minimus, fugacissimus.* If you have the opportunity, mention it in your *addenda*.

Monsieur Allioni, following Haller, has done very well to make a distinct species of *Primula acaulis*.[500] Having done it myself in my notes, I did not dare to reproduce it in my catalogue. But I would say that it is closer to the *elatior* than to the *officinalis*; and I would not quite call it acaulis, but subacaulis, as I have observed that its scape is not always simple

~

[497] The lengthy passage appears as a discussion of *Pedicularis gyroflexa* Vill., **Hist. Pl. Dauph.** 2: 426-430 (1787).
[498] There is considerable room for confusion here: Evidently Allioni did not realize that Haller, **Hist.**, no. 317 = *Pedicularis comosa* L., perhaps misled because Haller did not cite the Linnaean name as a synonym. Chaix then erred by giving the number 323, which is = *Pedicularis palustris* L., for which Bauhin's *P. bulbosa* is indeed a synonym. Allioni, meanwhile, had equated Haller's no. 323 to *Pedicularis tuberosa* L.
[499] Chaix refers here to a passage in the forthcoming flora, **Hist. Pl. Dauph.** 2: 471 (1787), where Villars cited these species in distinguishing between *Aretia, Dispensia*, and *Androsace*; following which he published several transfers.
[500] Linnaeus gave his *Primula veris* three varieties: *officinalis, elatior*, and *acaulis*. Whereas Haller gave those varieties distinct specific rank, his number (**Hist.**) 610, 609, and 608 respectively.

and uniflora, but sometimes multiflora with a very short peduncle.

If there is an annual *Cerinthe minor* [L.], as Monsieur Allioni designates it, our perennial in our mountains must be a distinct species, as the said Allioni has separated it with the name *Cerinthe maculata* [L.].[501]

So we have lost Monsieur Guettard![502] May the God of mercy have clasped him to His bosom! . . .

Monsieur Gilibert's preface appears interesting to me, but it should have been edited and purged of some mistakes in Latinity. That was part of his own responsibility in regard to his **Chloris Grodnenois** and his **Flora lithuanica inchoata**, but he has only been editor of the rest. I am annoyed that your **Flora delphinalis** appears there with so many errors. I shall not repeat to you here the criticisms I made to you elsewhere regarding this matter.[503]

Has the **Théologie de Lyon**, printed in 1784 and authorized by the archbishop, raised any storm in Grenoble? According to a critical pamphlet sent to me from Embrun (although anonymous, I think I know its author and the place of publication), this **Théologie** favors the five propositions of Jansenius, condemned by the Church. Our bishop, to whom it had been proposed, declined for very good reason to have it taught in his seminary in Gap. Tell me something about it.

I have had the two books that I requested from you earlier, **Voltaire parmi les ombres** and the **Institution divine des curés**, sent from Lyon. I should like you to read the first one.

93
Les Baux, 29 January 1786

I had a letter posted to you yesterday to respond to you regarding *Pedicularis tuberosa* etc. As you might need Monsieur Gilibert's first volume, I hastened to read it and to take some notes from his **Flora lithuanica**, as well as from the **Nomenclator** by [J. D.] Leers. I am sending them back to you with much gratitude. . . . As you have indicated to me that you have the complete [J. J.] Reichard,[504] I am retaining them to be read at leisure. If Monsieur Gilibert had limited himself simply to perfecting his **Flora lithuanica** and had given us only this piece, we would be under more obligation to him than for all the rest of his production. The few notes that Monsieur [Claret de Fleurieu de] La Tourrette gives us in his **Chloris [lugdunensis]** makes one hope for the publication of his flora. I catch a glimpse there of a much more distinguished botanist.[505]

Here are a few observations I am addressing to you:

1. *Epilobium roseum* [Schreber] in Gilibert, **Lith.** p. 64, I have seen and collected from

~

[501] *Cerinthe minor* L. can evidently be either annual, biennial, or perennial; and *Cerinthe maculata* L. has been reduced to synonymy.

[502] Jean-Etienne Guettard (1715-1786), physician and naturalist, grandson of François Descurain (1658-1740); Médecin-Botaniste to the duc d'Orléans. Guettard had visited Chaix in 1775.

[503] See letter 90. Gilibert, trained in medicine at Montpellier, had been invited to practice at Grodno by Stanislas Poniatowski, King of Poland (1764-1795). He founded a botanical garden at Grodno before moving to Vilna when the university was transferred there. He removed to Lyon in 1783 after his criticisms of Polish-Lithuanian medical services won him many enemies, becoming an affiliate of the local Hôtel-Dieu.

[504] Which title of Reichard is unclear, but it may have been his new edition of **Caroli a Linné, Systema plantarum** (1779-1780), issued in four parts.

[505] The anticipated flora of Lyon was never published. His two more limited works were **Botanicon pilantense** (1770) and **Chloris lugdunensis** (1785).

marshy ground at St.-André near Embrun, a very distinct species.[506]

2. *Epilobium purpurascens* in Gilibert, ibid. I have met this at Chaudun in the gravel on the [River] Buëch. Only a variety of *Epilobium montanum* [L.]?

3. D. Villars, **Flor. delph.** [Distinguish between *Lichen parientinus* L. and *Lichen candelarius* L., one morely likely to adhere to trees, the other to rocks.]

4. *Lichen pallescens* L., frequently seen on fruit trees, is leprous and crustose; but *Lichen stellaris* L., abundant on said trees, is foliaceous and imbricate.

5. D. Villars, **Fl. delph.**, replace *Lichen stellaris* L. with *Lichen omphalodes* L.? I have often found it in cool woods on the bark of trees. . . .

6. *Byssus botryoides* L. grows on the damp walls of the church in Les Baux.

7. *Agaricus clavus* L. is around Les Baux on the rotting roots of willows and poplars amidst debris.

8. *Boletus fomentarius* L., *inaequalis*, *obtusus*. Mine is Haller, **Hist.**, no. 2283.[507] *Boletus annulatus* All., **Fl. Ped.**, [no. 2746][508] On trunks of aged rotting trees in woods. *Boletus igniatius* L. is smooth, always hard; grows on walnut trees.[509]

9. *Boletus agaricum* All. [= *Boletus officinalis* L.], on larches; rare here.

10. I do not see *Agaricum coriaceum* Latourr., **Chlor. lugd.** in your **Fl. delph.** It is Haller, **Hist.**, no. 2263, on dead trees in our woods and not at all infrequent.[510]

Write to me when you can. I am impatient to know whether Monsieur de Jussieu has met your expectations. I have paid the shipping costs for the accompanying books to Pellissier.

94

Les Baux, 24 February 1786

I cannot give you adequate expression of the overwhelming feeling of gratitude I have for the gift of the first volume of your history that you have made to me, in which you reveal me to the public quite beyond what I merit, and in which you sketch me with your own hand in the most affectionate possible terms to a friend. I have read it with an indescribable pleasure.[511]

The preface is very well wrought, adorned with very interesting incidents. The dictionary is the most thorough of its kind. Your method appears to be the shortest way to introduce botany, and you will have to be permitted some irregularities, which you have been obliged to introduce, in order to preserve the natural families. The general idea you

~

[506] This species, although not recognized by Villars, was within his range.

[507] Haller's no. 2283: *Polyporus sessilis, convexo planus, anulis versicoloribus, poris albis tenuissimus.*

[508] Allioni also cited Haller's no. 2283 as a synonym.

[509] Haller's no. 2288: *Polyporus sessilis, convexo planus, durissimus, cinereus, inferne albidus.*

[510] Haller's no. 2263: *Agaricum coriaceum, squamis ellipticis, imbricatis, superne villosis & diversicoloribus.* Villars, quite aware that Chaix had little experience with cryptogams, did not acknowledge any of this material in his flora.

[511] "At first sight, [Chaix] gave promise of being a man full of merit and candor, who, beneath a pensive and cold exterior, combined notable talents and solid judgment with the rarest and most estimabale qualities of soul. Monsieur Chaix had the same passion for botany as I. He was made to help and encourage me: incapable of letting me experience the slightest dissaffection, and quite beyond those failings or petty jealousies to which botany sometimes makes men susceptible. We were meant to be bound by taste and sentiment, to share our troubles and our successes, and to support one another." Vill., Hist. Pl. Dauph. 1: x (1786). The preface was dated Grenoble, 10 January 1786.

provide about the virtues of plants is sufficient, since you reserve the details for the individual species; your genera are very good; and if anyone criticizes you for having based them on those of Linnaeus, could you have stepped off Nature's path that this celebrated naturalist was the first to follow?

As for your lists, they bind you to many repetitions.[512] Most [of the plants] are without comment; but those you have accompanied with notes are disfigured by the great number of mistakes in your use of the Latin language, very unfortunately employed. As a rule, you ought to have had them reviewed by a Latinist, or have sent them to me. Your friends who will read them will groan. Your enemies will make fun of them. Those indifferent to you will have a horror of them. They are too pointed to attribute them to the incompetence of the printer. I shall call your attention to some of them below, but I think it is not inappropriate for you to compose an *Errata*. Henceforth follow my advice: Do not risk writing anything in Latin in your two remaining volumes, which will be the core of your work--with the exception of your plant descriptions that, being composed entirely of ablatives, are more familiar to you and less susceptible to gross mistakes. You will be obliged, however, to express yourself differently. See my attached criticism:[513]

pp. 113-114: As the number of stamens does not correspond absolutely in Classes 3 and 10, it should not be necessary to assert it as a general rule. *Crocus* with 3 stamens comes as a surprise after the assertion of 6 stamens.

p. 163: *Carex*: You say flowers are sometimes hermaphrodite. I do not know of any. That would put them in Linnaeus's Triandria. I only know of androgynous spikes.

p. 198: *Taraxacum*: According to me, the pappus is not plumose, but of simple hairs.

p. 211: Class 7: [Plants with seven stamens,] *or with seven stamens* [united by their filaments]. You must eliminate *seven*.

p. 220: *Philadelphus*: Why call it Persian lilac when *Syringa persica* bears that name?

p. 236: Walnut trees. Did you make an error in separating this species from those bearing catkins and joining it to the hermaphrodites? its flowers being, as you say, monoecious.

p. 237: *Fraxinus*. Beyond being polygamous, it is dioecious. In this region, there are completely sterile stalks, others being fertile.

p. 307: *Antirrhinum genistifolium* [L.]. You say *radix repet*, whereas I say *radix non repet*, which provides the principal separation from *Antirrhinum linaria* [L.] that creeps conspicuously. At Le Bourget and in Les Florins, I have observed that *A. genistifolium* does not creep, or very little.

While you can be more attentive to Latin spelling, even so I repeat: Do not describe anything in Latin in the future, for you express yourself very well in French. Perhaps the prodigious number of mistakes in **Flora delphinalis** Gilibert came from your manuscript done in great haste. I say this to you alone to advise you about the matter once and for all. Your combination of two adjectives into one is not always good usage, but I have fallen into

⁓

---

[512] A substantial portion of the first volume was given over not only to Villars' itineraries, but to lists of plants collected on them. A particular plant could be collected on more than one trip, hence the reference to repetitions.

[513] On the following list, I omit all spelling and grammatical corrections as no longer of interest, but include everything related to plant descriptions.

the same error: *Juncus alpino-pilosus*[514] and *Juncus alpino-articulatus*.[515]

Whether because of jealousy or because of personal interest, Monsieur de Jussieu did not render unto you the justice he owed to you. I hope that the public will [be just].[516] You suggest that I write to him on your behalf to cause him to be less severe in the future. But would my letter have any effect upon his thinking? Can I protest against his report? I could only complain about having been praised beyond my merit and to call myself one of your first teachers of botany, something he can infer from your preface.[517] Consequently, as far as he is concerned, I ought to remain silent. But, if you would be willing, as I hope, to publish my supplement, which I am sending you on a separate sheet, at the end of your terminal volume, the public will be made aware of the truth about this issue through my short preamble. I have put in a few notes there that could be of some value.[518]

Monsieur Deleuze, who has just written to me in response to a copy of my **Plantae Vapincenses** just given to him, charges me to remember him to you while offering his respectful and friendly sentiments.

Madame de La Roche, to whom I carried your first volume, wanted to pay me the [entire] 24 livres for her subscription.[519] I have had that sum remitted to Pellissier to transfer it to you. When Messieurs Gaude and Martin give me their allocations, I shall also forward them. I could have made them to you in advance on their behalf if I had believed that would accommodate you better.

You will find three cryptogams enclosed, about which please give me your opinion when you have the time.

Reichard was not right to refer *Bromus versicolor* Poll. to *Bromus asper* L. fil.[520]

∿

[514] *Juncus alpinopilosus* Chaix, Pl. Vap. 14 (1785); Chaix in Vill., **Hist. Pl. Dauph.** 1: 318 (1786) [= *Luzula alpinopilosa* (Chaix) Breistr. (1947)].

[515] *Juncus alpinoarticulatis* Chaix, Pl. Vap. 14 (1785); Chaix in Vill., **Hist. Pl Dauph.** 1: 378 (1786). *Juncus alpinus* Vill., ibid. 2: 233 (1787) becomes a synonym.

[516] Villars republished (unpaged at the opening of his first volume) the report signed by Geoffroy, Jussieu, and the abbé Tessier. "Extrait des Registres de la Société Royale de Médecine," Séance du 13 Janvier 1786. It reads as a very fair summary of the first volume, but was quite properly critical of Villars' new method: "According to [Villars], this method is easier, simpler, and a more appropriate way to conserve the natural families. On this matter, it should be pointed out to him that, as the number of stamens is not uniform in many plant families, they can be used jointly in his method only by noting numerous exceptions, rendering the method less perfect and more difficult. We should also add that he has only reduced, not perfected, the Linnaean system. . . . But the art of simplifying a system of Botany lies not in the reduction in the number of its classes, but in making them clear, precise, well-characterized, and distributing them according to a methodical order based upon well-chosen general characters. From which it follows that it will be more advantageous to multiply these classes, when the general characters lend themselves to it, since this multiplication reduces the number of sections within each [class].

"Except for his method, the critics believe that the work will be useful to botanists, above all to those in Dauphiné, and recommend approval of the work by the Society and to be published under its privilege."

[517] Jussieu, evidently the principal author of the report, had noted that Villars' first volume concludes with a catalogue of plants from the region around Gap and Embrun by Monsieur Chaix, the curé of a parish in that canton, "a very learned botanist and one of Villars' first teachers. The list is extensive, done with care, and arranged according to Linnaeus's natural orders. Several critical notes regarding the determination of some questionable species indicate the good observer and an unpretentious man."

[518] Villars would respond to the criticism himself in the preface to his second volume, defending his mixed-method at length.

[519] The volumes were priced at 9 livres each, or 24 livres for subscribers to all three volumes.

[520] Correct: it is a synonym of *Bromus arvensis* L.

How could the judicious Ray have not understood that his *Crista galli* [*montana*], p. 770, no. 7, was the same as *Pedicularis foliosa* [L.]? How did he dare insist that his plant was very different from the latter? I shall bow, however, to your opinion.

As for the paper meant for the competition, which you mentioned to me, . . . I shall do what I can to put myself at your disposal.

I pray the Lord to preserve your health. My poor nephew, Joseph Chabot of Les Baux, whom you know, is wasting away from one day to the next thanks to a very weak stomach that can barely retain any food. May the will of God be done! I baptised another daughter for him last Sunday. He will leave his wife in great trouble.

P.S. It is certain that the author of the **Théologie de Lyon** advocates the errors of Jansenius. A piece of writing much discussed!

~

Villars' "Tableau des Classes,"
Hist. Pl. Dauph. *1: 113, 116-125 (1786)*
(Modified Sexual System)

Definite		MONANDRIA I
		DIANDRIA II
		TRIANDRIA III
		TETANDRIA IV
		PENTANDRIA V
		EXANDRIA VI
		OCTANDRIA VIII
		DECANDRIA X
		DODECANDRIA XII
Indefinite	UNITED BY THEIR FILAMENTS	MONADELPHIA VII
		DIADELPHIA VII
	ABOVE 12 INSERTED ON THE CALYX	ENEANDRIA IX
	ABOVE 30 INSERTED ON THE RECEPTACLE	ICOSANDRIA IX
		POLYANDRIA XI
Invisible		CRYPTOGAMIA XIII

(Natural Families by Class, modified for Dauphiné)

I. None shown, being highly unusual.

II. Second Class
    1. Orchidaeae

III. Third Class

2. Gramina	3. Cyperi	4. Typha
5. Junci	6. Liliaceae	

IV. Fourth Class

7. Dipsaceae	8. Rubiaceae
9. Labiatae	10. Personatae

V. Fifth Class

11. Borraginae	12. Ombelliferae	13. Cynarocephalae
14. Chicoraceae	15. Corymbiferae	

VI. Sixth Class
    16. Cruciferae

VII. Seventh Class

17. Malvae	18. Papilionaceae or Leguminosae

IX. Ninth Class
    19. Rosaceae

X. Tenth Class
    20. Caryophilleae

XI. Eleventh Class

21. Ranunculi	22. Amentacei	23. Coniferae

XIII. Thirteenth Class

24. Filices	25. Musci	26. Algae
27. Fungi		

~

The Linnaean Natural Orders of 1764 as modified by Chaix for Dauphiné, dated 5 January 1785[521]

[After each family name, several representative genera are added, as not all the names will be recognized by modern readers.]

1. Piperitae: *Arum*
2. Calamariae: *Sparganium, Typha, Scirpus, Carex, Cyperus.*
3. Gramina: *Triticum, Secale, Avena, Bromus, Festuca, Poa*
4. Tripetaloideae: *Juncus, Triglochin, Alisma*
5. Ensatae: *Crocus, Gladiolus, Iris*
6. Orchideae: *Orchis, Satyrium, Ophrys, Cypripedium*
7. Spathaceae: *Allium, Narcissus, Leucoium, Colchicum*
8. Coronariae: *Asphodelus, Ornithogalum, Veratrum, Lilium*
9. Sarmentaceae: *Uvularia, Convallaria, Asparagus, Aristolochia*
10. Oleraceae: *Atriplex, Chenopodium, Herniaria, Polygonum*
11. Succulentae: *Tamarix, Sempervivium, Sedum, Saxifraga*
12. Gruinales: *Linum, Oxalis, Geranium, Tribulus*
13. Inundatae: *Potamogeton, Myriophyllum*
14. Caliciflorae: *Hypophae*
15. Calycanthemae: *Epilobium, Lithrum*
16. Bicornes: *Rhododendron, Pyrola, Erica, Vaccinium*
17. Hesperideae: *Philadelphus*
18. Rotaceae: *Lysimachia, Gentiana, Swertia, Hypericum, Cistus*
19. Preciae: *Primula, Androsace, Aretia, Soldanella*
20. Caryophylleae: *Dianthus, Saponaria, Silene, Lychnis, Spergula*
21. Trihilatae: *Acer*
22. Corydales: *Fumaria, Utricularia, Pinguicula*
23. Multisiliquosae: *Aquilegia, Aconitum, Delphinium, Ranunculus*
24. Rhodeae: *Chelidonium, Papaver*
25. Luridae: *Verbascum, Digitalis, Solanum, Hyoscyamus, Atropa*
26. Campanacae: *Convolvulus, Campanula, Phyteuma, Jasione, Viola*
27. Contortae: *Vinca, Asclepias*
28. Vepreculae: *Daphne, Stellera, Thesium*
29. Papilionaceae: *Medicago, Trifolium, Astragalus, Genista, Vicia*
30. Lomentaceae: *Polygala*
31. Cucurbitaceae: *Bryonia, Cucurbita, Cucumis, Momordica*
32. Senticosae: *Dryas, Geum, Sibbaldia, Potentilla, Rosa, Rubus*
33. Pomaceae: *Spiraea, Ribes, Sorbus, Crataegus, Amygdalus, Prunus*
34. Columniferae: *Malva, Althea, Tilis.*
35. Trioccae: *Euphorbia, Mercurialis, Buxus*
36. Siliquosae: *Draba, Lepidium, Alyssum, Iberis, Thlaspi, Erysimum*
37. Personatae: *Pedicularis, Orobanche, Scrophularia, Veronica*
38. Asperifoliae: *Borrago, Asperugo, Lithospermum, Myosotis*

∽

[521] In Villars, **Hist. Pl. Dauph.** 1: 310-311.

39. Verticillatae: *Thymus, Origanum, Hyssopus, Lavandula, Salvia*
40. Dumosae: *Rhamnus, Euonimus, Viburnum, Sambucus, Rhus*
41. Sepiariae: *Jasminum, Olea, Fraxinus, Syringa, Ligustrum*
42. Umbellatae: *Daucus, Ligusticum, Laserpitium, Angelica, Carum*
43. Hederaceae: *Hedera, Vitis*
44. Stellatae: *Sherardia, Galium, Asperula, Rubia, Cornus*
45. Aggregatae: *Globularia, Scabiosa, Valeriana, Lonicera*
46. Compositae: *Arctium, Centaurea, Cichorium, Hieracium, Aster*
47. Amentaceae: *Salix, Populus, Fabus, Juglans, Quercus, Betula*
48. Coniferae: *Pinus, Juniperus, Taxus, Equisetum*
49. Scabridae: *Ficus, Urtica, Morus, Ulmus, Canabis, Humulus*
50. Miscellaneae, dubiae: *Plantago*
51. Filicis
52. Musci

~

95
Les Baux, 18 March 1786[522]

You are no doubt convinced about my good intentions from everything I have said in the preamble of my supplement. I believe I have now removed everything you suggested to me that might make trouble for you. In redoing it, I have added even more plants that a reading of your table recalled to me. But I have not excluded the five or six already mentioned in your *Errata* in order to keep all the plants that concern me grouped together. If you deign to publish my enclosed supplement, would you please be willing to have some copies run off separately so that I can supply them to those to whom I gave copies of my **Plantae Vapincenses**, most notably to the Monsignor Bishop. I am more than aware that, in all of this, I have no right other than to depend upon your unceasing kindness to me on all occasions.

So, you want me to point out to you still more mistakes in your first volume! I turn to that:[523] In regard to *Taraxacum*, it will be easy for you to correct the mistake about its pappus by saying [simple] hairs instead of plumose. Similarly, it will not be difficult for you to put the division after the walnuts rather than before them. . . . You must exercise great rigor in order to spell names precisely; watch the editing carefully and do not rely on the skill of your printer. I should have preferred that the initial letter of trivial names be capitalized, as in Linnaeus, when they are used as nouns in order to distinguish them from adjectives. . . .

Are you quite sure you have *Amaranthus oleraceus* L. growing naturally in Grenoble? I do not believe any botanist locates it in Europe. Last year, I asked the abbé Pourret for seed from it; and, under that name in his shipment, I only raised *Amaranthus hybridus* [L.].

~

[522] This is the first letter that Chaix addressed to Villars with the *s* in his name. At the head of the first sheet, and in Villars' hand, is the following note: See if Barrelier, Icones, 95, 1, represents our *Melica ramosa* from Le Buis. [= *Melica minuta* L., a quite variable species.]
[523] Once again, his spelling and grammatical corrections are omitted here.

I would be very much obliged to you if you got me seed from *A. oleraceus*.[524] (Is the *Chenopodium arvense* on your lists actually known to you? I do not find it in any of my authors.)[525]

You do not have *Lonicera periclymenum* [L.] on your lists. I may be mistaken myself by confusing it with *Lonicera caprifolium* [L.]. But my plant does not have connate leaves, as the figures in Daléchamp represents them for *Lonicera caprifolium*.[526]

Should we not have *Santolina chamaecyparissus* [L.] in Lower Dauphiné? It grows near Sisteron. Also, *Gnaphalium stoechas* [L.] [= *Helichrysum stoechas* (L.) Moench (1794)], found near Aix? I can assure you that I have collected *Gnaphalium sordidum* [L.] between Le Buis and Mollans, now called *Conyza sordida* [L.] [= *Phagnalon sordidum* (L.) Reichenb. (1831)], the same plant sent to me from Narbonne by Monsieur Pourret; and *Pinus pinea* [L.] between Le Buis and Pierrelongue (at least I took it for that). But your lists do not cite them.[527]

You have *Serapias xylophyllum* L. fil. [= *Cephalanthera longifolia* (L.) Fritsch (1888)] at Grenoble. I have an example of it from Monsieur Liottard. Does not *Rhamnus zizyphus* [L.] [= *Ziziphus jujuba* Mill. (1768)], which provides the hedgerows in the Comtat, extend up to Orange? Should you not see *Rhamnus infectorius* [L.] [= *Rhamnus saxatilis* Jacq. ssp. *saxatilis*] somewhere? We have *Rhamnus saxatilis* [Jacq.] here that yields seeds like those in Avignon, and I have cited it. Do not forget our *Conyza* or *Inula bifrons* [L.].

You do not have the genus *Momordica*. I have certainly collected *Momordica elaterium* [L.] growing naturally in the village of Clamensane, but I do not know whether it had escaped from some garden. Gérard also lists it, so I presume it is to be found around southern villages.[528]

I am sending you 16 livres by Pellissier from Messieurs Gaude and Martin in payment for your first volume. Please be good enough to acknowledge receipt when you next write to me. Monsieur Gaude, who remains very fond of you, has recently had many crosses to bear. His sister, the widow Mesclé, who was serving him, became ill and returned to her son in Montmaur. Her daughter-in-law came down with the same malady, then gave premature birth to a child with the aid of the summoned Dr. D'heralde and died thereafter. A young niece, who, meanwhile, had gone to replace her aunt at Oze, also fell sick. After returning home to her mother, the widow Gaude, she, too, died a few days later. The poor curé, so imbued with religion, seeks consolation in God. Do you owe some response to Monsieur Martin? He asks me to ask you if you received a letter from him in response to one you had written to him. He is troubled by the matter, fearing that his many have been lost.

I am quite concerned about your health, which you tell me has been very disturbed by a cold. Keep me informed, if you please, about it, and give my regards to your dear family. . . .

~

[524] The genus is mostly introduced, and *oleraceus* is not among the species recognized as naturalized.
[525] Chaix was right, and the species did not reappear in Villars' second volume.
[526] Chaix was correct, and Villars added the species in his second volume.
[527] The last two species were added to Villars' later volumes.
[528] A cucumber, now usually *Ecballium elaterium* (L.) A. Richard (1824).

P.S. Gilibert, or rather Reichard, has just given incised leaves to *Chrysanthemum atratum* [Jacq.]. In that case, it would be my *Chrysanthemum coronopifolium* [Vill.]. But the serrate leaves in Linnaeus had suggested the former to me, which may be your *montanum*.[529]

At the moment, poor Joseph Chabot is a bit improved. It seems that some of his food is passing through the normal channels. He is improved by the regimen that you prescribed for him: olive oil taken by mouth and the enemas have relieved him. Four or five days ago, I though he was about dead.

96
Les Baux, 28 March 1786
(In care of Monsieur Mondet of La Roche)

Monsieur Mondet, who is going to Grenoble, provides me the opportunity to respond right away to your letter of the 24th that I received yesterday.

My poor nephew, Joseph Chabot, died on the 23rd, imbued with the greatest religious feelings, just as he had always lived. His long illness had afflicted me very much; and I shall so remain for a long time because of the sad situation left for my niece, his poor widow, and her young children. . . .

I praise your prudence for wanting to assure yourself once more about the character of the pappus in *Taraxacum*. It has always appeared to me to be simple hairs. I do not have in my notes what Haller said about it. But Allioni, **Fl. Ped.**, no. [758], agrees with me: *pappus simplex* he says in describing this genus.[530]

Figure 95, l, in Barrelier, well fits a *Melica* given to me by Monsieur Danthoine labeled *Melica imberbis foliis setaceis rigidis, panicula pyramidali,* Danth. *Culmus bipedalis, basi nudus, infra medium foliis involutis rigidis vaginatus etc.* Perhaps you yourself gave me another one that I have, with a very simple spike, which I believe is *Melica minuta* L., **Mantissa** [p. 32].

In your preface, you do not distinguish between the trivial name and the specific name. That is not the way I understand Linnaeus. Read **Philosophia botanica** beginning on p. 207. According to our great teacher, the specific name is the definition or description that characterizes a plant to distinguish it from all other species: something the trivial name cannot always do, as it often indicates only one character, albeit the most striking character.[531]

I learn with much pleasure that Monsieur [Pajot] de Marcheval[532] has appreciated your first volume, and that he has been touched by the expressions of gratitude and affection that you set down in the most moving manner. Your lists, which indicate the species of plants pell-mell [for each botanical field trip], are less to my taste than the one you

---

[529] *Leucanthemum coronopilifolium* Vill., **Prosp. Fl. Dauph.** 32 (1779) = *Chrysanthemum coronopilifolium* Vill., **Flor. delph.** 98 (1786) [= *Leucanthemum atratum* (Jacq.) DC. ssp. *coronopifolium* (Vill.) Horvati (1935)]. The leaves are incise-dentate.

[530] Villars, **Hist. Pl. Dauph.** 3: 72 (1788), would call it "a simple pappus." He was following Haller in segregating *Taraxacum* from *Leontodon.* Evidently neither Villars nor Chaix knew of the segregation made by Weber in Wiggers, **Primitiae florae holsaticae** (1780).

[531] For them, the trivial name was what we now call the specific name. Their specific name was the *generic* name plus the *trivial* name.

[532] The former intendant who had been Villars' steadfast patron.

have provided, in alphabetical order, for plants within two leagues of Grenoble. If the work should be reprinted, I would recommend providing only one alphabetical list for all the plants, indicating the native location of each one. This would eliminate the many repetitions, and the reader would gain much from it.

Assurance of the friendship you have for me is what gives me the liberty I am taking to give my opinion as openly as I am doing. I know that you take everything I say to you without offense. True friends hide nothing from each other because of the mutual good that can devolve upon them. I do hope to come to embrace you during the month of June. For the moment, I do it from my heart.

P.S. If you are not mistaken in giving *Taraxacum* a plumose pappus, then Linnaeus, Allioni, and I are all mistaken, believing to have seen it with simple hairs. *Dens leonis, pappo simplici seu capillari, gaudet.* Linnaeus, **Genera**, p. 402.

97
Les Baux, 16 April 1786

I received your letter of April 4th only yesterday afternoon. As for icon 7 in Barrelier, the generic word *Gramen* was omitted in the descriptive phrase accompanying the figure; but it is cited in the **Observatae**, p. 110, no. 1209: *Gramen miliaceum aquaticum. Brizae locustis semine rufo. Gramen paniculatum aquaticum fluitans* Tourn., **Inst.**, [521]. *Gramen fluviatile* Tabern., [Icon 216]. All these synonyms belong to *Festuca fluitans* L.; and Gouan, **Hort.**, quite appropriately, refers Barrelier's synonymy to it, even though the spikelets on said icon 7 in Barrelier approach those of *Poa eragrostis* icon 442. They are larger and do not form an open panicle. But what proves that this figure is certainly that of *Festuca fluitans* L. [= *Glyceria fluitans* (L.) R. Br. (1810)] is that its lateral culms are bent horizontally and throw roots from their nodes, something that I do not see in the specimens of *Poa eragrostis* [L.] [= *Eragrostis minor* Host (1809)] that have been sent to me, and which are not native here. That is all I can say in response to your question.

Monsieur Thouin has not yet sent me anything this year, although I sent him, as customary, a small collection of seeds, along with a list of what I want, which he told me he would have done. If I receive nothing more, I will get over it. Monsieur Pourret has taxed me for a long time with his great compliments and his fine promises about botanical materials and about Weis and Weber that he has solemnly promised me on two or three occasions. I now know the man, and I have made him aware that I still have some feeling. He will write to me when he finds it to be appropriate. We are now quite distant from each other. The discontinuance of these important correspondences will leave me greater leisure to occupy myself with eternity about which it is high time I think more seriously than I have done in the past.

Take care of your health and extend my fond sentiments to your wife and you lovable daughters.

P.S. I have some notion of having seen pappus in *Taraxacum* that is more than simple hairs, but I would not want to say *plumose* as you do. Can you not do better by taking a term midway between these two extremes? I hope to read your middle expression if God prolongs my life. *Interum vale.*

98

Les Baux, 16 May 1786

Monsieur Martin and I have made the necessary arrangements for our trip to Grenoble. We count on arriving there the 30th or the 31st of this month.

You tell me in a short note accompanying Monsieur Pourret's letter that you wrote to him about confiding the Weis and the Weber he means for me to have to the intendant in Paris. Please make it your business to see that they reach you, whether through the intendance or by the ordinary messenger service. You will spare me writing to Monsieur Pourret until I shall have received said books.

You had remarked to me some time ago that Monsieur de Lamarck had presented, as new, many plants known and named by other authors, and that you had become resolved to make no mention of this botanist. In order to call attention to him properly and to be useful to botany, I would favor, on the contrary, adding his trivial name to the string of synonyms, as Monsieur Allioni has done, and to whom should you not be obliged to give an accounting of your intention?

The small teeth that the microscope has enabled you to see on the pappus rays in *Taraxacum* do not authorize you to call it plumose. This would be to misuse the term knowingly. This pappus is truly of simple hairs, seen with the naked eye or with the ordinary magnifying glass. I also do not approve your failure to distinguish the specific from the trivial name, . . . contrary to Linnaeus's usage.

If Allioni's *Danaa*, with leaves like columbine and seeds like coriander, is my *Ligusticum austriacum*,[533] I would burn my botanical books and renounce this study forever. It may well be that my *Ligusticum austriacum* is not *L. austriacum* Jacq., for I have never seen the one in Allioni, no. 1323, tab. 43; but it is even less like *Danaa aquilegifolia* All.

I should also like to merge *Ligusticum ferulaceum* All., which is the one in Séguier, **Veron. 2**: 41, tab. 13, with *Ligusticum pyrenaeum* Gou., which you have called *Ligusticum seguierii* [Vill.], and with Lamarck's *ferulaceum*; not having seen Allioni's plant in nature, but having compared figures of one and the other.[534]

If you do not have *Anemone halleri* All., no. 1922, in your notes, you will surely not omit it after Allioni's testimony, locating it in the Queyras [= *Pulsatilla halleri* (All.) Willd. (1809)].

I have made an observation about the flowers of the ash [*Fraxinus*] that is perhaps new, being different from that in Linnaeus, **Genera**. Around my village I find some individual trees with flowers absolutely male without any pistil; but others with bisexual flowers where I do not see any of solitary sex. (On these two points I am in agreement with Linnaeus.) But still others, finally, with female flowers, amongst which some bisexual flowers are found (and I do not see this in Linnaeus). Take a look at the examples enclosed and make use of this observation if you will.[535]

◠

[533] See letter 89.

[534] See letter 90. Allioni had properly separated his *Ligusticum ferulaceum* (1774) from Gouan's *Ligusticum pyrenaeum*. But the latter species, plus *Ligusticum seguierii* Vill., are considered synonyms of *Ligusticum lucidum* Mill. (1768).

[535] Villars did not choose to utilize Chaix's observation, but a modern description of the Oleaceae encompasses what Chaix reported.

My *Iberis aurosica* [Chaix] is not a variety of *Iberis umbellata* L. I shall bring you an example of it that is about to flower in my garden. It is a biennial.

Monsieur Deleuze has often remarked to me that *Salvia verticillata* [L.] is dioecious; and that, having only one individual in his garden with female flowers without any sign of stamens, he has never seen it brought to fertility.[536] The ten stamens in *Geranium pyrenaicum* [Burm. fil.] are certainly fertile. Therefore Gérard, whom Murray followed, confused two plants.

Lamium hybridum Deleuze, *forte natum e L. purpureum* [L.] *et L. amplexicaule* [L.]. *Valernae abundans; differt foliis floralibus inaequaliter et argute incisis, acutis, non cordata, crenatis, obtusis; item colore obscure virescenti. . . .*[537]

Monsieur Martin, who is here today, offers you his compliments.

99
Les Baux, 20 June 1786

The woman in my parish, about whom I consulted you in Prémol,[538] had died at the end of the week I left Les Baux [for Grenoble]. Otherwise, I am very pleased about my trip, during which you gave me proof of the close friendship with which you have always honored me. I hasten to send on to you some botanical clarifications obtained during an interview with Monsieur Deleuze, who spent between three and four days here, and who asked me to remember him to you.

The *Ephedra*, of which I brought you a male specimen, and which I believe is *E. distachya* L., Barrelier, **Icones** 731, grows near Valernes on the Montagne de Gache. You will likely have collected it during your botanizing in our province.[539]

The flax from Grenoble, which you suspect is [*Linum*] *perenne* [L.], is certainly not the *perenne* I have previously cultivated from seeds from Paris. The latter rises more than three feet, the root perennial, calyx obtuse, petals very large. Yours appeared to me to be an annual and does not have those other characters. I thought it was either a variety of *L. usitatissimum* [L.] or *L. humile* [Mill.] in Reichard.[540]

Your *Sisymbrium asarifolium* from the Lac d'Annecy, which is now in flower in my garden, is nothing else than a small variety of *Sisymbrium nasturtium* [L.] [= *Nasturtium officinale* R. Br. (1812)].[541]

The honeysuckle from Prémol is indeed *Lonicera periclymenum* L. Daléchamp's figure shows an imbricated flowerhead very well, different from his other flowerheads. So, ours is *Lonicera caprifolium* [L.],[542] but the one that Danthoine had given me with that name is a very noticeable variety. Monsieur Deleuze had another variety of it with a short

⌒

[536] For sexual dimorphism in the Labiatae, see Tutin et al., Flora Europae 3: 126.

[537] Villars had already published *Lamium hybridum* Vill., Hist. Pl. Dauph. 1: 251 (1786); but he would later disavow it, ibid. 2: 385 (1787), making it a synonym of *Lamium purpureum* L. This letter intervened.

[538] The Chartreuse de Prémol (a convent of Chartreusines) south of Grenoble.

[539] The specimen was seen by Grenier and Godron, who named it *Ephedra villarsii* Gren. & Godr. But Verlot, who saw it shortly thereafter, could not distinguish it from *Ephedra nebrodensis* Tineo ex Guss. [= *Ephedra major* Host (1832)].

[540] While Villars did not change his opinion, Chaix may have been either right or close. Verlot thought it more likely to be *Linum angustifolium* Huds. [= *Linum bienne* Mill. (1768)].

[541] Vill., Hist. Pl. Dauph. 3: 340 (1788), accepted the verdict, but gave the location as Lac du Bourget in Savoy.

and regular corolla from below Sisteron at Valonne. It remains to be seen if it is [*L.*] *sempervirens* [L.]. He has promised to examine it closely.

I had taken the tall sorrel from the woods in the mountains to be *Rumex arifolius* [All.], but now I have been corrected regarding it. That species, which is Allioni 2: 204, does not have panicles, but rather a branched spike. And Monsieur Deleuze has given me *Acetosa fagopyrifolia* Boccone [**Musea di piante di Sicilia**] 1: 126. Instead of *Acetosa arifolia* ibid. 1: 125, it is [= *Rumex acetosa* L.].

The potentilla that I took for *Potentilla argentea* [L.] is a grayish variety of *Potentilla* [rodent damage].[543] I am sending you a small specimen of *P. argentea* [L.], of *P. hirta* [L.] from Paris, and *P. hirta* from Provence, not to teach you anything, but to present to you what I can about [the matter].

Can you not make a species out of this small astragalus from the subalpine pastures? I am unable to refer it either to *Astragalus arenarius* [L.] or to *Astragalus glaux* [L.].[544]

If I had seen *Arenaria tenuifolia* [L.] earlier, which Monsieur Deleuze has just given to me, I would not have called Vaillant's *Arenaria saxatilis* by that name, from which I have always been put off by the characters Linnaeus and Haller gave to it. I do not doubt that you know it, but I had never run across it. It is indeed different from *Arenaria saxatilis* Vaillant and from *Alsine mucronata* L.[545]

From seeds sent from Paris, I have Vaillant's variety *altissima* of *Crepis biennis* [L.], as well as that tall *Picris hieracioides* [L.] from Valgaudemar, which may be a different species.[546] *Valeriana dentata* [L.] [= *Valerianella dentata* (L.) Pollich (1776)] is also found there with *Valeriana olitoria* [L.] [= *Valerianella locusta* (L.) Laterrade (1821)]; but one would be wrong to make one only a variety of the other.

Lotus hirsutus [L.] [= *Dorycnium hirsutum* (L.) Ser. in DC.] certainly grows on the slopes at Ventavon, and *Rosa eglanteria* [L.], with yellow flowers, on a slope between Le Poët and Sisteron. *Centaurea calcitrapoides* [auct.] [= *Centaurea torreana* Ten. (1830)], from Ventavon and raised from seeds, has not degenerated in my garden.

Instead of printing my *supplenda* and *amendanda*, I should much prefer that you take the trouble to make my corrections in the notes for your species, as I perceive you have already done in the case of Gérard's figure for *Festuca spadicea* [L.] [= *Festuca paniculata* (L.) Schinz & Thell. (1913)]. If you think otherwise, I will gladly concur.

Neither Monsieur Deleuze nor I can reconcile to *Lichen saxatilis* [L.] the lichen you know by that name. When you can turn to the matter, perhaps your reflections will lead to a different opinion.[547]

~

[542] Verlot believed they had failed to recognize a novelty that came to be *Lonicera etrusca* Santi (1795).

[543] Probably *Potentilla cinerea* Chaix in Vill., **Prosp. Pl. Dauph.** 46 (1779). Vill., **Hist. Pl. Dauph.** 3: 566 (1788), no longer recognized it, publishing in its place *Potentilla opaca* Vill., non L., that remains a synonym.

[544] It is evident from Villars' observations, **Hist. Pl. Dauph.** 3: 460 (1788), that Chaix had wanted to call this small plant *Astragalus exilis*, but that Villars could not distinguish it from *Astragalus onobrichis* L.

[545] Vill., **Hist. Pl. Dauph.** 3: 633 (1788), has a long discussion of *Arenaria tenuifolia* L., describing three varieties, the third of which remains as *Minuartia hybrida* (Vill.) Schischkin (1936).

[546] There are numerous subspecies of *Picris hieracioides* L.

[547] The following words were added by Chaix between the lines of this paragraph, written after receiving books sent by Pourret: Now I shall be able to examine what Weis and Weber say about the matter.

P.S. Having completed the above letter, I was brought your package enclosing the Weis and the Weber, which the abbé Pourret sent to me, in addition to your amiable letter in which the feeling of your heart toward me can be read, expressed with increasing vigor. I am very touched by it. One could not be more so.

Our return from visiting you was favorable except for an inconvenience to both of my feet caused by a blister on each of them, which made me suffer considerably from Laffrey to La Mure. There, having met Monsieur Martin's brother, I took advantage of the horse he brought to us to go on to spend the night at Les Terrasses. That evening I lanced the two blisters. The next day I was less inconvenienced by them; and a whole day of rest at the home of my companion's father healed me completely.

Smallpox is spreading in our neighborhood. If God wills that my niece not catch it naturally, I am resolved to put off the innoculation until after August 14th on account of the indispensible tasks during this season and fear that the children of the village will catch it from her, their parents becoming more distressed by that possibility in this most critical time of the year. You have no doubt learned of the nearly sudden death of poor Garnier's wife in Gap, he being your relative, which occurred around Pentecost. According to what I have been told, the grief of this family has been pitiable. God deigns to bring His consolation to them. Everything warns us to keep ourselves ready for this important passage.

Last year, I gathered *Iberis amara* [L.] in abundance, grown from seeds sent from Paris. It has reappeared this year in my garden, having seeded itself. Therefore, I am putting your specimen aside for Monsieur Martin. You will give me pleasure if you will dry again two specimens of *Potamogeton crispus* [L.] for me from the La Troche reservoir. I thought I had brought back three of them; but I found only two, one of which I had to give to Monsieur Deleuze. Try to get me seeds from your fine *Myagrum perenne* L. [= *Rapistrum perenne* (L.) All. (1785)],[548] from that small marjoram from Monsieur de Rochechina's garden, and from that fine variety of *Antirrhinum majus* [L.].[549] I shall obtain some fritillary bulbs for you.

Take care of your health that is so dear to so many people, above all to me who will be devoted to you until death with the greatest affection.[550]

100

Les Baux, 30 June 1786

I am sending you enclosed some bulbs of our *Fritillaria meleagris* [L.]. I have not separated any of them from their stalks, because I have learned from experience that the Liliaceae, if their stalks are cut before they dry out, will not flower the following year. If you need any *Bulbocodium verum* [L.] or any *Ornithogalum luteum* [L.] [= *Gagea lutea* (L.) Ker-Gawler (1809)], I can provide them for you. I only have one bulb of *Erythronium dens-canis* [L.] in my garden, given to me earlier by Monsieur Liottard, which has flowered only once

---

[548] Vill., **Hist. Pl. Dauph.** 3: 277 (1788), having cultivated this perennial, knew it was not native to Dauphiné and distinct from *Myagrum rugosum* L.

[549] *Antirrhinum majus* L. var. b.: *Antirrhinum majus alterum folio longiore*, a cultivar native to Russia and Iran, Vill., **Hist. Pl. Dauph.** 2: 343 (1786).

[550] Joyeux and Dejarnac, "Le Médecin Dominique Villars," p. 130, indicate that Villars had suffered from acute, and then chronic, malaria in 1786 and 1787.

over the past several years, because worms have eaten it considerably. If you could at least get me a second one, I should have less fear of being deprived of it entirely.

I am depending on your care for the seeds I asked you for in my last letter. You would also give me great pleasure if you would add to *Potamogeton crispus* [L.] two or three dried specimens of *Mentha [spicata* L.] var. *rotundifolia* [L.], very abundant at Grenoble, which I have not been able to find anywhere in flower since my trip. Consequently, what I have reported about it is quite incomplete. See what I can do from my side to help you.

I no longer doubt that your lacunary lichen, which we have only on forest trees here, is *Lichen saxatilis arborum* Weis. For the one you want to call *Lichen pallescens* [L.] that is very common, even in the orchards, that is surely *Lichen stellaris* [L.] according to Weis, just as Monsieur de La Tourrette believes. *Lichen pallescens*, instead of being in the section of the imbricated, belongs with those of scaly scutella; and it may only be, as Weis says, a juvenile form of *Lichen subfuscus* [L.]. I have previously indicated to you my opinion about *Lichen candelarius* [L.] and *L. parietinus* [L.], and I hold to it: The former decorates nearly all the stones in my garden and in our cemetery; the latter is ordinarily on the branches of trees, but more common in Champsaur. The lichen you have designated as *L. amorphus*: should it not be *L. candidus* Weber, p. 193?[551] Weis has omitted the section of the *Lichen umbilicati* [in Linnaeus]. That is a shame. It is not unusual to notice serious mistakes in Latin grammar by this author. . . . Weber was quite right to say that most of the sections within the *Lichen* really constitute genera. Who could not be convinced that the *Lichen leprosi* constitute a single genus, or the *Lichen filamentosi*? I doubt whether I shall be able to muddle through my cryptogam authors any farther into the *Bryum, Mnium, Hypnum*, or *Jungermannia*, the most part of which do not have any fruit in our area. But this study is only some superrogation. The principal work is that of salvation. *Fear God and keep His commandments; for this is the whole duty of man.* Ecclesiastes 12:13.

Our venerable Madame de Colvin, superior of the Couvent de la Charité in Gap, has just died.[552] I shall miss her very much. It will be difficult to replace her. The prior of Furmeyer [Jean Faure] has resigned from his parish in favor of our good friend Monsieur Gaude, and the date of it must already have been fixed in Rome. I learned the news upon my return from Grenoble, and Monsieur Gaude got word of it to me yesterday through a letter he had written to me. I do not yet know the circumstances of this resignation. It can hardly increase the worthy confrere's income.[553] But it becomes advantageous for me given the pleasure I will have, so long as God permits, in having a friend with whom I am closely associated in our most critical duties closer at hand.[554]

I am again enclosing here a specimen of *Centaurea* calcitrapoides,[555] grown in my

~

[551] *Lichen amorphus* does not appear in the flora, but *L. candidus* Web. does. Vill., **Hist. Pl. Dauph.** 3: 967 (1789).
[552] Marguerite de Colvin, known as the Soeur de Colvin de l'Incarnation, had been the superior for 36 years. A native of Corps, she died 27 June 1786, age 69. She had encouraged and developed the botanical interests of both Chaix and Villars. See Manteyer, **Les Origines de Dominique Villars**, pp. 168-170.
[553] Gaude was not formally named to Furmeyer until October of 1786. As he did not resign his former parish in Oze until May of 1787, he would have drawn a double income during that interim. See Timothy Tackett, **Priest and Parish**, p. 111n.
[554] The move cut the distance between the two priests roughly by half.
[555] See letter 99.

garden. Note its difference from *Centaurea calcitrapa* [L.]. Plus a specimen of my *Hieracium auricula* [L.]. If this is not the *H. auricula* of Linnaeus and Haller, it is a new plant that has not degenerated into *Hieracium cymosum* [L.].[556] It has been in my garden for a long time. I shall send you seeds from these two plants if you want them.

101

Les Baux, 28 July 1786

The figure in Barrelier, 469, does not have the characters of your *Pedicularis gyroflexa* [Vill.] that you detail for me in your last letter. Its stalks are erect, glabrous, leafy; the spike long, without corolla, but well supplied with capsules; bracts linear; the roots are not represented as tuberous. But Jussieu calls it *Pedicularis alpina asphodeli radice, flore alba. . . . flores modo purpurei, modo albi* in Barrelier, p. 22, no. 210, with the synonym from Ray, **Hist.**, p. 771, no. 10, where Ray, describing the roots of his plant, *principio angusto, senoim extuberantes, multum filipendulae radicibus cedentes*, means our *Pedicularis comosa* [L.], the one we want to make the true *P. tuberosa*. Either Barrelier or Jussieu may well have confused the two plants when alleging the flowers to be purplish or white; for I do not believe that our *Pediculari comosa* ever has purplish flowers.[557]

If *Rhaponticum angustifolium* Lobel, with rumex-like leaves, has been found at Charousse, I am quite comfortable with that. I already consulted you about that plant.[558] Our plant in the Gapençais is *Rhaponticum folio helenii incano*, C. Bauhin, **Pin.** 117, with inula-like leaves, crenate rather than denticulate on the edges, whose petioles are sometimes furnished with some appendages of alternate leaflets, which one could call lyrate, as is sometimes seen on those of *Berardia*. Therefore, I believe that ours is different from Haller, **Hist.**, no. 160, which you say you have from Charousse.[559] [= *Leuzea rhapontica* (L.) J. Holub. ssp. *helenifolia* (Gren. & God.) J. Holub. (1973)]. Ours does not have sterile florets on the circumferance as Haller says his does. The large leaves on ours are one cubit, about eight inches in diameter, oval in form. The stem bears only one large round head of an inch and a half.

Tussilago frigida [Vill., non L.][560] is so abundant in the quarter of La Grangette called Gravasson, at the foot of the crag of Bure, that the sheep fatten on its leaves in the months of August and September when the leaves reach maturity. Earlier, the leaves are insipid to them. What criticism are you preparing for Haller, who makes a *Petasites* of his [Hist.] no. 141 and not a *Cacalia*, never having seen the plant radiate as we have. Linnaeus in

~

[556] *Hieracium* x *hybridum* Chaix ex Vill., **Hist. Pl. Dauph.** 3: 100 (1788) [*H. cynosum* L. x *H. peleteranum* Mérat]. Villars held to *H. auricula* L. but indicated here Chaix's inclination to call the plant *hybridum*. A specimen was found in the Chaix herbarium. See letter 101.

[557] Villars incorporated these observations into his treatment of *Pedicularis gyroflexa* Vill., **Hist. Pl. Dauph.** 2: 426 (1787), and his treatment of *Pedicularis tuberosa* Vill., non L., ibid. 2:431 (1787). He indicated that the latter species, based on a specimen give to him by Dr. Charmeil in 1775 [formerly a surgeon in Château-Queyras, and initially published in **Flor. delph.** 64 (1785)] had great affinity with *Pedicularis comosa* L. and might only be a variety of it.

[558] Based upon *Rhaponticum angusto folio incano*, C. Bauhin, **Pin.**, 117.

[559] Haller, **Hist.**, no. 160 = *Rhaponticum alterum angustiori folio* Lobel, **Ic.**, p. 288. Villars, following Linnaeus, put these plants in *Centaurea*, suspecting that the above plant was really a synonym of the other.

[560] Both *Tussilago frigida* Vill. and *Tussilago nivea* Vill. have been made synonyms of *Petasites paradoxus* (Retz.) Baumg. (1816) by Burnat, **Flore des Alpes-Maritimes** 5: 268.

Mantissa, 469, gives assurance that it is radiate and floscular, and we must rely on his opinion.[561]

.... [What we have called] *Picris echioides* [L.] from Valgaudemar is not really a different species from the common picris [*Picris hieracioides* L.], but only a noticeable variety: a more erect stalk about 3 feet high; its branches apart in panicles; its leaves flat, lightly toothed, and not deeply toothed or crispate. I do not see any other differences. One should not use *Picris pyrenaica* as it was abandoned in the later editions of Linnaeus, being nothing else but *Hieracium pyrenaicum* [L.].[562]

For several years when I have had *Hieracium auricula* [L.] in my garden, even cultivating it in pots, it has not degenerated. But as it is not stoloniferous, I have come to believe that it is not either Linnaeus's or Haller's *Hieracium auricula*, but rather a hybrid plant, perhaps the offspring of *Hieracium pilosella* [L.] and *Hieracium cymosum* [L.]. Therefore, I should call it *Hieracium hybridum*:[563] *foliis ovato-lanceolatis pilosis, caule nudo piloso, ramis uni-bifloris, planta tota pilosa; pili ramei et calycini tuberculo nigro innascitur. Nulli prodient stolones. Caulis nudus, firmus, unum aut alterum superne edit ramum, bractea suffultum, uni-biflorium.* Found at Les Baux near the Bois du Devès, from which come the *Hieracia pilosella* and *cymosum*. You will want to mention it in your treatment of species.

The generation of the mushrooms, the algaes, etc., is a mystery that botanists have not yet explained, no matter whether they specify seeds or stem suckers as the primary cause. Weis maintains, following the experiments of his teacher, [C. W.] Büttner, that mushrooms are nothing but abodes for insects. I do not believe that he is right, and you no doubt do not share his opinion. *O altituds!* The learned Monsieur Allioni cites the species of the cryptogams for us with their synonymy without giving a single word to instruct us. Such works, which only repeat descriptions, only hamper and annoy. They are useless.

Well, if they continue to publish the same repetitions, the land will soon be inundated by them without contributing the slightest degree to science. Botanists will increase their libraries at great expense, but they will not be more instructed. You will not fall into this deficiency since you are applying yourself to the description of all the plants that you cite. I have just reread your preface with renewed pleasure, and I still find it well worked. I hope that the public will avenge you for your censor's severity.

You must have known for a long time what use the Brothers of Charity have for you. I have known for a long time that you have been neither liked nor respected by those monks.[564] Few people esteem true merit. What is to be done? You must keep brazen-faced against such malignity, always continuing to fulfill your duties, not breaking with them

~

[561] Vill., Hist. Pl. Dauph. 3: 175 (1788), put Haller's no. 141 as a synonym of *Tussilago frigida*, see note above.
[562] Vill., Hist. Pl. Dauph. 3: 148 (1788), gives the picris from Valgaudemar as *Picris pyrenaica* Gou., but noted that it might be a variety of *P. hieracioides* and remarked about the blackish involucre. [= *Picris hieracioides* L. ssp. *villarsii* (Jord.) Nyman (1879)].
[563] See letter 100.
[564] We noted much earlier Villars' brief comment, in his short biographical piece written in 1805, that he had found considerable rivalry between the monks who administered the Hôpital de la Charité in Grenoble and the physicians and surgeons of the city. On 4 August 1782, Villars, already an army physician, had been also elected to be physician to the poor in the Hôpital de la Charité by the mònks in their community convent. This letter indicates that the association had not been gratifying. See Manteyer, **Les Origines de Dominique Villars**, p. 155.

openly, and not giving vent to your feelings. But never soften: make yourself feared but without vanity, and make yourself obeyed even if without authority. I know that your prescriptions for your patients are not always carried out, but resist abandoning this post. Your desertion would give jealousy a victory. A little more firmness will defeat the arrogance of your adversaries. The Lord protects his own.

My niece was innoculated for smallpox by Monsieur D'heralde at the beginning of this month. She was quite sick the three or four days that preceded the rash; but that eruption was not extensive, and she has begun to recover her health quite well.

I thought to render her a great service by having her spend the time with me, because she is on the verge of quitting me through a marriage contract that is being negotiated. If the marriage takes place, the eighth and last daughter of my late brother will become my fourth governess. These changes are hardly comfortable for me, but reason requires acceptance. My entire confidence is in God.

In response to the news that the prior of Furmeyer had resigned [his parish] in favor of Monsieur Gaude, a priest related to said prior, who aspired to that parish, transmitted his sorry complaint to him on the subject. In the meantime, the prior became aware that 12,000 livres in gold, which he kept in a money-box, had been stolen from him. The crime is dreadful, and the public has not been slow to reach its suspicions. Its author, who merits the gallows, will probably only meet the judgment of God. This recalls in particular the teaching of our divine master to the ecclesiastics: *Lay not up for yourselves treasures upon the earth, where moth and rust consume, and where thieves break through and steal: but lay up for yourselves treasures in heaven where neither moth nor rust doth consume, and where thieves do not break through nor steal.* **Matthew** 6: 19-20. The honorable prior still has enough resources to subsist decently.

You promise me specimens of *Potamogeton crispus* [L.] and *Mentha rotundifolia* [L.]. Can you not get me the seeds of *Antirrhinum majus* [L.], *flore rubro; Myagrum perenne* [L.]; *Anchusa tinctoria* [L.] [= *Alkanna tinctoria* (L.) Tausch (1824)]; *Origanum majorana* [L.], which are in your public gardens, which I also asked you for,[565] but about which you do not say a word? If I am not to be sent anything, each one will henceforth keep what he has, I mean this in regard to your famous director of gardens [Liottard].

102

Les Baux, 2 August 1786

Although I wrote to you last week by way of the intendance, allow me to cover the postal expense for this one in order not to burden said office so frequently. I have some later observations that could be of some use to you.

You tell me that you have found *Cineraria pratensis* L. in the Lans marsh [southwest of Grenoble]. No doubt you meant to say *C. palustris* L. [= *Senecio congestus* (R. Br.) DC. (1838)]. I have never seen it.[566] Plus, in the woods of said Lans, a superb rose *with red leaves* (you use the same expression in your letter to Monsieur Serre), when you perhaps have

⁓

---

[565] See letter 99.

[566] Apparently Villars decided that he had not seen it either, publishing instead *Senecio paludosus* L. from that site. **Hist. Pl. Dauph.** 3: 232 (1788).

meant to say *with red petals*. Well, this could really be my *Rosa montana* [Chaix][567] that I have recently observed again at Chaudun. Monsieur Deleuze assured me he has seen a rose with yellow flowers on a slope between Le Poët and Sisteron. This would be *Rosa foetida* [J. Hermann], found in All., Fl. Ped., no. 1792, a variety of *Rosa eglantaria* L.

Saxifraga hypnoides [L.], which I have found among the cassias on Pic de Bure, is quite viscous; radical leaves 3-4-5-fidus, none being entire; cauline leaves 3-fidus, the upper one entire; corolla white; is truly different from *Saxifraga caespitosa* [L.].[568]

Galium saxatile L., in Vill., Hist. Pl. Dauph. 1: 302, 307, *foliis obovatis, margine subscabris, apice spinula terminatis*, according to me. Why does Haller, Hist., no. 718, say *follis ellipticis* without *spinulis*? Should not this latter character, along with *pedunculis brevissimis vix de floribus eminentibus*, better fit *Galium hercynicum* Vill., Hist. Pl. Dauph. 1: 271? and, to the contrary, Haller's *Galium*, no. 717, *foliis subrotundis aristatis, petiolis ramosis*, does it not fit our said *Galium saxatile* [L.]? Tell me your opinion. [Rodent damage.] According to my understanding, *Galium hercynicum* [Vill.] is not different from [*Asperula*] *saxosa* Chaix, Pl. Vap. [57 (1785)] [= *Galium saxosum* (Chaix) Breistr. (1948)].[569] I collected *Galium uliginosum* L. about a month ago at Bayard, where I had the opportunity to compare it with *Galium palustre* L. to see the difference between them.

Bupleurum junceum L. [= *Bupleurum praealtum* L. (1756)] is Gérard, Fl. Gallo-prov., p. 234, no. 9, *tripedale, involucra 3-phylla quidem*, but not like Linnaeus *longitudine umbellae, sed duplo, etiam triplo breviora; radii plerumque 3, raro 4, inaequales, flores laterales solitarii*. This came up in my garden this year.

Bupleurum gerardii All. [1774], Fl. ped. no. 1356,[570] *pede brevius, involucrum 4-5-phyllum; radii plerumque 6, inaequales; involucella 5-phylla; floribus paulo longiora*. My specimen sent by Monsieur Danthoine.

Bupleurum petraeum [L.], *caulis plerumque omnino nudus; involucra 3-4-5-phylla* as in Gouan, Ill. 8, *ab involucella non semifida dumtaxat ut Hallers, sed usque ad pediculum divisa ut Seguieris* 3: 221, *laciniae saepe*.

Your dictionary of botanical terms, which I reread after your preface, appears to me superiorly well done and rendered very well in our language. The public that compares it to what has been said by authors who have written in Latin will judge you to have a true knowledge, no matter the indifference of your critic. I shall count myself as indeed fortunate and advantageously recompensed if, by my small notes, I have been able to contribute in some manner to the perfection of a work that will immortalize your memory among the learned, above all in this province.

Tomorrow I am going to see our respectable prior of Furmeyer, where I hope to find our dear friend, Monsieur Gaude. You will certainly not be forgotten there; and I shall learn about the circumstances of the theft that he recently suffered.

P. S. I hope to spend the festival of my St. Dominique with these gentlemen, which is also yours.

∾

[567] *Rosa montana* Chaix, Pl. Vap. 42 (1785); Chaix in Vill., Hist. Pl. Dauph. 1: 346 (1786); subalpine hill-and-mountain slopes. From Rabou and La Grangette. A specimen found in the Chaix herbarium.
[568] See letter 103.
[569] Villars did not publish either plant among his species, retaining only *Galium saxatile* L.
[570] Gérard, Fl. Gallo-prov., p. 233, no. 7.

103
Les Baux, 29 August 1786

Last Saturday, the 26th of this month, I received your letter of the 17th with all the more pleasure as I was eagerly anticipating it. Although I am delighted in advance of the imminent shipment of dried plants and of seeds that you promise me, I do not want to delay any further bringing to your attention some reflections that you can profitably use in time and place if they have some value.

The character you give to lavender, [Genus] 123, is incorrect. The corolla is not reflexed, nor is its upper lip three-parted or the lower lip two-parted. The contrary is true: the upper lip, erect and larger than the lower, is divided shallowly into two lobes; the lower lip is divided into three segments, short and separate. All of this is well marked in Linnaeus's **Genera**, no. 111. See there the correction you must make. . . .

The synonym from Barrelier, *Jacobaea montana polyanthos, flore aureo, foliis longis et integris, italica,* which you say belongs to your new *Cineraria pratensis,* is not no. 260, which is a *Calceolus* [= *Cypripedium*] from Canada, but rather icon no. 266, obs. no. 1089. . . .[571]

I see with pleasure that I was not far off when calling your *Saxifraga exarata* [Vill.] in Allioni [**Fl. Ped.** 2: 73], Tab. 88, fig. 2, by the name of *Saxifraga hypnoides* [L.].[572] But I am astonished that we do not have the real *S. hypnoides* on our high alps, reported by Linnaeus, Haller, Allioni, as well as Gérard; [but that you report it] rather from Vienne and Toulon. I shall be pleased to get your specimen of it.[573]

I am sending you three species of *Lathyrus* that you may determine for yourself following the notes that I made about them. Please tell me subsequently your opinion about them. Monsieur Deleuze has still not found the true *Lathyrus angulatus* [L.], *seminibus angulosis* in Tournefort and Gérard. If you still have the copy of my supplement, you will be able to consult it about these three species and to make use of my notes regarding some other plants. In my various letters, I have mentioned many plants to you that I feared you might have omitted, . . . hoping you might not be obliged to make a supplement, or at least to make it shorter.

A pretty shoot of *Chenopodium urbicum* [L.] has come up for me from seed, which I am caring for in a pot. I shall give you information about it. I am cultivating *Chenopodium purpurascens* Juss. for the second time. I should like very much to know from where they got it in Paris. Tell me whether you want specimens or seeds of it. I could also send you *Anchusa sempervirens* [L.] [= *Pentaglottis sempervirens* (L.) Tausch ex L. H. Bailey (1949)] for planting, and a hesperis that may be *Hesperis sibirica* L. [= *Hesperis matronalis* L. ssp. *m.*] sown this year from seed collected in Digne.

Monsieur Martin wrote me that he had found *Silene inaperta* [L.] at Clausonne [southeast of Le Saix]; but not having seen his specimen, I cannot vouch for its reality. His

∽

[571] Vill., **Hist. Pl. Dauph.** 3: 225 (1788), would actually publish this as *Cineraria pratensis* Jacq., making it Variety B of *Cineraria integrifolia* (L.) Murray [= *Senecio integrifolius* (L.) Clairv. (1811)]. And it could well be ssp. *integrifolius.* If so, Variety A may be ssp. *capitatus* (Wahlenb.) Cuf.

[572] Allioni gave his no. 1538 to *Saxifraga hypnoides* L. and his no. 1539 to *Saxifraga exarata* Vill., indicating their affinity.

[573] Chaix had listed both *Saxifraga exarata* and *S. hypnoides* in his catalogue, indicating he had collected the latter at Sigottier above Serres.

claim would acquire some probability if you have found this plant growing naturally in Dauphiné.[574]

I do not believe I ever ran across your *Cacalia hybrida* [Vill.] [= *Adenostyles alliariae* (Gouan) A. Kerner ssp. *hybrida* (Vill.) Tutin (1973)] from Sept-Laux, or at least paid no attention to it. It may be that the true *Galium saxatile* [L.] is in our mountains; and, in that case, I have not been able to distinguish it from *Galium pusillum* [L.]. In the future, I shall be on the watch for that *Galium* and for the above said *Cacalia*. I have not seen the seeds of *Galium uliginosum* [L.] and cannot characterize them. I have no species within *Splachnum*. Therefore, either your *Splachnum ampullaceum* [L.] from Prémol or your new one from Sept-Laux would be very welcome.

If Monsieur de Jussieu does not want to authorize publication of your response to his criticism, the public will understand without fail that there is some [bad] temper in his proceedings.[575] Monsieur de La Tourrette, who gave you the idea, thereby gave you proof of his esteem and affection. I bear the greatest gratitude to him in my heart. Beyond that, you already know, as a philosophe and a true Christian, how to extract advantage from [Jussieu's] severe relations. The blows that knock down those of weak spirit become glorious for those well-born souls by purifying their merits: *For my power is made perfect in weakness.* II **Corinthians** 12: 9.

After the death of the curé de La Bâtie-Vieille, the grand vicars nominated Monsieur [Isidore-François] Robin, the bishop's chaplain, to that parish. But that nomination is prejudicial to the right of those with [academic] degrees, unless Robin, a Parisian, is himself a graduate and had earlier given proof of his degree, something not to be presumed. Therefore, as among the graduates who are claimants for the benefice, your brother has seniority, not to speak of being very worthy and having influential patrons, it is said that the bishop will choose him in preference, inflexibility being inappropriate for [appointment] to parish benefices.

104
Les Baux, 12 September 1786

While rereading your first volume on the plants of Dauphiné, I made corrections of mistakes I observed there on the copy you gave me. I hasten to send you a copy of the outcome on the enclosed separate sheet of paper. You will be able to make some use of it in the subsequent volumes, either in your *errata* or in your second edition should you produce one in the future. I believe I have noted the principal mistakes. As for the description *utraque ad basin magis coalitae quam vexillo*, p. 290, l. 4, it appears to me to be eligible for correction. But, in order to correct it, I have to wait for you to tell me what meaning you meant to give it. Then I can recommend according to the rules of Latinity.[576]

Monsieur Danthoine has sent me a vetch that I have never run across in my

---

[574] Villars did not list *Silene inaperta* L. for Dauphiné. But he did give a place to *Silene polyphylla* L. that did not belong there, leaving reason to believe that *S. inaperta* was actually the plant he had found. See Grenier & Godron, **Flore de France** 1: 212.

[575] Villars' response to Jussieu's criticisms *would* appear within the preface to his second volume in 1787.

[576] This refers to his description of *Trifolium thalii* Vill, published earlier in **Prosp. Pl. Dauph.** 43 (1779). When it reappeared in the third volume of the flora, the description would be in French.

botanizing, and which is not in Linnaeus. As you yourself may have found it, or as it may at least grow in Lower Dauphiné as well as in Manosque, I am sending you a fragment of it, along with everything I know about it, in asking you to give me your opinion of it: *Vicia minima cirrhis hexaphyllis; foliolis linearibus; pedunculus longis, paucifloris; leguminibus cylandricis* Danthoine. *Vicia angustifolia* Chaix. *Vicia parvum genus polis longo* J. Bauhin, Chabr. 148. Ray, **Hist.** 903, no. 18, *caulis cubitalis, angulosus, ramosus, fere glaber, stipulae lineares, basi calcarata, folia 3, juga; foliola linearia, acuta, longa; pedunculi longi, pauciflori, corolla parva* [rodent damage] *affinis Viciae craccae* [L.]. . . . What do you perceive from this?

If you can send me a pinch of the wheat-grass called *moutin blanc* in Grenoble,[577] you would give me great pleasure. I would sow it in my garden along with some seeds of *Triticum polonicum* [L.][578] sent to me by Monsieur Thouin. Upon seeing these seeds, I realized that our *regagnon* around Gap is actually *Triticum turgidum* L.;[579] and that our *froment* is *Triticum aestivum* [L.], a biennial, of which our *tremois* is only an annual variety. Is that what you have decided about them in your work?

Has the young man from Lausanne, a printer, who spent a year here with relatives, presented himself to you? I directed him to you for help in finding work in his profession. Can you give me any news about his situation? Although he is a Protestant, I liked him for his good qualities.

Our region is in a sorry state because of the excessive dryness for the past two and a half months. The north wind blows every day. Planting is impossible, for, if one sows, the wheat blows from the ground. *God, Who covereth the heavens with clouds, Who prepareth rain for the earth.* **Psalms** 147: 8. Most of the trees have already lost their leaves on account of the great drought.

Monsieur Serre, who dined here yesterday, sends you his respects and hopes to preserve your friendship with him. He has quite enough work in his profession, succeeding with his operations and with his treatments. Those who have employed him are very satisfied with them. Take care of your own health. I pray God to preserve it a long time for you, both for His glory and for the public welfare.

105

Les Baux, 30 October 1786

You tell me in your last letter of September 16th that, on p. 290, line 4, you meant to explain that the wings and the keel are more adherent to each other at their base than at the calyx. As you have three parts of the corolla in mind in this description, it is necessary to replace *utraque* with the words *hae* and *illa*, or *hae tres sibi ad basin magis coalitae quam vexillum* instead of *vexillo*. . . .[580] I have not yet been able to understand how Linnaeus uses *corolla resupinata* to characterize the genus *Lavandula*. The corolla of the lavenders that I

~

[577] *Triticum turgidum* L., see Vill. **Hist. Pl. Dauph.** 2: 156 (1787).

[578] A Spanish wheat-grass; elsewhere a cultivar or weed. The author of the specific name forgot that there are two Galicias in Europe.

[579] Vill., **Hist. Pl. Dauph.** 2: 156 (1787), said that *regagnon* was *Triticum maximum*, which is obscure, or possibly *Triticum polonicum* L.

[580] See letter 104.

know, my *officinalis*,[581] my *spica*,[582] *stoechas* [L.], and *multifida* [L.], is not bent backwards, the ventral part up, but rather ascending, the upper lip directed upward; and below, the three segments are thrust out nearly horizontally. That does not constitute a resupinate corolla. This epithet has led you into error.

I thank you for the wheat-grasses you have sent to me. I have sown them in my garden along side the *Triticum polonicum* [L.] of which Monsieur Thouin had sent me nine or ten seeds, and which have come up. If I had had the latter last year and your *moutin blanc* from Grenoble, I would not have called our *regagnon* or *gros blé* the *Triticum polonicum* in my catalogue. So, our *regagnon* or *gros blé* is a variety of *Triticum turgidum* [L.], and your *moutin blanc* is the true species of it as defined by Linnaeus. Our *froment* in Gap, or *prime annone*, is a variety of *Triticum aestivum* [L.], even though sown in the fall; for our *fromois* wheat, which is sown here in the spring, does not have really distinct characters. That is how, given time, the difficulties were smoothed out.

It may well be that your *Ligusticum gmelini* [Vill.] hardly differs from *Ligusticum austriacum* [L.]. But why did Linnaeus give it very entire leaflets? when they are incised and serrate![583] And why refer *Sesseli montanum cicutae folio glabrum* C. Bauhin to it? which seems to me more suitable to *Ligusticum* [rodent damage].

The *Chenopodium urbicum* [L.] that Monsieur Deleuze had found at Le Poët, and which he had sent to me both as a dried specimen and in seed, has come up for me and matured this year. Here is the note that I made on it: *Planta annua, sesquicubitalis, tota viridis; folia ampla, triangularia; racemi nudi, stricti, sursum recta spectantes; cauli paralleli.* For the second time, I have sown and cultivated the *Chenopodium rubrum* [L.] that Monsieur Thouin had sent to me previously. (It may be *Chenopodium purpurascens* Juss.) It is surely a species different from *Chenopodium murale* [L.]. Here is my description: *Statim ab radice ramos emittit sumi jacentes, purpureus, ramoso; folia triangularia, dentata, obtusa, crassa seu pulposa; racemi subfoliosi, rubescentes, ex foliorum axillis; summus caulem terminat.* Whereas *Chenopodium murale differt ramis horizontalibus, imis diffusis, foliis crasse oribus, obtuse oribus.* Linnaeus was not mistaken nor inappropriately multiplying species inasmuch as the synonyms and the figures in Daléchamps and J. Bauhin, which are the same, belong to *Chenopodium murale* [L.]. And *Chenopodium rubrum* L. remains a species distinct from *C. murale.* I am sending you specimens of these three species, and you can verify them yourself.[584]

I had written the above notes by the time I received your letter of the 13th instant on the 21st, included within the valuable bundle of plants you have sent me. The packet you added to it, coming from Vienne by the messenger service, was from Monsieur Pourret, sent from Narbonne. It contained 100 dry specimens, of which only 30 proved to be new to me. The most interesting among them for us is *Lepidium iberis* L., about which I speak to you on a separate sheet of paper on which you will see some leaves and a branch I have

⌒

---

[581] *Lavandula officinalis* Chaix, Pl. Vap. 51 (1785); Chaix in Vill., Hist. Pl. Dauph. 1: 355 (1786), on barren hillsides at Les Baux and elsewhere. [= *Lavandula angustifolia* Miller (1768) ssp. *angustifolia*].

[582] *Lavandula spica* Chaix, non L., Pl. Vap. 51 (1785); Chaix in Vill., Hist. Pl. Dauph. 1: 355 (1786), southern exposures on the mountains, Le Buis and elsewhere. [= *Lavandula latifolia* Medicus (1784)].

[583] *Ligusticum gmelini* Vill. [= *Pleurospermum austriacum* (L.) Hoffm.] has lobes pinnatifid, coarsely crenate-dentate.

[584] Verlot, Cat. Pl. Dauph., pp. 286-287, did not accept *Chenopodium rubrum* L. for Dauphiné, nor had Villars.

detached from it. The example of this *Lepidium iberis*, of the above-mentioned *Chenopodium rubrum*, and of *Myagrum perenne* L., teach us that we have to tremble when we undertake to correct or, in particular, to reduce the species of the incomparable Linnaeus.[585]

I next turn to your shipment in detail in order to give you my opinion about it as you request. Please understand that I agree with you about those plants that I do not mention in my critique.

Your *Rosa rubrifolia* [Vill.] [= *Rosa glauca* Pourret (1788)] is exactly the same as my *Rosa montana* [Chaix].[586] Its new shoots and the petioles are usually red, but the leaf blades do not always have that tint.

Your *Salix hastata* [L.] is the same as my *Salix spadicea* [Chaix][587] but does not have [rodent damage] either glabrous or sessile leaves. I do not believe it is *Salix hastata* L.

Your *Salix amygdalina* [L.] differs from our *Salix triandra* [L.] because of the glaucous color on the underside of its leaves. But these leaves, being medium-sized and not quite large (*permagna*), are not in my opinion those of Haller, [**Hist.**], no. 1636 [*Salix amygdalina*].

Your *Sphagnum alpinum* [L.] is my *Sphagnum palustre* [L.].

Your *Phleum perenne* Bellardi has a very great affinity to *Phalaris phleoides* L.[588] The only difference I see is the numerous hairs on the glumes.

I believe that your *Erigeron atticum* [Vill.] is a variety of *Erigeron alpinum* [L.].[589]

Your *Hieracium andryaloides* [Vill.] is *Andryala lanata* L. because of its dentate, nearly lyrate, leaves. The much taller plant in Haller has completely entire leaves.[590]

Centaurea alba [L.]? *C. amara* [L.] in my catalogue: *non est C. alba calcilibus integris, mucronatis; sed est C. amara squamis ciliatis, lauris.*

Orchis laxiflora Lam. I believe is only a variety of *Orchis morio* [L.].[591]

Your *Arenaria striata* [L.] from Mont Néron is *Arenaria liniflora* [L.], Murr., **Syst.** Haller should never have linked this plant to *Arenaria larcifolia* [L.] as he did with *A. striata*.[592]

<hr>

[585] Chaix in Vill., **Hist. Pl. Dauph.** 1: 349 (1786), had listed *Lepidium iberis* L. (1753) as good "for the reason that it is not separate from *Lepidium graminifolium* L." (1759). Whereas, Villars, ibid. 3: 286 (1788), would correctly make *L. iberis* a synonym of *L. graminifolium*.

[586] *Rosa montana* Chaix, **Pl. Vap.** 42 (1785); Chaix in Vill., **Hist. Pl. Dauph.** 1:346 (1786). Subalpine hill or mountain sides, from Rabou and La Grangette.

[587] *Salix spadicea* Chaix, **Pl. Vap.** 69 (1785); Chaix in Vill., **Hist. Pl. Dauph.** 1: 373 (1786), from St.-Julien-en-Champsaur and Chabottes. Several botanists in the 19th century thought this might be *Salix caprea* L., but that was a species Chaix had recognized. A probable hybrid: *Salix lanata* L. x *Salix hastata* L.

[588] Published as *Phleum phalarideum* Vill., **Hist. Pl. Dauph.** 2: 60 (1787) [= *Phleum phleoides* (L.) Karsten (1881)].

[589] Both species remain good today.

[590] They were quite comparable. But *Hieracium andryaloides* Vill., *H. lanatum* Vill., and *H. liottardii* Vill., **Prosp. Pl. Dauph.** 35 (1779), are all considered to be part of the *H. lanatum* group today. Villars regarded *Andryala lanata* L. to be a synonym of *H. lanatum* Vill.; and the leaf description given by Chaix suggests *H. liottardii* Vill.

[591] Both species are held to be good today.

[592] *Arenaria liniflora* L. = *Minuartia capillacea* (All.) Graebner in Asscherson & Graebner (1918). *Arenaria striata* L. = *Minuartia laricifolia* (L.) Schinz & Thell. (1907). Villars, **Hist. Pl. Dauph.** 3: 630 (1788), thought that *striata* and *laricifolia* might be varieties of the same species.

Your *Pulmonaria latifolia* from Berne is identical to ours from Manteyer [i.e., *Pulmonaria officinalis* L.]

Your *Euphorbia paralias* [L.] is surely *Euphorbia esula* [L.], not the *E. paralias* sent to me a long time ago by Monsieur Pourret: *folia habet arctissime sursum imbricata, acuta, pungentia, caulem occultantia.*[593]

I extend a thousand thanks to you for all the fine plants in your shipment. But you have forgotten [my request for] *Mentha rotundifolia, Potamogeton crispum, Saxifraga hypnoides, Chaerophyllum cicularia* Vill., and the seeds of *Myagrum perenne* L., *Anchusa tinctoria, Lunaria rediviva,* and *Antirrhinum majus* with purple-white flowers.[594] You had also promised me a clarification about a small vetch from Manosque.[595] Your subscribers are asking me for news about your second volume. When will you be able to deliver it to the public?

Your brother has taken possession of the parish of La Bâtie-Vieille. My own family established three domiciles on the 17th of this month. My nephew from La Grangette married a widow from Rabou. His son married that widow's daughter. And my niece, who lived with me and whom you know, married the son of the same widow, the heir of his late father. In consequence of which I now have the eighth and last daughter of my late brother as governess of my small household. So goes the ebb and flow of circumstances. God unites us all in the bosom of His glory!

I think that it will not probably be very difficult for you to find the public announcement in Grenoble in which the king's edict of last September 2nd was published, which orders and fixes the income of the curés at 700 livres.[596] Please send it to me with Monsieur Serre when he returns home. I need it so that I can take the steps prescribed by the sovereign that will enable me to enjoy this established benefit.

106

Les Baux, 5 December 1786

It is nearly a routine to have written each other every month. Now you have raised the bid, the recent month having passed without me doing my part. I have received two letters from your side, to which I shall respond below, but not before informing you of an accident that very much afflicted me and caused me concern during the entire month of November.

My niece, who came to my house to replace her sister, whose marriage I announced to you in my last letter, has not experienced her menstrual discharges for a year and a half, beginning with an extremely cold temperature that she endured in a critical circumstance at about the age of seventeen. On October 29th, she was seized by a violent pain in her knee, accompanied by a high fever. Not knowing what course this malady would take, I deferred having her bled for four days. The inflammation and the swelling, however, became greatly increased. It was a true erysipelas that neither the bleeding done on two

---

[593] It appears that Villars, **Hist. Pl. Dauph.** 3: 826 (1788), put the plant under *Euphorbia seguieri* Scop. [= *Euphorbia seguieriana* Necker (1770)].

[594] See letter 101.

[595] See letter 104.

[596] The royal declaration of 2 September 1786 that raised the *portion congrue* from 300 to 700 livres a year.

separate occasions, nor the application of hot, moist elder, nor the diet alleviated. That was the time of Monsieur Serre's trip to Grenoble and Lyon. By the time of our young surgeon's return, which I had been anxiously awaiting, the injury gave the appearance of a running sore. After two days of maturative plasters, he made an opening with a lancet, releasing much pus, reducing the swelling little by little. The patient is now beginning to be active, but I took care of her for more than three weeks when she was in bed. I am now hopeful for her imminent recovery, but I remain troubled about the resumption of her menstruation. What must I do to obtain that result?

I received *Tussilago frigida* L. [= *Petasites frigidus* (L.) Fries (1846)] from you right away, with roots fresh for planting. I welcome it only because it comes from your hand. It is very abundant at La Grangette in the vale of the Gravasson at the foot of the great perpendicular crag of the Pic de Bure. I have mentioned it in my catalogue; and you yourself must have trod upon it, having on one occasion, when coming to see me in Les Baux, crossed the very difficult pass of Le Foulet coming from St.-Etienne-en-Dévoluy and through La Grangette.

If *Seseli montanum cicutaefolia glabrum* C. Bauhin belongs to *Ligusticum austriacum* [L.] in Allioni, why did Bauhin say it has leaves like conium?[597] And why did Clusius compare its leaves to those of *Seseli peloponnesence* [Matth.] [= *Ligusticum peloponnesiacum* L.], which does not fit *Ligusticum austriacum* in Allioni? I do not understand this outcome.[598]

I would refer my *Salix spadicea* [Chaix], your *S. hastata* [L.], to Haller, no. 1655 rather than to no. 1654, because of its slender, not oval, catkins.[599]

Do you know *Rumex fagopolifolius* Boccone from Monsieur Deleuze? Its spike is always simple, never paniculate. Monsieur Deleuze has sent it to me from his mountains.[600] He has promised me a specimen of *Hypericum coris* [L.] that he collected beyond Digne with Monsieur Danthoine, which differs particularly from my *Hypericum hyssopifolium* [Chaix][601] or *Hypericum galloprovinciale*[602] with which I had earlier confused it following an opinion given me by Monsieur Danthoine in a letter. Do you know this *Hypericum coris* [L.] well? The same Monsieur Deleuze had sent me a euphorbia found at Digne, with solitary, axillary flowers that appeared to me not far from *Euphorbia pepoloides* [Gouan], Allioni no. 1032. Do you know it?

As for your *Euphorbia paralias* [L.][603] from Le Valais, it is certainly the same plant that we have at Le Poët: *umbellae multifidae; involucella subcordata; folia linearia, sparsa, rara; petala integra, obtusa.* It is the same as Gérard, Fl. **galloprov.**, p. 540, no. 18, Linnaeus, **Mantissa**, p. 324, that I believe to be *Euphorbia esula* L. But why did Linnaeus, who gives

$\sim$

[597] Allioni was simply following the synonymy given by Linnaeus, Sp. Pl., p. 250.
[598] Some authors have called *Ligusticum peloponnesiacum* L. an enigmatic plant. See *Molopospermum peloponnesiacum* (L.) Koch (1824), and letter 105.
[599] See letter 105.
[600] No *Rumex* survived in the Chaix herbarium, and this brief description does not well fit *Fagopyrum esculentum* Moench (1774).
[601] *Hypericum hyssopifolium* Chaix, Pl. Vap. 25 (1785); Chaix in Vill., Hist. Pl. Dauph. 1: 329 (1786), from Les Baux, Rabou, in open woods. Still good as ssp. *hyssopifolium*.
[602] Villars explained that this was the name used by Brother Gabriel, a capuchin friar and learned botanist in Aix-en-Provence, for *Hypericum hyssopifolium* Chaix.
[603] See letter 105.

two-horned petals to *Euphorbia esula*, refer it to Gérard who sees only entire petals in his [euphorbia]? Is it to make troubles for botanists? In any case, I doubt that Monsieur Villars could have seen the true *Euphorbia paralias* in Switzerland, even though it may be Haller, Hist., no. 1055, as it is a maritime plant with leaves so appressed that they are imbricate. Do consult the figures in J. Bauhin [Chabrey, **Stirpium icones**], Gérard, p. 538, no. 9, and the fragment enclosed taken from a specimen that the abbé Julien gave me.

I am very much obliged to you for the copy of the king's declaration concerning the *portions congrues* that you have sent me. Our temporal lot is in the hands of men. May divine mercy dispose of our eternal lot! *Fiat, fiat.* The detail you have given me about your response to Monsieur Jussieu's criticism has given me real pleasure. I shall not be the only one to applaud it when reading it at the beginning of your second volume. You delight me doubly by informing me that we may expect it in the near future. Monsieur Deleuze, who did subscribe, has not received your first volume, and he wants it very much. Could it still be in your brother's possession, the curé de La Bâtie-Vieille? I learned that he had two copies of it. Whatever is the case, you should see to it that this deserving botanist, who has such a high opinion of you, is served according to his desire and to his pledge.

107

Les Baux, 19 December 1786

What you tell me about being bound for Corsica pleases me on the one hand, because your reputation will become increasingly established there, and your interests would be served there. But, on the other hand, [the news] upsets and afflicts me, as I fear for your health that is so precious to me, and because I fear losing much of the friendly association that I have had with you for so long. If the good Lord has so decreed it, then His will be done and His holy name be blessed.[604]

The more I read your first volume, the more I find it learned and well done. Your botanical dictionary is a perfect piece. Your method, which, like all the others, has defects, is indeed the easiest for beginners as you claim. Beyond the critical list of mistakes I had observed that I earlier addressed to you, I have noted still others on rereading the work. . . .[605] I think that these errors will have been repaired in your forthcoming volume on species, but they may explain what Monsieur Jussieu meant in his report by the words "the author has often forgotten to make additions or changes."

I am very familiar with *Tussilago frigida*[606] with its triangular leaves, very white below, which fills our vale of the Gravasson and the torrents therein on the gravel, and more of the same between La Grangette and the mountain in a place called the Pranons. I have also found *Tussilago alba* [L.] and *T. hybrida* [L.][607] in the Loubet woods, in the Gravasson, but in moist places, and which can be differentiated only by their sex. And when the thyrse has disappeared, the leaves do not present a distinctive character, their circumference being

∾

[604] Villars did not mention a possible move from Grenoble to Corsica in his brief autobiography, although we already know of tensions and conflicts within the medical and clerical communities in Grenoble that must have been a constant temptation to move elsewhere. His eventual move would not occur until after Chaix's death.

[605] A substantial list of errors follows, many of them spelling or typographical errors, some of the minor errors in plant description or plant sequence.

[606] See letter 106.

[607] All of these species are now in *Petasites.*

reniform, angular, their surface downy and dirty-white.

If your *Euphorbia paralias* [L.] from Aigle [Switzerland] is Haller's plant, it is then not Linnaeus's, to judge by his description of it.

Your plates should be very good, judging by the one you sent to me. The natural habit of the umbel in *Allium narcissiflorum* [Vill.], my *Allium grandiflorum* [Chaix], is more nodding. I regret this defect in your figure. I cannot imagine to what narcissus you are comparing its flower.[608]

Two pretty mosses surround the roots of *Tussilago frigida*. I am sending you specimens enclosed in the hope you will determine them for me. I already have them in my collections.

108

Les Baux, 29 December 1786

Long letters cost the abbé Pourret nothing. The Asian style is much his own, especially when it comes to extending himself with compliments and promises. The results of them are more limited. He has written you a seven-page letter, to which is joined a list of plants he wants from you. He has written equally at length to me, also adding a list on three large pages, three columns each, to tell me about a large herbarium he is assembling at Brienne.[609] It will be arranged in 210 atlas-sized containers using large-sized paper from Annonay, with a synonymy of his choice, a description and usage, the ink from China, no luxury to be spared. But this superb botanical theater, which perhaps will never see the light of day, reminds me of cadavers enclosed in superb mausoleums where they are not provided shelter from putrefaction or from worms. *Putreds et vermes haereditabunt illum.* **Ecclesiastes.** I shall see what I can do to fill his list. As for his authority in botany, that is not clear in my mind. I need more witnesses about the truth. . . .

Regarding *Lepidium iberis* [L.] [= *Lepidium graminifolium* L. ssp. *graminifolium*], it still remains that we only have *Lepidium graminifolium* from our botanizing, so surely *Lepidium iberis* is a different species.

In six years, Linnaeus will no longer be mentioned in Paris. That is jealousy or absurd, overstated, by whoever has launched that prediction. The vision of such geniuses reminds me of ants undertaking to beat down the walls of the Louvre.[610] Even Monsieur Allioni, the most moderate of men, has not been able to resist expressing his opinion of the strange Monsieur Lamarck. Let us pray God that He gives us the grace to know ourselves.

～

---

[608] *Allium grandiflorum* Chaix, Pl. Vap. 16 (1785); Chaix in Vill., **Hist. Pl. Dauph.** 1: 320 (1786) [= *Allium narcissiflorum* Vill. (1779)].

[609] Pierre-André Pourret (1754-1818), abbé de Saint-Jacob in Provence, specialized in the flora of the Pyrenees. The brothers Loménie de Brienne [Etienne-Charles de Loménie de Brienne, archbishop of Toulouse and soon to be head of Louis XVI's *conseil des finances*; and Louis-Marie-Athanase de Loménie, comte de Brienne] put Pourret in charge of their plant collections at Brienne-le-Château east of Troyes. Pourret would emigrate to Spain in 1789.

[610] This would appear to be some Parisian botanist's prediction that Linnaeus would soon be eclipsed by the imminent publication of A.-L. Jussieu's **General Plantarum** (1789).

The *Euphorbia pilosa* [L.] in my **Pl. Vap.** is really *Euphorbia carniolica* [Jacq.] as in All., **Fl. ped.** 1: 287, no. 1048.[611]

I learn with real pleasure that your son is giving you all the satisfactions that his age permits. Formed from your blood, he can only have its characters. My nephew in La Grangette has confided the education of one of his sons to me. He is beginning to mumble the Latin primer and to scribble short phrases. He is still young. Although without a notable talent, he will be able to succeed if he persists and if he is encouraged.

P. S. Have you read your dissertation on the conservation of forests in the province to your academy? Has the prize been awarded? In that case, who has been crowned?[612]

109

Les Baux, 29 March 1787

I received the four copies of your second volume with the four folders of plates that relate to them, which you have had the kindness to send me. You tell me that you have added a fifth folder for Monsieur Martin, but I assure you that I only found four of them. Therefore, either that fifth folder was inadvertently not sent, or it was stolen en route. I am close to suspecting the latter; for the package, contrary to your usual practice, was poorly tied-up, badly bent, without any address, and without closure or seal. When I had Madame Garnier recompensed for the shipping costs she had paid for me, I had her asked who had brought said package to her. Whenever he may see them, I am quite persuaded that Monsieur Martin will not be deprived of your plates. I delivered Madame de Flotte's copy and folder to her (last year I sent you her payment for the entire work); Monsieur Serre's copy to him (I think you are making him a gift of it); and also Monsieur Gaude's to him, for which I send you the price, that is to say, 8 livres, for which please acknowledge receipt. I thank you very much for the copy you have given me. I have wanted it very much for a long time, and I have already read a part of it. Everything seems well done to me and quite correct. I think that the connoisseurs will be pleased with it. I shall draw up a list of the small number of mistakes that I have noticed and will send it to you in the future.

I thank you and Monsieur Liottard for the *Saxifraga crassifolia* L. that you sent me with Monsieur Serre. In return, I am sending you and Monsieur Liottard *Hesperis hieracifolia* Chaix. *Hesperis sylvestris hieracii foliis hirsuta* C. Bauhin. Ray, Hist. 791, no. 5. Gérard, **Galloprov.** 365, no. 1.[613] Exclude the Linnaean synonyms. *Caulis viviradix, cubitalis, ramosus, pilis raris instructus; folia lyrato-dentata, crassa subhispida; flores spicati,*

~

[611] A plant given to Chaix, probably by Deleuze, it having been collected in Upper Provence across the Durance at Vaumeilh. This species is, in fact, an Asian plant; and Villars, Hist. Pl. **Dauph.** 3: 832 (1788), remained uncertain about Chaix's suggested change.

[612] Villars, "Liste et observations sur les arbres de la province du Dauphiné." **Mémoires de la Société littéraire de Grenoble** (Grenoble: Allier, 1787), pp. 174-244. His observations were on the trees and shrubs of Dauphiné, not on the question of deforestation or reforestation that was the subject of the prize sponsored by the society. Villars merely supported those other members of the society committed to conservation by providing an extensive and useful catalogue.

[613] Gérard, in fact, thought this plant was *Hesperis sibirica* L. [= *Hesperis matronalis* L. ssp. *matronalis*]. See letter 103.

rari, cernui; calices subvillosi; petala purpurea, reflexa, semiuncialis, etiam longiora, obtusa.[614]
I am also sending *Hieracium auricula* Chaix, which may only be an unnatural offspring of *Hieracium cymosum* [L.]. But where should *Hieracium auricula* be placed? as I have not yet satisfactorily separated it from *Hieracium dubium* [Vill., non L.].[615] More of *Anchusa sempervirens* [L.]. I presume that you will cite the above mentioned *Hesperis hieracifolia*, because, as it was found at Digne, it probably inhabits the southern part of Dauphiné. I am sending in addition about fifteen packets of seeds for Monsieur Liottard. And I am asking him, whether for planting or seeding, for *Antirrhinum majus* [L.] *flore rubro*, *Anchusa tinctoria* [L.] [= *Alkanna tinctoria* (L.) Tausch], *Myagrum perenne* [L.], *Chaerophyllum alpinum* Vill. [= *Anthriscus sylvestris* (L.) Hoffm. var. *tenuifolia* DC.], and *Saxifraga umbrosa* [L.].

The great Monsieur de Buffon has greatness alone in his usefulness and his eloquence. His hypotheses contrived to refute the sacred text are merely reveries, absurdities, and contradictions. By what right should he be called the French Pliny? I am not surprised that his confrere, Monsieur de Voltaire, has said that Buffon's literary output comes out of very small houses. But then the one is hardly worth more than the other! *Nescis quo modo nihil tam absurdi dici potest, quod non dicatur ab aligus philosphorum.* Cicero, **Divinatione**, *lib.* 11. Must one be a philosophe in order to talk nonsense?

P.S. If *Lathyrus inconspicuus* L. comes up for you, please examine it in order to talk about it. It grows among the wheat-grasses in the southern region.[616] Our *Lathyrus aristatus*, which you perhaps refer to *Lathyrus angulatus* L. (though its seeds are not angular), having come up for me only in the fall even when sown in the spring, is doing well in my garden. I will be able to send it to you either dried or in seed.

110

Les Baux, 27 April 1787

Under the intendant's cover, I sent you a package enclosing some plants and the money from Monsieur Gaude for your second volume. I also talked to you in the accompanying letter about the folder of illustrations for Monsieur Martin that I had not received. Was that package delivered to you? What annoyances have they created for you lately? Your brother told me on Tuesday at our conclave in Gap that you have been made uneasy. On this miserable earth, the peaceful are often in unsteady boats. I am not surprised about it. The sovereign prince of peace endured such inconsistencies.

I am sending you herein some specimens of incomplete plants along with a list of a few mistakes that slipped into your second volume, plus a short catalogue of a few species that pertain to it. All of this might be of some use to you if you have to add a supplement at the end of your third volume.

---

[614] Villars, **Hist. Pl. Dauph.** 3: 317 (1788), published this species as *Hesperis hieracifolia* Vill.; but it appears to be *Hesperis laciniata* All.
[615] *Hieracium cymosum* L. var. C (*Hieracium spurium foliis ovato-oblongis pilosis, caulo subnudo, pedunculis sparsis inaequalibus.* Chaix manuscript.) Vill., Hist. Pl. Dauph. 3:102 (1788). [= *Hieracium x spurium* Chaix ex Froelich in DC. (1838). (*H. cymosum* L. x *H. pilosella* L.)].
[616] It had been collected at Rosans and Le Poët by Chaix and Deleuze.

111
Les Baux, 1 June 1787

Since that time in the month of March when you sent me copies of your second volume, I have not had one word of news from you. But since then, I have written to you on two occasions, each letter accompanying a small shipment, sent under the cover of the intendant. Are you ill? Have I offended you in some way? Do you not have a moment to yourself to give me two words of response? In the first case, I would share greatly in your ill state and pray to God for your recovery. In the second case, my fault being purely inadvertent, I would then be able to mend my ways after having made my apologies to you. In the third case, I would be content with a single word from you. Whatever it may be, I shall not remain silent in your regard until such time that your silence convinces me that I must become silent.

I enclose here a second catalogue of a few mistakes in your second volume that you may add to the first one I sent you, in addition to a few observations Monsieur Pourret made to me about certain astragalus, plus a small specimen of *Allium ampeloprasum* L. that I have cited from around Gap. You made no mention of it, not risking to report it on my word alone. As it has quite the appearance of *Allium arenarium* [L.], you could have confused it with the latter if you have not seen it in fructification, and if you have not dug up its onion, which is always surrounded by bulblets, as the authors say and as J. Bauhin's figure depicts it.[617] Also a small leguminous specimen found by Monsieur Martin in the La Bâtie de Montsaléon preserve that I do not know but should like to know. Also, I can assure you that Monsieur Martin did find *Crucianella angustifolia* [L.] in said preserve, having compared his specimens with my own that I have often cultivated.[618]

The week before St.-Jean, Messieurs Martin, Serre, and I must go to botanize around Le Buis to snare, if we can, some of the plants you indicate from there.

112
Les Baux, 13 July 1787

All the brothers are not brothers; and the name, Charity, is not always reinforced by the virtue of charity. What a stroke of good fortune that you perceived the underground fire before it broke into the open, and that you were able to smother the flames before you were surrounded by them. All of those guns had been deployed to knock you down. I bless the Lord for being the declared protector of the innocent. While we have true friends to whom we may express ourselves for consolation, we also have dangerous enemies against whom we must always be on guard. We owe propriety and decency to everyone, but we owe to ourselves the preservation of our legitimate rights and those of conscience. . . .

You were too preoccupied by your troubles to respond to all the items on my list from Le Buis, for which I am sorry.[619] You tell me that you will have it inserted at the head of

∾

[617] Chaix was correct that *Allium ampeloprasum* L. ought to have been included in Villars' second volume for Dauphiné. Their *Allium arenarium* L., however, was *Allium scorodoprasum* L. See Verlot, **Cat. Pl. Dauph.**, p. 318.

[618] Another good species omitted by Villars from Vol. 2.

[619] This indicates a missing letter sometime between 1 June and 13 July 1787; but this letter discusses the plants collected around Le Buis that had been sent to Villars previously.

your *errata*, but I cannot agree to that. It is neither complete for the botanizing in that region nor interesting for the reader. I only sent it to you for any advantage you might be able to draw from it. If you are to print anything by me, it should be my corrections or additions to my **Plantae Vapincenses**, which I would send to you at the appropriate time.

I turn next to your botanical clarifications:

1. Pardon me if I wrote *chloris* for *chlora*.

2. I gladly believe that my *Cistus dubius* is *Cistus salicifolius* [L.], but why is it called willow-leaf, as its leaves are usually only three times longer than broad?

3. My *Ervum monanthos* [L.] [= *Vicia articulata* Hornem. (1807)] is exactly the same plant from Monsieur Martin that I had already sent you another time.

4. As the *Geranium* you call *purpureum* [Vill.] is found with *Geranium robertianum* [L.], I believed *purpureum* to be a variety of it.

5. If my *Lathyrus caeruleus* [from Le Buis] is the same as *Orobus* Gérard, **Galloprov.** 493, no. 4, where it is named a synonym of J. Bauhin [2: 326], Gérard was wrong to make an orobus of it. For my plant is certainly a *Lathyrus*, the style flat, villous, and enlarged at the end. It is certainly not *Orobus angustifolius* L., *flore lutes.*[620]

6. Our melica from Les Rochers-de-Ste.-Euphémie is not *Melica lobelii* Vill., which I have never yet found; but rather *Melica ramosa* Vill. that Lamarck calls [*M.*] *pyramidalis* following Barrelier, **Icones** 95, no. l: *gramen avenaceum angustifolium panicula pyramidali.*[621] Whereas *Melica lobelii* Vill. is *gramen avenaceum latifolium minus sparsa panicula*, **Icones** 94, no. 2 [= *Melica uniflora* Retz. (1779)]. But *Melica minuta* L. differs from both one and the other.

7. *Ophrys monorchis* [L.] [= *Herminium monorchis* (L.) R. Br. (1813)]. See my note attached to the enclosed specimen.

8. *Rubia tinctorum* [L.] from Le Buis. I cannot say whether this differs from the one around Grenoble [*Rubia peregrina* L.] that I have never seen, but I believe that it is more likely *Rubia lucida* L. [= *Rubia peregrina* L.]. Its leaves are in divisions of four or five, but they are not linear as in *Rubia peregrina*, nor do they have prickles on their central nerve as in *Rubia tinctorum*; they are elliptical and smooth except on the margins. The flowers are paniculate, five-petaled, pentandrous, *ex speciminum autopsia*. Thus, it appears to be the one you describe in [**Hist. Pl. Dauph.**] 2: 313.

9. *Lysimachia thrysiflora* [L.]. See the written note. If I am mistaken, provide me with the name of this plant. It is a find by Monsieur Meyer [vicar of La Bâtie-Neuve], and he should be honored for it.[622]

10. *Rhododendron intermedium* Chaix.[623] I have never noticed these silvery dots on the peduncle and the corolla of *Rhododendron ferrugineum* L.

11. *Triticum maximum* Vill., which is our *regagnon* [or *gros bled*], appears to me to be a glabrous variety of *Triticum turgidum* [L.].[624] Last year in Grenoble, I collected the latter,

∽

[620] Villars had already named this species twice: *Orobus vicioides* Vill., non DC., **Prosp. Pl. Dauph.** 41 (1779); *Orobus angustifolius* Vill., non L., **Fl. Delph.** 75 (1786), the name he would retain in his third volume. [= *Lathyrus filiformis* (Lam.) Gay (1857)].

[621] Both *Melica ramosa* Vill. and *M. pyramidali* Lam. = *Melica minuta* L., a variable species.

[622] The fragmentary specimen, seen by Verlot, **Cat. Pl. Dauph.**, p. 236, suggested *Lysimachia nummularia* L.

[623] Unpublished.

[624] See letter 105.

its flowers pubescent, as Linnaeus describes it, and which is Villars 2: 155, no. 3. I expected to see it come up this year from seeds you had sent me labeled *moutin blanc*, but they only produced a more delicate variety of our *ragagnon*, with glabrous flowers, which may be the *blé fin* of Grenoble that I would call *Triticum optimum*.[625] I am not segregating these species as their calyx is the same, that is to say, without awn. Either you made a mistake when selecting seeds for me, or they degenerated after reaching me. When you can, please send me once again a pinch of this wheat with pubescent flowers so that I can test it once more. As for *Triticum polonicum* [L.], it is quite a different species. I have sent you a spike of it.

12. *Centaurea calcitrapoides*, which is now seeding itself naturally here, consistently retains its entire leaves and has not degenerated.[626]

Other *errata* from Vol. 2 by Monsieur Villars. . . .

Your "Mémoire [sur les maladies les plus fréquentes à Grenoble, suivi d'un essai sur la topographie de cette ville," 1787], which you say you sent to me in your letter of June 5th, having still not reached me, must have suffered the same fate as the folder of your plates meant for Monsieur Martin. Either it never left Grenoble, or it was destroyed on the road. I am quite annoyed about it, because nothing is dearer to me than anything that comes from your hand.

P. S. During the night of Tuesday to Wednesday of last week, Monsieur Vivian, the surgeon in Valserres, was murdered in his house, his head split, his own knife plunged into his chest, his nose and ears cut off, his personal effects scattered, and his money stolen. It is probable that his murderers pretended to want to have a tooth pulled, as the instruments for that operation were found laid out on the table. In which case, the surgeon, in good faith, was examining the tooth that he would have to remove from one of them when the other villain delivered the fatal blow. These mutilations indicate a diabolical vengeance beyond the criminal act. *Portentum ad tartara amandandum!* The authors are still unknown.

My friends, that is, your brother, his neighbor Meyer [vicar of La Bâtie-Neuve], the curé du Saix [Martin], and Serre, are coming to join me on Monday to go searching for plants around La Grangette. As I expect nothing new from this botanizing, I have not deemed it appropriate to defer this letter until after our small venture. Take care of your health.

113

Les Baux, 11 September 1787

You may judge with what pleasure I have read your paper on the most common illnesses in Grenoble and on the topography of that city, about which I had already heard much talk.[627] In my judgment, which is limited, the work is very learned. After the very

---

⁓

[625] This may have been the same as *Triticum compositum* L. in Vill., **Hist. Pl. Dauph.** 2: 157 (1787), the *blé de miracle*, a form of *Triticum turgidum* L.

[626] Villars, **Hist. Pl. Dauph.** 3: 55 (1788), concluded that this was not a hybrid, but rather *Centaurea calcitrapa* [L.] var. *b*.

[627] See letter 112.

favorable testimony about it rendered by the Royal Society in Paris, further praise is superfluous. I would have liked a bit more clarity and precision in your opening and in your line of argument, which is often complicated and metaphysical. Perhaps you would have done better either to be silent or to indicate more indirectly the means for salubrity, which, being burdensome for the province, have offended many people. I am permitted this criticism only by the friendship with which you honor me, but it does not go beyond the recesses of my heart. As it remains totally secret, make of it what you think it merits. I thank you for this valuable mailing. I have already had one of your copies read by a few of my friends, and it is in Durbon at the moment.

Although the use of the sexual system as published by Monsieur Gouan has nothing new to teach those with a profound knowledge of the Linnaean botanical works, it is very useful for beginners learning the sexual system and of some help to those of middling skills.[628] The chief merit of this work is the clarity and succinctness of its ideas, a quality that any author should keep primarily in his mind. Monsieur Gouan seems a bit reprehensible for declaring himself to an uncompromising partisan of his master, as if it would be mistaken in the extreme to deviate from the principles of the prince of botanists. I am returning this work to you with a thousand thanks. Your brother was the only one to read it after me.

I am sending you a *mantissa* that I have put together as a supplement and a correction for my catalogue published in your first volume. After you have read it, striking out from it anything that appears to you either superfluous or not in conformity with what you yourself will have stated, I would beg you to have it printed at the end of your third volume: either to free the readers of that catalogue from my errors, or to shore up the small reputation I may have acquired in botany when learned people [in the future] will see that I was aware of my aberrations. In sum, do not be afraid to remove anything that might shock you.

On still another sheet, you will see some *errata* relative to your two volumes, plus some notes, followed by a short list of plants I observed during my trip to Molines-en-Queyras. I had counted on getting higher into those mountains; but, in order to do that, I would have had to stop for a night in a shepherd's hut, something I did not have the courage to do. *Non idem possumus omnes.*

You ask me in your last letter of July 19th to give you my opinion about the canons [aphorisms] of Linnaeus. *Caracter non constituit genus, sed genus caracterem*, etc., **Philosophia botanica** no. 123.[629] It is true that human thought contributes nothing to the constitution of the object, and that nature [alone] has made the genera and the species. But as the boundaries she has set down for them are not perceived alike by all men, it follows from that fact that some of them extend, and others retract, those boundaries; and that the genera and the species, becoming arbitrary, are no longer, in that sense, the work of nature. Thus, Linnaeus sees a genus, and even a class, in *Poterium*, different from the genus and the

---

[628] Antoine Gouan, **Explication du système botanique du chevalier von Linné** (Montpellier: Jean-François Picot, 1787).

[629] *Scias characterem non constituere genus, sed genus characterem, characterem fluere e genere, non genus e charactere.* **Phil. bot.** no. 169. Know that the character does not make the genus but the genus the character; that the character derives from the genus, not the genus from the character.

class of *Sanguisorba*.[630] Haller merged the *Phaca* from Linnaeus with the *Astragalus*, and so on. Who would attribute a crime to the botanist who would transfer *Oxalis corniculata* [L.] from decandria to monadelphia?[631] More than once, Linnaeus himself has followed this course: witness *Poterium*, the violets, the grasses, the polygamous, etc. We have defined the genus in the abstract, an abstract rank applicable to all the species. That is to say, all of its attributes must fit all the congeneric species. The latter differ from each other through their special characters. Thus, Linnaeus has sometimes made the genus depend upon the character, contrary to the formation of his own canon. That seems obvious to me.

Monsieur [Jacques-Henri Bernardin] de Saint-Pierre says in Vol. 2 of **Etudes de la nature** [1784] that the peloria reproduces itself by seed.[632] How unusual this famous observer is in so many things! In order not to talk nonsense about the matter, I would rather believe Linnaeus on this subject, who, in giving assurance that the fruit of this marvelous plant always aborts, would certainly have given that assurance only after many proofs of it. Have you any more recent clarifications about [this phenomenon]?

Was your trip to Mont Cenis pleasant?[633] If it was, I do not doubt that it was both agreeable and fruitful for you. Please tell me something about it when you have the leisure. If you can, send me some seeds from your garden, for example, *Chaerophyllum alpinum* [Vill.] [= *Anthriscus sylvestris* (L.) Hoffm. (1814)], and *Chaerophyllum cicutaria* [Vill.] [= *Chaerophyllum hirsutum* L.], which I have never found, and so on. You would give me a great pleasure. I am sending you some specimens from my own culture, pressed in your book.

P.S. [From Dr. Serre to Dr. Villars]. I beg you, Monsieur and dear friend, to buy a pair of convex eyeglasses for me, meant for an uncle on whom I have operated for a cataract. I received the books that you had the kindness to send me through Madame de Calvière. I will benefit from them and send them back to you after reading them. I enclose the money for the glasses. I wish you a health as perfect as my own.

114
Les Baux, 18 September 1787

Last week I sent you your book by Monsieur Gouan with my short *mantissa* and a few notes on mistakes in your second volume. I think you will have received them by now. As you recommended to me in your last letter in the month of July that I read your books closely and report to you the mistakes that I might observe in them, I have done so; and I hasten to send you enclosed what I believe must still be brought to your attention so that,

⌒

[630] Whereas Villars, **Hist. Pl. Dauph.** 2: 306-307 (1787), had found no legitimate separation between the two genera; and Chaix in Vill., ibid. 1: 375, indicated the closest affinity between the two, while being uncertain where they belonged.

[631] *Oxalis* has ten stamens that are monadelphous. Within the Linnaean system, *Oxalis* is found under **Decandria Pentagynia**.

[632] *Peloria*: the particular state of certain flowers that, normally irregular, become regular. Linnaeus regarded them in that state as monsters.

[633] Villars, **Hist. Pl. Dauph.** 3: xvii (1788), described a trip into Savoy in 1787 through the Maurienne country as far as Mont Cenis. The region had already been botanized by Allioni, Bellardi, and Molinelli from Turin.

as the printing of your third volume proceeds, you will receive it in time for practical use if you find the remarks appropriate.

Enclosed, I am also addressing you a short notice on rural husbandry that you will please take to Monsieur Giroud for insertion in his periodical if he finds it suitable.[634] I believe that many people in the public will be able to gain advantage from it. Here in Les Baux, we are finding that the practice works very well. It may not be new, but it is possibly neglected too frequently or even known too infrequently by most people in the country. In the event that Monsieur Giroud publishes it, please ask him for a copy of his sheet for me.

Enclosure: **Economic Notice**

In Volume 2 of **Etude de la nature** by Monsieur Bernardin de Saint-Pierre, he recommends digging up the bulbs of autumn saffron [*Colchicum autumnale* L.] from the meadows in order to destroy the moles that feed on their roots. The remedy would be worse than the disease, as it would be necessary to turn over an entire meadow. There exists a much simpler and even more efficacious method that has worked very well for me. Make a decoction from the roots of white hellebore, *Veratrum album* L. Let a handful of wheat seeds steep in this decoction. Put a pinch of these soaked seeds into the most recent of the mole tunnels, immediately reclosing the openings you make for fear that the mole, which does not want any daylight in its apartments, will hasten to repair any breaches made, pushing out the seeds along with earth; and also for fear some fowl will eat them. Soon the entire garrison, which will be regaled by this new booty, will exchange the colic for the sleep of death in the underground corridors. One can also use this method to get rid of rats that are a pest in houses. Thorn-apple, *Datura stramonium* L., possibly could produce the same effect, but it is less available to the country people, especially in the mountains.[635] Chaix, curé des Baux près de Gap

115

Les Baux, 5 October 1787

When I sent you my *mantissa*, I believe that I urged you to use it as if it were your own. If I did not explain myself clearly enough, I again beg you to proceed as I have just said. In all of my opinions about botanical facts, I have only sought to contribute, as much as I can, to the perfection of your work. I do not value any of this for myself, for all my happiness in this world lies in the success of your productivity.

If my notice on moles is deemed appropriate for publication, I mean for the editor to put it into the form that suits him, whether it be the choice of words or a question of style. A countryman can observe, but one must not expect purity of language from him.

I thank you for your copy of **Voyage au Mont Blanc**,[636] for your description of Mont Cenis and your pretty plants, but principally for *Gentiana glacialis* Haller,[637] *Gentiana*

<hr />

[634] Alexandre Giroud, the printer in Grenoble.

[635] Thorn-apple, in Solanaceae, has properties similar to belledona. It was not native to France but had been naturalized from America in some areas.

[636] Horace-Bénédict de Saussure of Geneva, physician and naturalist, who studied the geology and the botany of the Alps, had reached the summit of Mont Blanc in 1787, guided by Jacques Balmat of Chamonix, who had been the first to climb the mountain on 7 August 1786. Saussure published attractive accounts of his trips, but had earlier been one of Albrecht von Haller's important correspondents.

brachyphylla Vill., *Saponaria lutea* L., and *Saxifraga muscoides* All., none of which I had ever seen. The restraint in your works, once completed, will bring me the satisfaction of many additional items in the future, if God grants me life.

P.S. I see so little difference between your *Gentiana brachyphylla* [Vill.] and *Gentiana bavarica* [L.] that I am not surprised that Haller only reported one species, just as he also did not make two species of *Gentiana verna* [L.] and *Gentiana pumila* [Jacq.].[638] Moreover, I believe that Allioni, through his *Gentiana nana* [Jacq.] means your *Gentiana brachyphylla* [Vill.], as his *Gentiana nana* is a perennial.[639] *Gentiana glacialis* Hall., you tell me, ought to be *Gentiana nana* All.; but *Gentiana glacilis* is an annual, is close to *Gentiana nivalis* [L.], and I believe that you call it *Gentiana minima* in Vill., [Hist. Pl. Dauph.] 2: 528. . . .[640]

Saxifraga muscoides [All.] does have some affinity to *Saxifraga androsacea* [L.], but I have not see the teeth on its leaves that the latter has. According to the synonym from Séguier, cited by Linnaeus, this should be *Saxifraga sedoides* L. Instead of citing Séguier 1: 450, t. 9, f. 3, Monsieur Allioni should have cited Séguier 3: 205, t. 5, f. 3, where Séguier revises his figure and provides a more exact description.[641]

Monsieur Linnaeus, as we have previously mentioned, has a great number of genera in which the species are maintained by only one character, *Hyssopus nepetoides* [L.], for example, by its elongated, exserted stamens, which I have before me at this moment, and which otherwise would be a *Nepeta*.[642]

Silene acaulis [(L.) Jacq.]. I have observed it to be bisexual here. I do not know whether it can also be dioecious.[643]

Salix fragilis L. Monsieur Villars would be mistaken if he takes the slightly red variety of *Salix alba* [L.] for *Salix fragilis* L. The former is planted more commonly in Les Baux, as I remember having noted in your manuscript on the trees of Dauphiné.[644] *Salix fragilis* L. is really the one for which I gave a short description in my **Cat. Vap.** 1: 372: *Salix folio lato longo splendente, fragilis.* Ray, **Hist.** p. 1420. Linn., **Lapp.**, no. 349, t. 8, f. b. Haller, Hist., no. 1638. This is indeed the fragile willow of the English at Cambridge.[645] Ray only mentions the fragility of the small branches, *ramulis fragilibus*, so that his statement does not run counter to the flexibility of the young shoots. In reality, among all our tree-like willows, we have no others whose branches are so breakable (*dissilientes* L.) at the

~

[637] This was the name that Vill., **Hist. Pl. Dauph.** 2: 532 (1787), gave to Haller, **Hist.** 1: 290, no. 652. Allioni did not recognize that number.

[638] All four species are still considered to be good.

[639] Allioni did cite *Gentiana brachyphylla* Vill. as a synonym, Fl. Ped. 1: 99, no. 360.

[640] Verlot, **Cat. Pl. Dauph.**, p. 242, believed this species to be *Gentiana tenella* Rottb. [= *Gentianella tenella* (Rottb.) Börner (1912)].

[641] Chaix was right about *Saxifraga androsacea* L. sometimes having toothed leaves. But the segregation of these close, but still good, species is based on more fundamental characters.

[642] This was probably an American form of *Hyssopus officinalis* L.

[643] Villars, **Hist. Pl. Dauph.** 3: 613 (1788), would say that the species was often dioecious in the mountains around Grenoble.

[644] *Salix fragilis* L. was commonly known in Dauphiné as red willow.

[645] Part of the confusion here stems from the fact that the description from Ray in Chaix's catalogue fits Haller's no. 1636 (not Haller's no. 1638), as Villars had detected. And Haller's no. 1636 is *Salix amygdalina* L., which Chaix did not recognize.

points where the branches are inserted, which Linnaeus called *genua* (knees), than *Salix fragilis*; we have none whose leaves are so acute at their tips, so shiny above and so obviously serrulate on their edges; we also have no other that equals the height of the white willow (*arbor procera* L.). *Salix alba est species maxima cum Salice fragile*, L. **Sp. Pl.**, p. 1021. If Haller has said that this is not the fragile willow of the English, that is true if he is speaking of the first shoots put forth after a cutting. For they are then so flexible that they can be very easily twisted to make withes, and thus the fragile willow of the English will be *Salix triandra* L., and the above said *Salix fragilis* would make a flexible and pliant willow for them. Haller, indicating *stamina fere quatuor*,[646] is in agreement with me, even though most of the scales are diandrous as Ray has seen them. This seems so clear to me that I no longer have any doubt about it. Beyond which, *Salix phylicifolia* L. could well be the same species, as Haller does not make any particular mention of it, no matter that our *Salix fragilis* very probably exists in Switzerland, where it is perhaps not indigenous as in our region. Should *Salix alba* be called indigenous here? I do not believe so.

 Ranunculus lanuginosus [L.]. I have found it at Durbon, the flowers double, cultivated by the monks. I have not seen it again in any other place. It is acrid like *Ranunculus acris* [L.], but much more pubescent, much larger in all its parts. The lobes of their leaves are not distinctly cleft as they are in *Ranunculus acris* with their notches. *Lobi non soluti, sed oris lateralibus imbricatio.* Note: The Montagne des Sept-Laux has no rare plants if one excepts *Arenaria apetala* Vill.[647] *Geranium pratense* [L.], from the same place, if it has not been examined carefully enough, might only be *Geranium sylvaticum* [L.].

 Note: *Tussilago frigida* [L.] [= *Petasites frigidus* (L.) Fries (1846)] is abundant in the mountain stream near the foliate rock of Durbon, along with *Tussilago petasites* [L.] [= *Petasites hybridus* (L.) Gaertner, B. Meyer & Scherb. (1801)]. The former is very abundant in the gravel on the Berthaud property of the former Chartreuses.[648] The latter I knew only from the meadow belonging to the comte de La Roche.

116
Les Baux, 17 October 1787

 I had written my note on your umbels before receiving your letter of the 10th instant, brought by Monsieur Serre. You will pardon me if I send it to you on a separate sheet enclosed rather than transcribing it here in this letter. In truth, I am embarrassed for having wearied you for so long and on so many different occasions with my mailings of *errata* or critical notes on your edition. Attribute it to my lack of discretion, because I ought to have proceeded with more perception, with greater brevity, with better order, and put the entirety of it in one summary. I indeed beg your pardon. Having read and reread your two volumes, I understand how much attention is required to make an edition correct.

 Monsieur Serre should not be criticized, without doing him an injustice, for having

               ~

[646] This is in Haller, **Hist.** no. 1638.

[647] *Arenaria biflora* L. is sometimes apetalous. Villars, **Hist. Pl. Dauph.** 3: 622 (1788), indicates that he found *Arenaria biflora* at "Alvar above Mont de Laus," presumably referring to Allevard north of the Montagne des Sept-Laux. One has to wonder, however, whether there was some confusion in their notes with the Mont de Lans, especially considering the frequent typographical errors in the flora.

[648] Chaix's birth place. In his catalogue, he located the plant at La Grangette.

made representations for the comtesse de La Roche. He was sent [to Grenoble] by Madame de La Roche, but also as the one chosen by his fellow citizens, and out of that affection that everyone must have for his country. The opposing party has spread the rumor throughout this region that Monsieur Serre has lost your good graces for having accepted that commission. Such a lie reflects upon its author, who, thereby, seeks to denigrate your sentiments and your probity. As for me, I have heard the rumbling of the storm, but as a timid dove. I have not emerged from the hole of my retreat. The unfortunate Icarus, by raising his flight too high, lost his wings, throwing him into the sea. The revolution, upon which this allusion bears, came to us precisely on the Sunday on which the Gospel concluded with these word of our divine savior: *For every one that exalteth himself shall be humbled.* Luke 14. 11.[649]

Those who are contemplating getting me to involve myself in such a situation do not know my incapacity. I shall always have my inadequacy as legitimate grounds for refusal. Your invitation to come to join you during the winter is very pleasing to me. I shall never have happier days than those I might spend with you; but, if not intending to be there this winter, I am not totally losing hope of it. Although the facility of the intendance is now closed for communication between us, do not fear to use the facility of the post to my address for letters, using the messenger service for your packages. Such expenditures will always be acceptable to me. I shall use the same means in your regard.

We are now in agreement about *Salix fragilis* L. I have characterized it in [**Hist. Pl. Dauph.**] 1: 372 [1786]. I will refer it, as you do, to Haller's no. 1636. What led me to refer it to Haller's no. 1638 was the number of stamens given by that author (*stamina fere quatuor*);[650] and this same number of stamens always diverted me from recognizing your *Salix daphnoides* [Vill.] under this no. 1638, which I only know with diandrous scales. The variety of *Salix alba* [L.] with red or blackish shoots, which we have more frequently here, cannot be *Salix amygdalina* L., as it never has stipules, and its leaves are not glabrous. Save for further clarifications about the above from you, I now think that *Salix amygdalina* L. does not differ from *Salix triandra* L., and that Haller's *Salix amydalina* is really *Salix fragilis* L.

I have cultivated *Ranunculus acris* [L.] *flore pleno*, *R. repends* [L.] *flore pleno*, and *R. aconitifolius* [L.] *flore pleno* for a long time in my garden. I am quite confident that Durbon has just produced *R. lanuginosus* [L.] *flore pleno* for me. In the mountains, I have seen *R. nivalis* [L.] and *R. glacialis* [L.] *floribus semiplenis*. I have not given up hope of finding *R. polyanthemos* [L.] *flore pleno* some place.

Saxifraga alpina muscoides, foliis superioribus oblongis, inferioribus rotundioribus et circumactis, Séguier, **Verona** 3: 203, t. 5, fig. 3. Similar to *Saxifraga alpina minima, foliis lingulatis in orbem circumactis, flore ochroleuco,* ibid. 1: 450, t. 9, f. 3. Very poorly engraved. [= *Saxifraga muscoides* All., Fl. Ped., no. 1528, from Mont Cenis]. . . . The petals in the figure are quite entire, obtuse (In all your specimens they are not all notched); the sterile stalks are one inch, covered with many leaves, taller than in your specimens. Also, Allioni's figure represents your specimens better than Séguier's do. In sum, I believe that both of the above are the same plant.[651]

∽

---

[649] This verse concludes: *and he that humbleth himself shall be exalted.* This somewhat opaque paragraph seems to give a hint of the political activity that would eventually lead Dr. Serre to the national legislature.
[650] See letter 115.

You promised me *Phellandrium mutellina* L. [= *Ligusticum mutellina* (L.) Crantz] from your trip to Savoy. I have it in the treasury of my hopes along with many others. Take care of your health and give me news of it when you can.

P. S. Monsieur Gaude, curé de Furmeyer, who went with me to Durbon, asked me quite insistently to send his respects to you. I never see this worthy priest, my true friend, that he does not speak of you.

I have not yet understood, and I cannot comprehend, what your adjective *brachyphylla* means when given to a gentian. I presume it is taken from the Greek. As I do not have a dictionary of that language, kindly explain it to me. Then I shall know on what feature the character of this plant rests.[652]

117
Les Baux, 17 November 1787

Monsieur Serre came last Thursday to show me the letter you wrote him concerning a trip to America. This project does touch me as it would deprive me of the society and the help of a true friend. But far from opposing it, I approved it very strongly. I have even hinted at additional advantages to be derived from it. If matters work out in support of his readiness, I hope they will be pleased by the commission with which he will be charged, and that he will enjoy the benefits of his undertaking. God will preserve him, his health being under the faithful care of his guardian angel.

Upon your arrival in Grenoble, you ought to have found my letter at home in which I made some observations about willows and your umbels. If you are following the the same course that you held to in your second volume, I foresee that you will not be able to include all the plants of Dauphiné, which you mean to treat, within the proper limits of your third volume. There remain genera loaded with many species: Beyond Linnaeus's syngenecious, for example, which I believe you already have set in print, such examples as *Geranium, Astragalus, Trifolium, Potentilla, Arenaria, Saxifraga, Ranunculus, Salix, Euphorbia, Polypodium, Bryum, Hypnum, Lichen,* etc. Consequently, I am nearly convinced that your third volume will not go beyond your 12th class. In that case, as you hinted to me when you were last here, you would need a fourth volume for the cryptogams, as well as for additions and corrections, which your subscribers would not be angry to pay an additional sum for, and which your readers would see with great pleasure. The typographical *errara* in the second and third volumes, or other mistakes that will have slipped in, could find a place there, too. You would also be able, if you should judge it to be appropriate, to enter my *mantissa* there. As I have sent you in several different letters the mistakes or the *errata* I have noticed in the first and second volumes, which might be difficult for you to recover and to put in order, I offer to make a general, coherent summary of them for you if that would be of use to you. In that event, you should make it known to me. I have ample time in the silence of my solitude; and whatever time I should employ to oblige you would be one of the sweetest times of my life. I would do the same thing for the third volume after it reaches me. Finally, it seems to me that it would be best to put all

∽

[651] Villars, **Hist. Pl. Dauph.** 3: 665-666 (1788), found *Saxifraga muscoides* All. on Mont Cenis but seemed to be uncertain whether it was different from *Saxifraga brioides* L.

the *errata* aside for placement at the end of the final volume.

[He adds a short list of grammatical corrections.]

P.S. Monsieur Allioni has rendered no service to his readers throughout his 12th class (the cryptogamy in Linnaeus), having simply repeated the authors' synonymy without adding a single note to it himself. At that price, it would have cost little to increase the volumes, but the readers now pay for a simple loss to themselves. You are quite far from following the example of our neighbor, otherwise so respectable, which leads me to believe that you will need a fourth volume for that part of your history.

If Monsieur [Serre] goes off to Paris, have the kindness to indicate to me under what conditions he is taking the trip and, with greater reason, the one to America. I am too concerned about everything that concerns him to treat this matter indifferently.

118
Les Baux, 15 January 1788

The literary society of Grenoble, having proposed the most interesting topic for the province as its program, has had the satisfaction of awarding its prize to Monsieur Achard's memoir, which has completely fulfilled the society's expectations.[653] It is to the glory of this group, so zealous for the public weal, to have used all its credit to recommend to the government those methods, so soundly established by the author of said memoir, for the conservation or for the maintenance of the woodlands in this province. If the Maîtrise continues to exercise its despotism in Dauphiné, in less than thirty years there will no longer remain any woods with full-grown trees.[654] Witness the superb forest of larch, cited by the author but unnamed by him, but which may be easily conjectured to be the Bois de Boscodon that no longer exists. Witness the Forêt de Loubet in our own community of La Roche and Les Baux, the most handsome in fir, and nearly unique in the Gapençais, which, being felled for the past few years under the fatal gavel of this odious tribunal, is already reduced to a pitiful state. What a reversal of order! The conservators of the forests for the state have become their tyrannical destroyers. Let reform take place, and our woodlands and fields will rise again from their [present] degradation. *Let the fields exult . . . the trees of the wood sing for joy.* **Psalms** 96. 12, and posterity will owe their recovery to the patriotic zeal of a society whose birth we may take pride in having seen.

You, Monsieur, have equally supported the goals of the society with your list of trees from the province.[655] I thank you for the volume enclosing all these important matters. Kindly allow me to make a few corrections here. . . .

∾

[652] In **Hist. Pl. Dauph.** 2: 528 (1787), Villars wrote that his plant had very short leaves, nearly round, compared to the leaves of *Gentiana verna* L.

[653] Alexandre Achard de Germane (1754-1826), an attorney affiliated with the Parlement de Grenoble since 1778, active in literary societies, and an opponent of royal absolutism. His prize-winning paper was entitled "Mémoire sur les causes du dépérissements des bois en Dauphiné et les moyens d'y rémedier." **Mémoires de la société littéraire de Grenoble** (Grenoble: Allier, 1787), pp. 29-131.

[654] Under the old régime, the country was divided into eighteen administrative districts called the *grandes maîtrises* that exercised jurisdiction over woodlands and forests. As this letter indicates, the royal desire for conservation was rarely well served by those courts.

[655] Villars, "Liste and observations sur les arbres de la province du Dauphiné." ibid. pp. 174-244.

When examining your list of plants from Lower Dauphiné, I notice some plants on it that belong in your second volume but that were omitted from it. I have drawn up a list of them here so that you can add those from your collections, which escaped from the volume, to your supplement: *Briza maxima* [L.]; *Euphorbia pilosa* [L.] is *Euphorbia carniolica* [Jacq.]; for *Euphorbia charassius* write *characias* [L.]; *Gentiana filiformis* [L.]; *Hyoscyamus albus* [L.]; *Ornithogalum narbonense* [L.]; *Phalaris aquatica* [L.]. As for plants belonging in your third volume, their fate is in your hands.

You would give me great pleasure, Monsieur, by informing me where you now are in the printing of your third volume, and whether you still believe yourself able to get all your remaining plants into it. Monsieur Serre told me about a critical response that you had prepared for your confrere, [Dr. J.-F.] Nicolas. He must have attacked you outrageously for you to be determined to mount a defense in writing. For you had appeared to me to have been inspired by Linnaeus's loftiness of soul (his letter to Monsieur Haller) in closing his eyes and ears to the satirizing from his opponents.[656] As for that, do not do anything without the advice of your best friends.

My neighbor, Bonnardel, not having had any news from his son for an infinite time, would like to know very much whether he is still in the service of Noyeret, or whether he has left the area of Grenoble. Kindly commission your servant-girl to get the information, and let me have a word about it in the first letter with which you will honor me. This poor father and his whole family will be obliged to you.

119

Les Baux, 29 April 1788

Sensing the burden of your occupations, I have not written to you since the month of January, fearing to become an annoyance to you. Some time ago I received the package containing five volumes of Latin letters to Haller (you have evidently retained the sixth volume, or you do not have it, as it is announced in the preface).[657] I have reached the end of the fourth volume in my reading of them. The materials treated in them have been of interest to me, with the exception of the anatomy about which I understand nothing. The Latin style of these scholars has delighted me above all else. I shall return them to you right away along with the two Gilibert volumes. I gave the other books in said package to Monsieur Serre, having recognized that they concerned him. He must have acknowledged their receipt to you. I also received your letters of March 15th and April 14th in good time. Your letters reached me without carrying charges. Can you tell me a means by which mine can enjoy the same privilege?

On two different occasions, I sent your brother a collection of notes to be passed on to you. In the last one, which you may not have received yet, following some observations

~

[656] Linnaeus to Haller in 1748: "If you will but listen to me as a friend, I would advise you to write no letters to Hamberger and such people. He is not on a par with you; and the more he is your inferior, the more consequence you give a man who would otherwise remain in obscurity, known only to those in his immediate circle. Our great example, Boerhaave, never answered anybody. I recollect him saying to me one day, 'You should never reply to any controversial writers. Promise me that you will not.' I promised him accordingly, and have benefited very much thereby." Wilfrid Blunt, **The Compleat Naturalist, a Life of Linnaeus** (London: Collins, 1984), p. 100.
[657] Haller published letters written *to him* up to 1775 in 6 volumes. He was primarily an anatomist, but with a dedicated interest in botany.

in Monsieur Pourret's **Chloris narbonensis**, I told you that *Ribes alpinum* [L.] is truly dioecious, about which I had some doubt in my **Pl. Vap.** There is no longer any doubt about it here.[658] Moreover, I told you that *Acer plantanoides* [L.] is polygamodioecious. That remains true, but also applies to *Acer monspessulanum* [L.]. It remains for me to examine *Acer pseudo-platanus* [L.] and our ayart.[659]

I am not surprised by the welcome given you in Lyon. Your merit, known far and wide, prepared the way for you. I congratulate you with the greatest satisfaction.[660] Already owing you so many things, I owe to you in particular the notice for which your learned society is proposing to honor me. This esteem is infinitely beyond my aspirations.

I have now written all I can say regarding the elimination of mendicity in this province. I shall have my scribbling sent on to you when you deem it to be appropriate. Perhaps I took a little too much liberty on some points, but I do not know how to conceal my sentiments. *Magis amica veritas.* I shall let Messieurs Gaude and Martin know about the proposition you have also made to them regarding this topic. . . .

120
Les Baux, 27 May 1788

I reply to your letter of May 6th that was filled with more and more new signs of your decency and your friendship for me. I only regret to find myself deprived of the means to prove my gratitude to you by deed, and to be obliged to bear that gratitude only in my heart. What to do about it? Someone must verify that sentence uttered by our divine savior in the presence of St. Paul: *It is more blessed to give than to receive.* **The Acts** 20. 35.

Although I do not have the brash presumption to aspire for the approbation of the Société littéraire de Grenoble, I have, nevertheless, directed my short work on the elimination of mendicity for its judgment. I shall be happy if some of its provisions merit attention. I have had it transcribed and sealed my name under my epigraph. I shall send it soon, carriage paid, to the secretary. How will I know just when the papers will be definitively judged?

I accept the charge of putting a herbarium of 1200 plants together as you propose, but on condition that I be provided the folders, the carriage be prepaid; and that the price be 2 sols per plant, which brings the total to 5 louis d'or. I am happy with that modest sum despite the opinion of friends to whom I have talked about it. I must know which botanical method I am expected to follow in the arrangement [of the plants]. If the person who is requesting this herbarium knows very little about plants, he would be indifferent to which species I should prepare for him. I shall not, however, introduce exotic plants, but will rather put in indigenous plants representing all the classes, and even all the sections that I can. I hope he will be pleased with my work. You must have the kindness to write me as soon as possible whether my conditions are acceptable, as I must set to work. It will be soon enough if I receive the folders during the month of August. . . . I am quite aware that a priest ought to be using his time for duties more analagous to his station; but I am poor, and I am obliged to obtain the necessities the best I can. I find authority in the

~

[658] It is *functionally* dioecious.

[659] *Acer opulifolium* Chaix, **Pl. Vap.** 29 (1785); Chaix in Vill., **Hist. Pl. Dauph.** 1: 333 (1786), common in Les Baux. [= *Acer opalus* Miller (1768)]. These *Acer* are usually monoecious, but can be polygamodioecious.

example of St. Paul, who, during his apostolic travels, worked in the home of Aquila at Corinth making tents: *And because he was of the same trade, he abode with them, and they wrought; for by their trade they were tentmakers.* **The Acts** 18. 3.

Your trouble in regard to the fate of Monsieur Deleuze was well grounded; and I myself became alarmed about him after what I heard from you about him, until I learned that he is very well in Paris, taking care of the children of Monsieur de Primini, a member of the Parlement, at a salary of 1200 livres. He was wrong to have put off writing to you for so long. I presume that, meanwhile, you have received news of him. Monsieur Pourret, according to what he has just written to me, has abandoned the publication of his flora of Narbonne and even his cystography. He is only working for his own satisfaction, or rather to concentrate his superior knowledge within the rich collection of Monsieur de Brienne.[661] Even if he has made up his mind to stick with this decision, he is a man of many talents beneficial to the public; and I shall overlook nothing to get him to change his mind.

Would you please take the trouble, Monsieur, to look in the bookstores for the **Mémoires en forme de lettres pour servir à l'histoire de la vie de feu M. de La Motte, Evêque d'Amiens,** in 2 volumes, 2nd edition published in Malines, 1785, and purchase it for me? I can either deduct the total cost from the price of the above said herbarium, if that comes to pass, or otherwise pay out of personal funds. A monk in Durbon lent me these volumes to read, and I have been so edified by the life of that holy prelate that I want to own the book myself to provide frequent reading for the salvation of my soul. Ah! if only the Church had among its ministers a number of pastors formed according to such a model, irreligion would be silenced, heresy would be overwhelmed, piety would be trimphant, and the souls would work out their salvation. I have no doubt that this incomparable minister of Jesus Christ will be publicly venerated as a saint through an official judgment of the Church. What more can the Carlo Borromeos and the François de Sales have done? Monsieur de La Motte has imitated their zeal in his holy ministry, to which he added the learning and the austerity of the Bernardines. Let us pray God to multiply such pastors for the care of His flock. *The harvest indeed is plenteous, but the laborers are few. [Pray ye therefore the Lord of the harvest, that he send forth laborers into his harvest.]* **Matthew** 9. 37-38.

121

Les Baux, 6 June 1788

As I have a small bundle to be sent to Monsieur Pourret, I am adding the two volumes by Monsieur Gilibert that I still have here, plus the five volumes of Latin letters to Haller. I thank you indeed for your courtesy in having given them to me to read. I did not get much use out of Gilibert, but I read the letters with pleasure. It must have been very costly to Haller's modesty to provide these letters for publication, filled as they are with such high praise for his talents, albeit highly merited;[662] for virtue fears display, as the prophet has

~

[660] On 1 April 1788, Villars read a paper before the Académie de Lyon entitled "Discours sur l'utilité de l'histoire naturelle." See Emile Callot, **Dominique Villars: Le naturaliste philosophe, le botaniste, le professeur, étudié à travers ses manuscrits inédites** (Gap: Société d'études des Hautes-Alpes, 1982), pp. 15-24.
[661] See letter 108.
[662] Haller was well-known for his reticense, dating from excessive shyness as a child.

said: *Ab altitudine diei timebo.* **Psalms** 55.... The letters in Vol. 5, pp. 242, 262, and 264, have so warned me against innoculation for smallpox that I shall never again advise it for anyone despite favorable results in the past and the defense of it made by other physicians in other letters. Do not bother to send me the sixth volume. I have had enough of these matters.

Please pay attention to remitting the package for Monsieur Pourret to the public messenger service. I have paid the transportation of this entire shipment, so it ought to be carried prepaid for you. I repeat my request about the **Mémoire de la vie de M. de La Motte** ... and also let me know about the herbarium that you proposed I make unless you have already sent your response. On Monday, I am going to spend two days with the curé de La Bâtie-Vieille [Villar] in response to his friendly invitation and to become reacquainted with your worthy mother.

122
Les Baux, 24 June 1788

Although I wrote to you a few days ago through the messenger Martin, when sending you your books, as I am pressed to learn whether I must prepare material to assemble the herbarium in question, and as the plants are disappearing day by day, please allow me to request an early response if you have not already done so.

Since, in the midst of so many serious occupations that carry off your time, botany is a solace for you, I bring several items to your attention here that will perhaps not displease you. Last week, I made a trip to Durbon with friend Serre; and it seemed as if Flora had been saving some of her gifts for me, as if to bring me out from the inactivity that has characterized me for some time. There is a monk in that pious monastery, the vicar of that house [Dom Grangier], who has, in addition to the most excellent qualities, a decided taste for the knowledge of plants, but for whom the retreat does not offer sufficient help. If he has just written to you, please, for my sake, do whatever you can for him.

1. In Les Baux recently, near the spring, I came upon the unfortunate decumbent fescue, previously unknown to me, whose discovery was very delightful.[663]

2. *Cistus apenninus* [L.], now blooming for me, given birth from seed, does not differ from *Cistus pilosus* [L.]: *caule foliis que viridalis, non incanis; foliis lanceolatis planis, non revolutis; petalis duplo minoribus, quatuor lineas non superantibus, albi tamen ut Cistus pilosus.* If you want to call it a variety, surely the variation is insignificant.[664]

3. *Valeriana dentata* [Poll.] [= *Valerianella rimosa* Bast. (1814)] comes from fertile land here and there in Champsaur, Les Baux, etc. *Valeriana coronata* [All.] [= *Valerianella discoides* Lois. (1810)] is around Gap, Veyne, etc. *Valeriana olitoria* [L.] [= *Valerianella*

---

[663] Possibly *Festuca decumbens* L. [= *Danthonia decumbens* (L.) DC. (1805)].
[664] *Helianthemum apenninum* (L.) Miller (1768) and *Helianthemum pilosum* (L.) Pers. (1806) are close but still separated.

locusta (L.) Laterrade (1821)] I have not yet met outside my garden. . . .

4. *Ranunculus sylvaticus* Vill.[665] (incorrectly referred to *Ranunculus auricomum* [L.]). Beyond the caracters in your clear description, evidently different from [*R.*] *lanuginoso* [L.], add *caule glabrius culo, fareto.*

5. *Ranunculus lanuginosus* [L.], found only within the woods around Durbon [rodent damage] *hirsuto, hirsutis que foliis.* The seeds not yet seen, but they may be in plants raised in my garden. One group is a double-flowered variety.

6. *Ranunculus bulbosus* [L.]: some dwelling in double-flowered groups. *Ranunculus acris* [L.] and *repens* [L.] now, and for some time, double-flowered in my own garden, with *Ranunculus aconitifolius* [L.] moreover double-flowered, of such a variant as would be a monster in nature, but a genuine botanophyllum if not very pleasant.

7. *Pinguicula alpina* [L.]: previously looked for in vain, but I finally found it at Durbon in a wet place near the cowshed.

8. *Chrysosplenium alternifolium* [L.]: not far from woods.

9. *Tozzia alpina* [L.], abundant near Durbon next to the torrents flowing through the woods, above the falls; young roots not mentioned by the authors, but this seems to me to be biennial, and I did not see any old remnants of stems.

10. *Daphne laureola* [L.]: coming from Durbon, here and there in the woods.

123
Les Baux, Undated
Between 24 June and 5 August 1788

You have probably been worried by my delay in responding to the shipment of your third volume, sent by way of the Dom vicar of Durbon, and to your letter included with the three folders meant for the proposed herbarium. The reason is that said volume, having been sent to the diocese without my address, allowed me to recover it only after a delay because of the absence of Monsieur Robin, secretary to the bishop; and I did not want to acknowledge reception of the folders without having fulfilled my duty in regard to your work. I have now completed reading it. I do not know how to express to you the eagerness, the pleasure, and the satisfaction I felt while reading it. I have admired the order of the materials; I have been struck by the choice of synonyms; I have applauded your learned and enlightening comments; and your definitions and descriptions leave nothing to be desired. I have been very touched by the honor you have done me by citing me on occasion (my only regret is not having been able to serve you better). How much work! How many evening did such a work cost you? to give it birth in the press of other occupations! The Lord has given you a rich talent; you have not hidden it. Bless Him for His gifts; and may the public render you justice through an eternal gratitude to you!

If the province values your works, its responses will not only open a career for you, but will put the seal of generosity on them. And I do not doubt that your censors will render you the testimony that is owed you, equitable and enlightened judges that they are. The editing of the third volume is much more exact than that of the two previous volumes as far as I can tell, and you will be able to judge that for yourself from my collection of the

~

[665] Published in **Prosp. Pl. Dauph.** 51 (1779), but not included in the third volume of the flora.

errata. On a separate sheet of paper, I have written down the mistakes I have noticed. Some of them are significant; some are perhaps too minute. It is for you to determine their suitability. I am also adding below a critical commentary on various items. As I presume that, following [publication of] your cryptogams, you will provide a short supplement or *mantissa*, my observations could be of some use to you.[666]

Critical Observations

p. 7. The root of *Carduus aurosicus* [Chaix] is not reddish, but whitish, rough or cracked all around.

p. 57. *Scolymus hispanicus* L. is improperly used [as a synonym] for *Scolymus perennis* Gérard, as the former is certainly only a biennial. That is how I have often seen it in my garden. . . . I have cultivated and noted the two species, which have given me the characters as given in Murray, Syst., 602.

p. 99. *Hieracium auricula* [L.]. I think that Linnaeus so-named this because of the resemblance of its leaves to those of *Primula auricula* [L.].

p. 104. Haller, Hist. no. 49 [= *Hieracium alpinum* L.], can only accommodate your *H. alpinum* and your *H. halleri* [Vill.] or *H. hybridum* [Chaix ex Vill.] as varieties.[667] Moreover, I would make your *Hieracium halleri* [Vill.], *H. valdepilosum* [Vill.], and *H. cydoniaefolium* [Vill.] into varieties of *H. villosum* [Jacq.].[668] Just as I would make your *Leontodon alpinum* [Vill., non Jacq.] a variety of *L. protheiforme* [Vill.] as you yourself suggest.[669] I am inclined to believe with Gouan that your *Hieracium prenanthoides* [Vill.] is the true *H. cerinthoides* L. because of the resemblance of its leaves to those of honeywort.[670]

∾

[666] Villars was responsible for creating later confusion about the publication dates of his third volume. In **Hist. Pl. Dauph.** 3: xviii (1789), he tells us that the third volume was ready for publication in 1788, but that the work on cryptogamy by J. Hedwig and G. F. Hoffmann had belatedly come to his attention. Their discoveries caused him to withhold publication until he could profit from the new knowledge about the reproduction of cryptogams. He *ought* to have said that he withheld publication of the cryptogams alone, for they were not published until the autumn of 1789 as the *second part* of the third volume. He enlarged the possibility for confusion by dating the preface to the entire third volume as 1 August 1789, consistent with the date 1789 on the title page. Moreover, the reader will also find an additional, undated frontispiece ahead of the cryptogams that offered several options about the placement of both that frontispiece and the plates in the third volume. The upshot was that later readers were easily confused about what had been the true date of Vol. 3, Part 1.

It is appropriate to remark here, therefore, that Chaix read 3 (1) in July of 1788, and read it in print. His critical notes in this letter ended with the *Euphorbia* on p. 832, which is the final page of 3 (1). The cryptogams and the plates, 3 (2), pages 833-1063, were not published until the autumn of 1789. Whoever provided F. A. Stafleu with information about the Chaix letters, "Dates of Botanical Publications, 1788-1792," **Taxon** 12 (1) January 1963: 82, misled him on two counts: (1) The critical letter above was not dated 4 August 1788. It bears no date at all, but has to have been written between 24 June and 5 August 1788. (2) Chaix did not give his Critical Observations merely for pp. 1-581 as stated, but for pp. 1-832. In sum, any new phanerogam published in the third volume should be dated 1788; any new cryptogam published in the third volume should be dated 1789. Finally, there never would be a supplement or *mantissa* as Chaix here anticipated, though certainly Villars had meant to publish one. Either he delayed publication for lack of funds, or his energies may have been diverted by the dislocation provoked by the French Revoluion.

[667] Both *Hieracium halleri* Vill., **Hist. Pl. Dauph.** 3: 104 (1788), and *H. x hybridum* Chaix ex Vill., ibid. 3; 100 (1788), are held to be distinct and good.

[668] All three are still considered distinct and good species.

[669] See *Leontodon hispidus* L. ssp. *alpinus* (Jacq.) Finch & P. D. Sell (1976).

[670] *Hieracium prenanthoides* Vill., **Prosp. Pl. Dauph.** 35 (1779) is still considered distinct and good.

p. 255. Since the leaves of *Achillea herbarotta* [All.] have teeth the length of their edges and principally at the top, I would remove the *basi* in the definition and, instead of *basidentatis*, I would rather say *oris serratis*.

p. 311. When you use *foliis enerviis* for your variety *B* of *Erysimum barbarea* [L.] [= *Barbarea vulgaris* R. Br. (1812)], you contradict both nature and the figures of former authors. When you link that description to *Sisymbrium barbarea* L., have you compared your plant closely with the latter of Linnaeus? Here is the description I made of *Sisymbrium barbarea* [L.] from a live plant in my garden: *fol. rad. spatulata, inferiora dentata, in petiolum decurrentia, semi-amplexicaulia, superiora amplexicaulia, insignita dentata; flores lutei pene Erysimum barbarea; siliquae patulo-erectae*. From around Corp. The figure from [Daléchamps] **Lugd.** [650] is not for it, but rather for *Erysimum barbarea folis latiore* var. *a.* L.; and your var. *a.* is var. *B.* L.[671]

p. 317. Monsieur Danthoine, who was here last week, assured me that *Hesperis hieracifolia* [Vill.] [= *Hesperis laciniata* All. (1785)], quite far from being born in [wheat] fields, is only found on rocks.

p. 342. *Sisymbrium erucastrum* [Poll.] [= *Erucastrum gallicum* (Willd.) O. E. Schulz (1916)]. Pollich, having copied Gouan's error, misled me with its whitish flowers; and I caught you up in my error. The plant of both [men] is the same, namely, *Sisymbrium murale* [L.] [= *Diplotaxis muralis* (L.) DC. (1821)]. So, Pollich's citation must be eliminated. I believed so strongly that my *Sisymbrium erucastrumn*, [**Hist. Pl. Dauph.**] 1: 351, differed from *Brassica erucastrum* L., [ibid.] 1:350 [= *Erucastrum nasturtiifolium* (Poiret) O. E. Schulz (1916)], that I put it in a different genus. Despite your observation, as you now present it as a variety [*B*] of your *Sisymbrium erucastrum*, which is *Brassica erucastrum* L., I am obliged to give mine another trivial name, *Sisymbrium neglectum* Chaix, as it is truly a very different species, an opinion shared by Monsieur Danthoine who brought me [the specimen] expressly to consult me about it. It is different in its simple, annual root, whereas the other plant is biennial or even perennial; by its smaller flowers, always whitish and never yellow; but the leaves are of similar form. It also differs from *Sinapis erucoides* L. [= *Diplotaxis erucoides* (L.) DC. (1821)], being smaller in all of its parts, and its branches being spread and open; its leaves are never entire with sharp teeth, but pinnatifid, obtuse; its petals are never perfectly white nor twice larger than the calyx, but rather white mixed with a little yellow and only one-third larger than the calyx; its siliques are not upright in the direction of the axis, but open nearly at right angles. Thus, I call it *Sisymbrium (neglectum) foliis obtuse pinnatifidis, petalis parvis albidis, siliquis patulis, radice annua.*[672]

p. 486. After *Trifolium angustifolium* [L.], you may add *Trifolium incarnatum* [L.], which is found at Pelleautier and Sigoyer near hedges at the edge of fields; as well as *Trifolium striatum* [L.] that I collected in abundance near Valernes with Monsieur Deleuze in a meadow bordering the mass of ice. In all probability, it must also be found across the Durance in Dauphiné. I believe it is annual or biennial, multicaulis, and a foot tall.

~

[671] Verlot, **Cat. Pl. Dauph.**, p. 20, concluded that Villars' var. *B* of *Erysimum barbarea* L. was *Barbarea praecox* (Sm.) R. Br. [= *Barbarea verna* (Miller) Ascherson (1860)].

[672] This passage suggests, despite its confusion, that Verlot, ibid., p. 38, was right to conclude that Villars' *Sisymbrium erucastrum* Poll. var. *a* was what we would name *Erucastrum nasturiifolium*; and that his var. *B.* would be *Erucastrum gallicum*.

p. 534. 2. This variety [*B* of *Prunus cerasus* L.] belongs to *Prunus avium* [L.]; for its fruit being sweet, the birds eat it; and it does not have the habit of the *griotters*.

p. 550. I fear that you have made too many species of roses.

p. 593. I have never seen *Dianthus scaber* [Chaix] at Aubesagne [in Champsaur],[673] but rather *Dianthus seguierii* [Chaix].[674]

p. 700. You had told me that my *Cistus medius* from La Bâtie-Montsaléon was *Cistus salicifolius* L., and I believe that after looking at Séguier's figure. But I do not know to what willow it has been compared. It is biennial: I planted its seed in the spring, and it has not given any indication of flowering this year.[675]

p. 709. The columbine found by Monsieur Charmeil [in Queyras] is *Aquilegia viscosa* L., **Mantissa** 77; Allioni, **Fl. Ped.** no. 1506 [= *Aquilegia viscosa* Gouan (1765)]. Monsieur Deleuze has found it also at Reynier.

p. 732. *Ranunculus phaeniceus miconi* [Daléchamps], **Lugd.** [1036] may indeed conform to the description of *Ranunculus gramineus* [L.], but the large leaves in the figure suggest *Ranunculus pyrenaeus* [L.].

p. 762. Having let *Salix vitellina* [L.] flower, I find its catkins appear no different from those of *Salix alba* [L.].[676]

p. 799. It seems to me that you showed me *Fraxinus ornus* [L.] by the side of the Bastille at Grenoble, the same that I believe I have observed in Le Buis. I did not see the flowers but reached that decision by comparing its leaves to those of *Fraxinus excelsior* [L.]. I am sending you a piece of both of them. Make a judgment.[677]

p. 823. *Euphorbia falcata* L. I do not believe that Monsieur Lamarck [could have found a worse name than] *Euphorbia acuminata* [Lam.] for this plant as all the others have that character. . . .

p. 823, bottom line. The plant that I gave to Monsieur Villars was not *Euphorbia terracina* L., but my *Euphorbia linifolia* from Gap and Le Buis. *Tithymalus annuus linifolio acuto*, Magnol, **Bot. monsp.**, 256. Ray, **Hist.**, 868. Barrelier, 7, no. 59, Ic. 742. *Semiped. annua, ramosa; foliis haud caducis, obsita inferioribus cuneiformibus, retusis, caeteris oblongis acutis; umbella 4-7 fida, involucellis cordato-acutis. Euphorbia autem terracina an est E. taurinensis* All., **Fl. Ped.** 1: 287? . . . *Euphorbia linifolium* Chaix.[678]

p. 824. *Euphorbia lathyris* [L.] surely is biennial.[679]

～

[673] *Dianthus scaber* Chaix, **Pl. Vap.** 27 (1785); Chaix in Vill., **Hist. Pl. Dauph.** 1: 331 (1786), from Reynier.

[674] This was mistakenly published as *Dianthus seguierii* Vill., **Prosp. Pl. Dauph.** 48 (1779), but later corrected to be Chaix in Vill., **Hist. Pl. Dauph.** 3: 594 (1788), from Rambaud, la Bâtie-Neuve near Gap, and Embrun.

[675] Villars had mentioned at the end of his section on *Cistus* that Chaix had found a *Cistus medius* very close to *Cistus aegyptiacus* L., adding that "As we have not had the opportunity to examine it, we shall be limited to simply indicating that collection here." Given the locality, it is possible Chaix had found *Helianthemum salicifolium* (L.) Miller (1768).

[676] The first of these was known as yellow osier, the second as white willow. See *Salix alba* L. ssp. *vitellina* (L.) Arcangeli (1882).

[677] The range of *Fraxinus ornus* L. was extended through frequent cultivation, and escaped cultivars have been noted. See Verlot, **Cat. Pl. Dauph.**, p. 237n.

[678] Verlot, ibid. p. 299, found indication in the notes Villars had been assembling for the supplement (that was never published) that he meant to replace *Euphorbia terracina* L. with *E. linafolia* Chaix. Villars' own description of the species led Verlot to believe that it was *Euphorbia segatalis* L.

[679] Villars had called it an annual. Chaix was right.

p. 825. *Tithymalus leptophyllos* [Titimalo leptifillo] Matthiolus, *involucellis acutis* is *Euphorbia exigua* L.[680]

p. 826. [You have a question about] *Euphorbia seguierii* (Scop.) All. Why not call it *Euphorbia gerardi* [from] **Galloprov.** 540, no. 18 [*Tithymalus umbella multifida, bifida: involucellis triangulari-cordatis, foliis superioribus latioribus*]. . . .[681]

p. 832. [You have a question about] *Euphorbia pilosa* [L.]. Yours is not *E. pilosa* L. Monsieur Danthoine suspects with good reason that it is also not *E. spinosa* L. On the other hand, I have compared each one with a specimen collected and sent by Monsieur Pourret. It is *fruticulosa, spithamaea, rigida, hirta; folia lineari-lanceolate, integerrima, reflexus hirsutie incana; involucra ovata, acuta; umbella 5-fida; petala integra*. Perhaps *Euphorbia langinosa* Lam., **Encyclopédie,** p. 436?[682]

124
Les Baux, 5 August 1788

When you proposed that I assemble a herbarium of 1200 plants for a person whose identity you did not reveal, I had the honor to respond to you that I could only do it for the average price of two sols per plant, which came to the total of five louis d'or; and that I should be supplied three folders with paper in which to put the dried plants, prepaid to Gap. You answered that my conditions were acceptable, although my work was considered a bit expensive. As a result, I set to work. Most of the plants are ready, and the folders have reached me. All that remains is to learn which method is desired for the arrangement of the plants. Kindly let me know as soon as possible.

Those who believe such a work to be expensive do not know the trouble and care one must expend in order to do it properly. The plants are not all in the first garden one enters, nor in the first meadow or the first field. One must go through the countryside and through the woods, climb mountains, to seek the items to be collected; bring them back with care, spread them out, check and recheck them for complete drying. Finally they have to be arranged according to a particular method in the folders, attached with an adhesive and labeled. An index has to be drawn up, and a short preface showing the arrangement of the work must be put at the beginning. Everyone to whom I have talked to about this has been astonished that I have taken this on at such a low price. I beg you, Monsieur, to make all this known to the person for whom I am working. If you prefer, show him this letter. I hope to have the work finished and have it off to Grenoble sometime during the coming October.

125
Les Baux, 14 September 1788

I am still working in haste to send you notes that I am making for you about your works. That is the reason I am tiring you with a multiplicity of letters instead of waiting to put a larger collection of them together. Please excuse my importunity. I am proceeding

---

[680] Chaix's observation was in connection with *Euphorbia leptophylla* Vill. [= *Euphorbia tenuifolia* Lam. = *Euphorbia esula* L. ssp. *tommasiniana* (Bertol.) Nyman (1881)].

[681] It had, in fact, already been done: *Euphorbia gerardiana* Jacq. [= *Euphorbia seguierana* Necker (1770)].

[682] Verlot, **Cat. Pl. Dauph.,** p. 298, believed the specimen to be *Euphorbia flavicoma* DC. (1813).

in this way, because I do not know when you will be printing your general *errata*. Here I am again on the same track:

1: 379, line 15: *lucens,* read *compressum.* For it was *Potamogeton compressum* [L.] that we found in a pond at Le Bourget, not *P. lucens* [L.], nor the *P. pusillum* [L.] indicated by Monsieur Villars in 2: 343.

2: 472, line 22: *teretibus,* read *planis subhirtis.* Its leaves [*Androsace diapensia* Vill. = *Androsace helvetica* (L.) All. (1785)] are not cylindrical but flat, *plana.* Therefore, on the following page, 473, line 5, for *cylindrical,* read *flat.* The cylindrical character gave Monsieur Danthoine trouble at Chaillol, and with reason.

Monsieur Serre's poa from the foot of the Pic de Bure, which you tell me is one of the varieties of *Poa angustifolia* [L.], should it not be Haller, **Hist.**, no. 1457? *spiculae triflorae, spadices virides, villosulae.* You may examine it at your leisure. I would call it *Poa frigida.*

2: 253: I do not believe that Monsieur Villars saw *Allium scorodoprasum* [L.] in the meadows at Durbon. He would have been misled by the name *rocambole* that is given to my *Allium palustre* [Chaix].[683]

3: 99: According to the letter Monsieur Villars wrote me on 2 September 1788, Haller, **Hist.** no. 53, must be referred to *Hieracium auricula* [L.] with some doubt.[684]

3: 429: *Ononis subocculta* Vill. [= *Ononis pusilla* L.]. The lower leaflets are nearly round to be sure, but all along the stems and branches they are obviously oval.

I must warn you that the names of *Hieracium hybridum* and *halleri* have been transposed on the plates, one for the other. . . .[685]

You really have the right to complain about your present situation, overwhelmed by the number of sick that the unusual garrisoning of soldiers gives you, and concerned about the monks who administer the hospital. When you have news, please do not deprive me of it, as I take great interest in everything that concerns you.

P.S. Kindly have some garden worker pick up some seeds for me, in particular those I have often requested from you.

---

[683] *Allium palustre* Chaix, **Pl. Vap.** 17 (1785); Chaix in Vill., **Hist. Pl. Dauph.** 1: 321 (1786), from swampy ground at Les Baux [= *Allium schoenoprasum* L.].

[684] It has also been referred to *Hieracium dubium* L.

[685] Chaix also called attention here to some errors in the pagination of the plates for the 2nd and 3rd volumes, which Villars would rectify following the frontispiece of Vol. 3 (2) in 1789.

Letters 126 to 170 cover the final decade of Chaix's life. While he remained concerned to correct errors in the three published volumes of the flora, his initial letter in this period, 13 October 1788, provides evidence that he was aware of, and reading about, the constitutional crisis. After the pending storm broke in 1789, both Chaix and Villars would have to cope with the course of political change and dislocation, as they both held titles, curé and médecin du roi, that eventually drew the attention of enemies of the old régime. The reader will find documents illustrative of Chaix's response, and ultimate plight, intruded among his letters to Villars where appropriate. [1788-1799]

126

Les Baux, 13 October 1788

I am finally sending you the herbarium in question. It is inconceivable how much trouble it takes for someone to arrange a great number of plants according to a method. I have done my best, and I have drawn up an alphabetical index. Having been dissatisfied with the 1200 plants we had agreed upon, I put 1300 in the herbarium, all in two volumes without any confusion. I still have the third folder, which I hope the person for whom I have worked will allow me to retain out of gratitude. But if said person does not wish to heed that voice from his heart, then I call upon him to let me have it as a matter of justice; for the 100 extra plants amount to a sum of 10 livres at 2.5 sols per plant, certainly more than the folder is worth. If, finally, said person wants a price for his folder, whatever it may be, I will send it to him. In which case you must then require an additional 10 livres from him over and above the sum agreed upon for the 1200 plants.

As I would presume that anyone would prefer to let me keep said folder rather than pay me an additional 10 livres, I beg you to have two similar ones made up for me, the same format and the same paper, to be sent to me by the messenger Martin, with the snippets that have served me very well as labels. I am going to occupy my time this winter putting my collection of plants into volumes; that is to say, one specimen of each species. So arranged, someone should be able to use them more advantageously after my death. Otherwise, if left in a pile, someone would probably use them to feed the flames. I am going to start work on this right away.

Do not send me anything out of the five louis d'or gained from the herbarium. A great part of this sum should be used as follows:

1. 6 livres 10 sols are due you for the book on the life of Monsieur de La Motte.

2. The cost of the two folders that you will have made.

3. A gold cross bearing the Christ in relief, with the heart, to be sent to me as soon as possible for my niece who serves me.

4. I should like a small picture that I have promised for the altar of the St.-Rosaire in our parish church. I have often seen images of the very holy Virgin painted on altar-fronts on leather. If someone wanted to remove such a piece from an altar-front, that would satisfy me. If one cannot be found in either Grenoble or Lyon, I beg you to have such a picture made in Grenoble: a small painting about one foot high and a little less wide, at the lowest price possible, representing the holy Virgin holding the infant Jesus in her arms, with some chaplets.

5. When I shall have received the final volume of your botanical work, I shall send all the volumes to you to have them bound along with some other books.[686] So, you see that I can spend my money in Grenoble without being there myself.

1. We read for the genus *Angelica*, which you define in your first two volumes, that *the marginal wings of the seeds are barely visible*.[687] Did you not mean to say *very visible?* *Angelica sylvestris* [L.] has marginal grains that are very apparent. See the enclosed specimen. Linnaeus said *semina margine cincta*; and Allioni, in relating it to *Selinum*, said *ala marginali magna*. Your character is more suitable to *Angelica archangelica* [L.]. See Allioni who put these two plants in separate genera.[688]

2. In 1: 211, line 5, remove *seven*. I do not remember whether I previously called this to your attention.

3. In 2: 621, you speak of an *Oenanthe* from Grenoble without characterizing it. The species that I have called *Oenanthe peucedanifolia* Poll. in 1: 359 is very well established, and it is abundant in marshy meadows around La Saulce. The abbé Meyer, who collected there this year, and who sent it to me to be determined along with many other plants, assured me he has seen a meadow at Vitrolles near the main road filled with this umbel.

I received your letter of September 30th. You complain of being obliged to be friendly to people who do not like you. The herbalist has acquired an illustrious name for himself, thanks to his [medicinal] herbs, that transcends the borders of the kingdom and will survive into future generations; while his adversaries, more circumspect during their lifetimes, will soon be forgotten. Jesus Christ offers you that portion of his chalice. After having endeavored to reject the bitterness in it, accept it, following his example, as having come from his hand. It is not in this world that virtue is rewarded.

Even in my solitude, I have not been totally ignorant of the excellent articles that have been appearing touching the affairs of our day. Your brother has had me read some of them. Madame de La Roche has also sent some of them to me.[689] I am very interested to learn what ruling will be approved for the Estates of Dauphiné. The parlement cannot avoid acknowledging the eagerness of the nation to support its restoration.

I am surprised that Monsieur Martin has still not returned your books. I saw him recently at Furmeyer. I shall let him know that you brought the matter to my attention. I thank you for the seeds of *Antirrhinum majus* [L.], *flore rubro, albo*. Try to get me the others that I have asked you for. May God preserve your health!

~

[686] As he indicated in an earlier letter, no. 123, Chaix anticipated that Villars would produce a fourth volume to encompass the cryptogams and a supplement for additions and *errata*.

[687] Villars, Hist. Pl. Dauph. 1: 192 (1786) and 2: 628 (1787).

[688] Allioni, Fl. Ped. 2: 7, 9 (1785).

[689] On 5 July 1788, an Order in Council had been issued requesting information on the method for convoking the Estates General. Before that issue could be resolved, an assembly of the Three Orders of Dauphiné met on 21 July 1788 at the Château de Vizille, led by the Royal Judge of Grenoble, Jean-Joseph Mounier. The assembly demanded the restoration of the provincial parlements (courts). Secondly, representatives of the Third Estate demanded an equal position to that of the first two orders in the Estates of Dauphiné, thus introducing the principle of doubling the Third and voting by head.

127

Les Baux, 24 November 1788

In response to your letter of October 28th, as you speak to me about the distinguished role played by the Brothers of Charity in the fashionable world of Grenoble, I must observe that if their founder, Saint Jean de Dieu, returned among them, perhaps he would not be received until there could be a new reform of morals and conduct.[690] But even in the soundest moral order, the greatest exaltation is founded upon the deepest humility: *For that which is exalted among men is an abomination in the sight of God.* Luke 16. 15. *The righteous shall be had in everlasting remembrance.* Psalms 112. 6.

Now that the yoke of oppression has been lifted, I can ask you whether you are not the author of the "Lettre aux Municipalités;" and you can respond to me freely. I thought I perceived your style in it.

Monsieur, will you kindly undertake the commissions I gave you regarding the money you received for the herbarium?

1. If you can tell me exactly the height and width of the painting (that you have not described for me with any precision) in order to be set within its frame, you will greatly oblige me; for the Reverend Father Dom vicar of Durbon, who is willing to make [the frame] for me, is waiting only for this exact dimension in order to start work.

2. I reiterate my request for two folders in the format and quality of the one left over for my own herbarium. I did not believe them to be so expensive, but no matter. I hope that these two folders, which should cost 18 livres, plus the one I already have, should nearly suffice to hold all the specimens of my plants species, with the exception of the umbels and the Cynareae, huge plants that I will put in folders that I will make myself; and also with the exception of the mosses and algaes for which I will have separate folders. Thus, my total expense will not be as considerable as it would be for 30 small folders at 4 livres each, which you have suggested to me, but which would amount to a total of 120 livres, something I cannot allow myself to do. I am using your method, and nearly your sequence, in the arrangement of my plants. You mentioned putting the rarer and exotic plants in the folder I now have. But what confusion! and what difficulty in going to find them far from their congeners! Besides, I like *Bursa pastoris* and *Carduus arvensis* as much as *Heliotropium peruvianum* and *Amaryllis formosissima,* because the omnipotence of God was necessary for the creation of both the former and the latter.

When will we have your cryptogamy? It is very much wanted. When you gratify me with it, please do not forget to send the frontispiece, the preface, and the index that you provided for your third volume. You know that my copy of it lacks all that.[691] I am adding below some further observations that could be of use to you if you provide a supplement or a *mantissa* in the manner of Linnaeus.

〜

[690] The Brothers of Charity, a hospital order of monks, had been founded in 1540 in Grenada by Saint Jean de Dieu and were brought into France during the reign of Henri IV.

[691] This statement confirms (see letter 123) that Chaix had read Vol. 3 (1) earlier in 1788, but without the title page that he would not receive until the following year, thus dated 1789, the actual publication date of only the cryptogams.

Critical Observations Relating to Our Botany

1. Monsieur, I think you ought not have corrected my name given to my specimen of *Anemone pulsatilla* [L.] [= *Pulsatilla vulgaris* Miller] in the herbarium I made for Monsieur Jourdan, changing it to *Anemone halleri* All. [= *Pulsatilla halleri* (All.) Willd.]. I have never seen the latter, its leaves simply pinnate. Mine has leaves that are hirsute and certainly bipinnate. Therefore, it is either *Anemone pulsatilla* or *A. pratensis* [L.], which differ very little from each other.[692]

. . . 3. My *Agrostis villosa* [Chaix] from Vallouise is only a variety of *Arundo calamogrostis* [L.]. My error has not been very great as *Agrostis arundinacea* [L.], from which I wanted to distinguish my plant, is the closest *Arundo* in Haller. Moreover, please tell me if *Arundo calamogrostis* is a branched plant around Grenoble, as it ought to be according the Linnaeus's description; for I have not observed a branched culm in the one that we have at Montmaur near the Buëch. If the culm in our region is always simple, we would be justified in making a separate species of it as I did in calling it *Agrostis villosa*.[693]

4. When you say that *Festuca duriuscula* [L.] has not shown you any crested glumes, possibly you have looked at specimens already somewhat aged whose awns had already fallen. I see its awns very clearly, as well as Haller did.

5. Since you say in 2: 210 that either you have never found *Carex elongata* [L.] or have confused it with *Carex remota* L., even though I drew your attention to this item on another occasion, I am describing the difference here, having the first species before me, collected at Prémol, as well as the second species collected on the Isère mort when I was with you in 1786. *Carex elongata* has leaves nearly as long as its culm; the leaves rise from the base; the culm, about a foot tall, is completely bare. Its spikelets are oval, sessile, and yellowish, rather compact, without leafy bracts, having only a bristle shorter than they are. Thus, the culm appears to be elongated, *elongata*. In contrast, *Carex remota*, which you have described very well, has very long bracts for each spikelet that surpass the top of the stem.

6. *Carex pallescens* [L.], with spikelets notably peduncled, would be placed better in your 4th division.

7. It appears to me that your *Carex frigida* [All.], which you say differs little from *Carex sempervirens* [Vill.], is really *Carex ferruginea* [Scop.], no. 2333 [in Allioni], and not *Carex frigida* All., no. 2334 (you have copied Allioni's error, no. 2344, when it should be no. 2334); for the *capsulis longe mucronatis* in Haller, [**Hist.** no. 1391] does not fit the plant that "differs little from *Carex sempervirens*." But it is very suitable to a specimen that you sent me labeled *Carex jacquini*, if I am not mistaken: leaves large; spikelets long peduncled, separated, blackish; scales very pointed. It had been sent to you by Monsieur Allioni. . . .[694]

8. I am sending you a blade of a uniflora grass that I am unable to determine, which Monsieur Deleuze sent me under an incorrect name.[695] He says it is abundant in the Alps,

~

[692] Chaix's complaint would seem to be justified, except that his plant could also have been *Pulsatilla montana* (Hoppe) Reichenb. (1832), which some botanists regarded as a form of *Pulsatilla vulgaris*.

[693] *Agrostis villosa* Chaix, **Pl. Vap.** 74 (1785); Chaix in Vill., **Hist. Pl. Dauph.** 1: 378 (1786), from wet fields or meadows in Vallouise. [= *Calamagrostis villosa* (Chaix) J. F. Gmelin (1791)]. The culms are unbranched, and the species stands as transferred.

[694] Haller, **Historia** 2: 195, no. 1391: *Carex spicis femininis termis quaternisque distichis, capsulis longe mucronatis. In summis alpibus.* Allioni cited this as a synonym for *Carex frigida* All.

[695] Villars has written *Aira* or *Juncea*? between the lines.

as you will see on his label; but I have never run across it and do not know where to place this specimen. Extract me, please, from the difficulty.

9. Your genus no. 19, *Arum*, 1: 158, has been quite properly omitted at the end of the Second Class in your second volume. But where have you put it? Did you forget it? or have I myself forgotten its place in the third volume? It seems to me that, according to your method, you ought to have placed it after *Ephedra*.

10. If someone is prepared to compete for the prize proposed by the Académie de Lyon regarding the family of the Stellatae [= Rubiaceae], it is surely you. I urge you to participate if you have not already resolved to do so.

128

Les Baux, 23 January 1789

You informed me in your last letter of December 6th that the painting in question would be completed by the following 15th or 20th. Having awaited it, as well as for my other commissions, for more than a month without having received anything, I am obliged to inquire of you once again. What is the cause for this delay? Has the person who was to pay for the herbarium not fulfilled his agreement? It is very unpleasant to have to wait such a long time for satisfaction. If a similar opportunity should present itself in the future, I would take other precautions. Or have the workers failed to live up to their word? or has the bad weather prevented the messengers from undertaking the shipment? If I had had my folders, I could have completed the arrangement of my plants during the severe months of the season when one cannot be busy in the field.

I thank you very much for having sent me your meteorological lessons.[696] You realize the pleasure I feel in seeing anything that comes from your hand because of the close tie that unites me to you. This work implies considerable knowledge of physics and chemistry on your part that I can only admire as being mostly above my sphere. I noted several sentences in the preface and in the preliminary discussion that could have benefited from a clear and more correct form. These defects in style mean nothing so long as your ideas are in harmony with the province of Nature. I cannot say anything more at the moment on this subject, having lent your copy to Monsieur Serre.

Our studious surgeon has just contracted a highly advantageous marriage. He has married an only daughter, the niece of Monsieur Chaix *notaire*, from La Roche, highly respected at the moment in this region, and who has great expectations of inheritance. Character responds to fortune. I have seen this alliance with the greatest pleasure. I wish a long life for them with all the blessings of Heaven.[697]

∾

[696] Villars, "Observations de météréologique et de botanique sur quelques montagnes du dauphiné." **Journal de physique** 22 (April 1783): 269-279, is one more indication that Villars had been broadening his knowledge of the sciences for some time.

[697] There is a *postscriptum* of some length entitled *Errata ulteriore ex Vol. 1, 2, et 3*, largely typographical, grammatical, and spelling errors.

129

Les Baux, 7 February 1789
By Express[698]

Four of my parishioners have been summoned to appear before the Maîtrise in Grenoble,[699] suspected of violating regulations, among whom is my nephew, Etienne Marcellin. He asked me for a letter of recommendation to you, and I have been unable to refuse him. For I do not believe him to be at fault in view of the certificate given him by the comtesse de La Roche. If, nevertheless, he finds himself in trouble, I beg you to grant him all the mediation you can before this court, whether by you directly or through your friends. I am the first to regret having to use you for such a matter, knowing that you will have to act on it as a consequence. But what can I do? In trouble, one can only have recourse to one's friends. As for the others, I am sorry for them if they are guilty. They have not made me party to their grounds for defense.

I have received your letter of January 27th, and I have sent the one meant for Monsieur Serre to him. I do not know, nor do I want to know, from where you got knowledge of Monsieur Serre's firmness in upholding his opinion. I was witness to a dispute in Durbon on a medical matter between Dr. Chevandier *père* and him. But someone has possibly given more importance to that disagreement than it deserved. Neither one of them said anything insulting. We have a saying that applies: *Peasants believe that we are fighting each other when we argue.* Said Dr. Chevandier died recently from apoplexy.

Dom Grangier, vicar at Durbon, has copied your entire history of plants from the copy belonging to the [monastic] house so as to have his own copy if the orders from his superiors should send him elsewhere. He has given to me those words and sentences that impeded him, which has helped me to construct a more complete *errata*. Following his recent careful search, I am again adding here some corrections to be made; and I strongly presume that I shall have no more to add on this subject in the future. . . .

I pray to God for your health and prosperity, and I await the shipment that you give me to anticipate.

P. S. Note: Has Haller adequately described the male flower of the alder? Séguier 2: 258, in giving each flower four stamens in a 4-lobed calyx, says nothing about a corolla. Pollich 2: 603, with his usual minute detail, indicates some scales, but says nothing about either corolla or stamens. Allioni does not sufficiently detail the parts of this fructification. J. Bauhin did not extend his research beyond that of his time. My other authors are either plagiarists or mute. This matter should require a definition by an experienced botanist.

~

[698] Villars made the following note at the bottom of this letter:
Remitted to Monsieur de La Valette 13 March 1789.

	One book	6 livres 10
76.10	Cross	20 livres
43.10	2 Folders	20 livres
——	Painting	*30 livres*
120		———
		76 livres 10

[699] The superintendence of waters and forests.

These catkins can now be found in bloom. I would like you to give them a moment of your time.[700]

130

Having occasion to write to Dom Grangier, I am taking advantage of the route he has offered me to have my letters sent on to you. I owe you a thousand thanks for the solicitude you showed for my parishioners, for my nephew in particular. You kindly took the trouble to support them in the difficulty into which their misconduct had led them. The Lord will recompense your charitable assistance.

I am very pleased with the painting and the gold cross. Keep an exact account of the total expense you have had for my commissions.

My reflections have led me to a conclusion about the mystery regarding M. . . . Enough of it! Let us draw a thick veil over a subject that a singular feature of your friendship has led you to mention, but about which your prudence has not permitted specificity. Count similarly on my discretion. . . . As for me, I shall always prefer my friends' corrections to their praises. The former will lead to my instruction, while the latter can only flatter and nourish self-esteem.

Although I have completed telling you what I can about your three volumes on the history of plants, allow me to remind you to correct the discordance that is evident between your description and figures of *Leontodon protheiformis* [Vill.] [= *Leontodon hispidus* L.]; of *Hieracium* [x] *hybridum* [Chaix ex Vill.]; and of *Hieracium halleri* [Vill.] in your *errata*. Has someone else before you given the same French name, Persian lilac, to *Philadelphus coronarius* [L.], 3: 529, line 5, and to *Syringa persica* L., 2: 7? Such misuse confuses two very distinct shrubs. I hasten to make those two additional observations, because you indicated to me some time ago that your cryptogamy was being printed, and I presume you will soon be working on your *errata*.

131

I have received with the greatest gratitude the sum of 120 livres, in money and in purchases, due me for the composition of the herbarium. You have given yourself much trouble and care on my behalf! And you obliged someone who can give you nothing in return, but who is overwhelmed by your good deeds. I am happy with everything; but the painting, no matter that a painter is to be indulged, seems expensive to me. And I see that the gilders have increased 20 sols each since the preceding ones [were purchased]. What is to be done? Everything is expensive in these times. I anticipate that I shall need still another such folder, which I shall have to ask you for in the future.

After I have assembled my herbarium in volumes, as I am putting in only one example of each species, you will realize that I shall be left with many additional examples. Therefore, try to get some advantage out of them for me. I would gladly give them up for

~

[700] *Alnus* Miller: Stamens 4; perianth 4 (-5)-merous.

2 sols each if it would only be a matter of labeling them. But for arranging them in a herbarium, as I recently did, I am resolved to accept nothing under 3 sols each. This takes into account the advantage Monsieur Serre has had over me in regard to his plants in the past, though most certainly I do not begrudge him. I would wish a sum one hundred times greater for him! So, do me still another favor, and you will increase your claims upon my gratitude.

Dom Grangier has the mind; he is hard-working and enthusiastic for botany; but he will still not be ready to make a compendium of Linnaeus. I would have scribbled something of the sort for the plants of Europe, with a diagnosis of two lines in Latin, had I not instead obtained some more plants for myself. Failing to do the above, and because of my advancing years, I am enclosing myself in my shell. When you will have completed your great work, you will be in a position to compose such a treatise for your amusement.

In a free and monarchical state, despotism has not failed to show its face from time to time, and the noble and ecclesiastical aristocracies have exerted their tyranny at the expense of their respective subordinates. May the Sovereign Judge of empires shatter the yoke of oppression at long last! *He raiseth up the poor out of the dust, And lifteth up the needy from the dunghill.* **Psalms** 113. 7.

I am quite convinced that, if indulgence has a place among the judges of the Maîtrise, your influence will work for the benefit of my poor parishioners, above all for my nephew who unfortunately finds himself involved. It belongs to humanity to extend its hand to the unfortunate one, to commiserate with him when the offense is not major and when commutative justice is not [thereby] endangered. . . .

Your kindness has moved you to invite me to visit you during the fine season. I am very touched by the invitation, and I would like nothing more if I did not see obstacles given my situation. I do not yet know what I shall be able to do. But will you not make a circuit through your native country this summer? You could come through Durbon to meet Dom Grangier personally and then direct yourself to my poor cottage. Take care of your health, and give me your news when you can.

P.S. Every time I see Monsieur Gaude, he asks me to pay you his respects. This good priest is entirely imbued with veneration and friendship for you. The death of his resigner has given him freedom, but the preceding circumstances caused him much bitterness. I think that his religion and the firmness of his soul have now pushed him beyond that situation.[701]

132

Les Baux, 3 April 1789

If I have written to you so often, it is sometimes to express my gratitude for your services, sometimes to ask you additional ones, often to unload difficulties upon you that had balked me, other times to solicit clarifications that I needed; but always because my heart's inclination committed me to write, besides being assured that my correspondence does not displease you. After this prelude, I beg you to repond about the matter I am going to raise here.

~

[701] See letter no. 100. The prior of Furmeyer had resigned from his parish in favor of François Gaude, curé d'Oze, in 1786, but not without unpleasant complications.

Monsieur [Pierre-Marie-Auguste] Broussonet, permanent secretary of the Société royal d'agriculture in Paris, has just written me and sent me the printed statement about the proposals of this worthy society. I have no doubt that they should have preferably been sent to you: **What are the surest ways to obtain new varieties of useful plants for the rural economy?** I do not know whether the society means by the term *varieties* what we understand it to mean in botany, or whether it means different species of plants. If you are certain about their usage of this term, settle the matter for me. If not, ask the said Monsieur Broussonet, or I shall ask him about it myself; so that, if I undertake to scribble something about the above, I will not deviate from the proposed topic.

Has the Société littéraire de Grenoble judged the papers on the elimination of beggary in Dauphiné, the topic proposed last year? If so, could you let me have a brief report on the results? If the session has not yet been held, will it be held soon?

The abbé Pourret has asked my opinion about your *Rubia peregrina* [L.] that he believes to be identical to the plant found on the mountains around Narbonne with oval leaves, which I brought back from Le Buis. (*Rubia tinctorum* [L.], he adds, grows naturally in that region.) I think that your *Rubia peregrina* with *large* and *oval* leaves is not Linnaeus's *peregrina* with linear leaves. I would have referred it to *Rubia lucida* L. if the keel of its leaves had been scabrous, and if Monsieur Pourret had not assured me he had brought *Rubia lucida* back from Spain, very different from ours in question. Therefore, I agree with him that the latter is a species that is not in Linnaeus. You will return to it before completing your *errata*.[702]

I say the same thing about your *Linum perenne* [L.] from Grenoble, which has always appeared to me to be an annual, and which I suspect is either *Linum humile* [Miller] [= *Linum usitatissimum* L.] or a variety of *Linum usitatissimum* [L.]. I have cultivated a very different species under the name *Linum perenne*; stalks upright, robust, three feet tall, with large blue flowers. I would prefer to refer to *Linum perenne* a perennial species we have here in our meadows and low pastures that I can hardly distinguish from our *Linum alpinum* [Jacq.] [= *Linum perenne* L. ssp. *alpinum* (Jacq.) Ockendon (1967)], but which is not the plant that I collected at Grenoble with you. At least, this plant from our low hillsides in not *Linum austriacum* L. I have cultivated flax sent from Paris under the incorrect name of *Linum austriacum*, which are only varieties of *Linum usitatissimum*.

I conclude in haste to take advantage of an opportunity to have this carried to Durbon.

133
Les Baux, 6 July 1789

Deprived of any news from you since April 20th, you may judge with what eagerness I received and read your last letter of June 28th written from Voreppe. In it, you give me a precise, but detailed, account of what happened in Versailles from June 17th to 23rd

---

[702] *Rubia lucida* L. is now held to be a synonym of *Rubia peregrina* L., its leaves in whorls, varying greatly from linear to ovate-elliptical.

inclusive.[703] I thank you for it, but I was not totally unaware of those events. Thanks to my friends in the vicinity, I am acquainted a bit with the interesting news of the time. The conflict between the two powers, royal and national, arouses my interest very strongly. Uncertain as to the outcome, I am pleased with my rural and inconspicuous lot. . . .

Can you not yet give me some news about the second part of your third volume on the plants of Dauphiné? Is it still not entirely printed? Your subscribers, I in particular, are very eager to have it. What is your current state relative to the Brothers [of Charity]? What is your situation relative to the new administration? How are the members of your family, especially your dear wife? Please extend my regards to her.

You are not unaware that our good friend, Dom Grangier, has left Durbon for La Verne near Toulon where, again, he will be vicar.[704] Although I had been expecting his transfer for some time, his departure wrung tears from me. He shed some himself, having to be removed from his association with you, and from mine I dare add. The ways of Heaven sometimes deprive us of appreciable consolations in order to purify our affections. The will of God be done! I have just spent two days with the holy monks of this Chartreuse. You are not forgotten there; and our good friend has been replaced by a worthy successor, a young man of 32 years, a native of Nîmes, who overwhelmed me with politeness and friendliness. I must pass on to you the affectionate greeting of my dear confrere, Monsieur Gaude, whom I saw on my way to Durbon.

Monsieur Serre remains quite close to me, and his reputation spreads more and more. It is he who obtained a great number of brochures about contemporary affairs for me to read. A few days ago, he passed on some greetings to me from you, and I entrusted him with a small piece of paper with two or three observations [for you].

134
Les Baux, 23 July 1789[705]

As you tell me in your last letter of July 10th that you are occupied in preparing the *errata* for your history of plants, I am sending you one or two more observations.

1. Your *Potentilla inclinata* [Vill.], which you have engraved very well with petals cordate or emarginate, is as I have always seen them. Why then have you said *truncate and not cordate* in your description in 3: 567? Remove those first three words.[706]

2. *Potentilla argentea* [L.]. I have cultivated it, and I have observed it in several places. The stalks have never appeared to me to be strict, as you describe them in 3: 571, but procumbent at first, later ascending, what Linnaeus has expressed as *caule erecto* (stalk

<hr/>

[703] That was the critical week in 1789 when the paralyzed Estates-General was converted to the National Assembly despite royal displeasure:

17 June: The Third Estate declared itself to be the National Assembly.

19 June: The First Estate voted to join the National Assembly.

20 June: The Tennis Court oath, a pledge to remain until a new constitution could be framed.

23 June: The king's opening speech; his Declaration concerning the Estates-General; his Declaration of Intentions; and his closing speech.

[704] The Chartreuse de La Verne was actually in the Massif des Maures, northwest of La Môle.

[705] Before this date, Chaix had habitually addressed Villars as *médecin du Roy à l'hôpital militaire de Grenoble, professeur de Botanique près du Collège royal*, with some variation. Beginning with this date, the addresses became strikingly shorter and stripped of political connotation: *A Monsieur Villars, Docteur en médecine, près le Collège.*

[706] Chaix was right. The petals are emarginate.

inclined at a high angle). He did not say, contrary to the truth, *caule recto* (stalk strict). Therefore, instead of stalk *strict*, read *raised* or *ascending*. You assign it cordate petals. They do not have that form but are simply obtuse and smaller than those of *Potentilla inclinata*. The two species grow near each other, along with *Potentilla intermedia* [L.], at Berthaud [La Crotte] near my sister's house and elsewhere.

3. In 2: 258, you have combined *Allium senescens* L. and *Allium angulosum* L. Following Haller and Scopoli, I believe that Linnaeus did not do wrong in separating them.[707] I have the first of these in my garden, which came from seeds sent from Paris under the name *Allium senescens* Rottboell, and which is a giant compared to *Allium petraeum* J. Bauhin [**Hist.** 2: 564] [= *Allium angulosum* L.] of our mountains. You give your *Allium narcissifolium* [Scop.] indiscriminately as native to Prémol, in Oisans, Champsaur, and Briançon. I have never found *Allium senescens* Rottboell in those last two places, and I do not have enough from Prémol and Oisans. Monsieur Allioni, in his nos. 1879 and 1880, has crossed the synonyms reported by Linnaeus. For he has described *A. angulosum* L. under the name *A. senescens* L., confirmed again in no. 1880 where he says that *A. angulosum* is the larger in all of its parts. That is false and contrary to the synonyms of both Bauhins that he cites.

[Chaix enclosed three dried specimens each of Potentilla argentea L. and P. inclinata Vill. in this letter.]

135
Les Baux, 12 September 1789

As I am [using] your method for the arrangement of my botanical folders, I have had the opportunity to compare plants I am unsure about with your descriptions. Do not be surprised, therefore, if I quibble with you from time to time. Here, again, are some of my cavils.

Filago arvensis [L.] [= *Logfia arvensis* (L.) J. Holub (1975)] in Vill. 3: 195: After having adopted *caule ramosissimo* from Haller, *floribus per caulem sparsis* from Tournefort, and *caule paniculato* from Linnaeus (and that is how I have collected it at Aubessagne, at Molines [-en-Queyras], and at Manse de St.-Julien [-en-Champsaur]), why do you describe it with a *stalk usually simple and a bit branched above*? Remove those words and substitute *stalk very branched, in the form of a panicle* for them. Could you have confused it with *Filago montana* [L.]? The latter does indeed have the leaves and the flowers you attribute to it; but its stalk, instead of *branching out immediately from near the root*, 3: 194, only divides dichotomously above. Also, has not Murray, **Syst.** (ed. 13), p. 662, substituted *subdichotomo* from Haller, **Hist.** no. 155, for the *diviso* in Linnaeus? I have collected it above the Forest de St. Julien....You will give me great pleasure if you can send me a specimen of *Filago pyramidata* [L.] and *Filago gallica* [L.] [= *Logfia gallica* (L.) Cosson & Germ. (1843)] that I do not have.

The seeds of *Sonchus plumieri* [L.] [= *Cicerbita plumieri* (L.) Kirschleger (1852)], which you sent me through Monsieur Serre, do not appear to be mature. I have recovered

---

[707] Villars had made *Allium senescens* L. and *Allium angulosum* L. varieties of *Allium narcissifolium* Scop. The latter species is now held to be a synonym of *A. senescens* L., while *A. angulosum* L. is still a good species.

from the loss as Monsieur Pourret has sent me a passable specimen of it from the Pyrenees. You wrote on the label *Sonchus or Lactuca*. If one makes a lettuce of this plant, the same thing would be necessary for *Sonchus alpinus* [L.] [= *Cicerbita alpina* (L.) Wallr. (1822)].

Hypochoeris uniflora Vill., [**Prosp. Pl. Dauph.** 37 (1779)]. Even though Monsieur Allioni's figure does not show the calyx bracts, well represented in your figure, his description is precise. For that reason, it seems to me you ought to cite him, as he preceded you in the publication of his work, although your prospectus was earlier.

The *Anthemis* from Paris, about which you speak in your footnote, 3: 252, has always been sent to me by Monsieur Thouin under the name *Anthemis nobilis* [L.] [= *Chamaemelum nobile* (L.) All. (1785)]. And I am quite led to believe that, even though it is an annual as opposed to the perennial character given by Linnaeus, and though it only differs from our *Anthemis arvensis* [L.] in its taller and more upright habit. It has even produced the double-flowered variety for me as reported by Linnaeus.[708]

I hasten to send these brief observations to you, because Monsieur Serre told me that the printing [of the remainder] of your third volume is nearly completed, and that we shall soon take delight in it. I have read the estimable letter you wrote to Monsieur de Jussieu with much pleasure and great satisfaction. I thank you for its communication to me on Monday the 21st. I hope to go to the home of the curé de La Bâtie-Vieille [Villar] to discharge the obligation of a visit that has been near to my heart for a long time. Perhaps I shall have the pleasure of seeing your adorable children there, a partial compensation for not likely finding you there.

P.S. More to be corrected and added in Vol. 3:
p. 416: *Dorycnium suffruticosum* [Vill.], *caulibus erectis*, stalks nearly crawling.
p. 417: *Dorycnium herbaceum* [Vill.], *caule diffuso*, stalks upright. Has not Monsieur Villars inadvertently transposed the description from one to the other? Could he have been misled by the specimens collected in Savoy and sent to me under the name *Dorycnium herbaceum*, when they are really *Dorycnium suffruticosum*? A word of response, please, regarding my observation of these two species.[709]

136

Les Baux, 6 October 1789

I never put my hand to pen more wholeheartedly than when I write to you. I then sense in myself an excitement that cheers and enlivens me. Thus, I seize every opportunity that presents itself to give myself this satisfaction. A few new observations provide me such an occasion, which will no doubt be useful to you if the printing of your *emendanda* is not completed.

1. In 3: 304, line 16, no. 2. Instead of *Biscutella didyma* [L.], this should be called *Biscutella laevigata* L., **Mant.** 255; because *Biscutella didyma* in **Sp. Pl.** has been divided

⁓

[708] *Anthemis arvensis* L. varies in height from 1-5dm.; it is an annual; and it may be erect, ascending, or spreading.
[709] *Dorycnium suffruticosum* Vill. = *Dorycnium pentaphyllum* Scop. ssp. *pentaphyllum* (1772); stalks woody, twisted and crawling; from Serres, Le Buis, and Die.
Dorycnium herbaceum Vill. = *Dorycnium pentaphyllum* Scop. ssp. *herbaceum* (Vill.) Rouy (1899); stalks nearly herbaceous, upright or ascending; from Savoy.

into several species, and as *Biscutella apula* L., **Mant.** 254, should be *Biscutella didyma* [L.], very different from your species because of its annual root, leaves noticeably serrate, flowers in a spike, and its scabrous silicles. You believe yours, instead, to be perennial, as do Allioni and I (and I speak knowledgeably here, because I have cultivated *Biscutella apula* and possess a good example of it): its leaves are only broadly dentate, the teeth obtuse; its flowers are in umbels; its silicles are smooth.

2. In 3: 305, line 4. Remove *An B. apula? Syst. III, 240* in view of the above.[710]

3. In 3: 306, line 7 [*Biscutella coronopifolia* L.]. For *siliculis hispidis* read *glabris* as with Linnaeus, **Mant.** 255. The silicles of this species are certainly glabrous on my specimens, which I am sending to you enclosed to convince you of it. You will then understand that your expressed difficulty on line 24 is unfounded. . . .

4. In 3: 296. Your observation on the fruit of the *Clypeola* and of *Alyssum minimum* [L.] [= *Alyssum desertorum* Stapf. (1886)], in order to distinguish these two plants, is very judicious. Without your comment about glabrousness and roughness, I should have been tempted to unite them.

5. You must have already received a note I sent to you recently ragarding *Filago montana* and *arvensis*.[711]

6. In 3: 342, line 2 [on *Sisymbrium*]. . . .[712]

7. In 3: 348, line 13, before *annual* add at St.-Bonnet, at Aiguilles in Queyras. [For *Sisymbrium sophia* L.]

When you next see Monsieur Liottard, extend to him my sincere affection and my gratitude. I have many thanks to give him regarding the plants he gave to Monsieur Serre for me according to your son's statement. I am a bit irritated at Monsieur Serre for not having told me that Monsieur Liottard had remembered me. What is more, I regard the plants that my worthy neighbor has at his home to be at my disposition, just as he has assurance that he has full right to anything I possess. This oversight on his part will have no effect on our association. Take care not to say anything about it either to Monsieur Liottard or to Monsieur Serre. I saw your lovable children at the home of the curé de La Bâtie-Vieille along with your beloved mother. They assured me that you had resolved to come to see me, but I am beginning to lose hope of that pleasure during this season. I strongly presume that your multiple occupations, and the uncertainty of public affairs, will not permit it. . . . When you mean to honor me with some response, and you do not have any convenient facility, use the regular post. The expense for the carriage of anything that comes from you is always very light for me.

137
Les Baux, 22 December 1789

Since you are willing to undertake my commissions, I am sending you the botany books I want to have bound; and I am enclosing here a separate note of directions for the

---

[710] This refers to *Biscutella longifolia* Vill. that Verlot recognized long ago to be only another form of *B. laevigata* L. [ssp. *laevigata*]. The species is considered to be extraordinarily variable.

[711] See letter no. 135.

[712] This paragraph reiterates why he wants *Sisymbrium neglectum* Chaix published in the *emendanda*. See letter no. 123.

bookbinder about how I want my books to be bound and entitled on their backs. Would you please urge him to pay attention to the order that I have prescribed for him to follow and not to transpose any leaves; and to bind everything properly and firmly.[713]

You note quite correctly, Monsieur, that the herbarium folders that you obtained for me last year contained more than three quires of paper each for the price of 10 livres each. For each one contained 210 leaves, that is to say, 105 sheets, which comes to four and more quires in each one. Therefore, will you please get me two more of them of the same quality and quantity of paper as those of last year, expecting me to pay the same price for them? .

The work by Monsieur de Jussieu [**General Plantarum**] is amazing in the great number of genera that he reports and in the sagacity of his proposals. What observations! What combinations! It is very well written in Latin, faithful to the style of this science. The author deviates as little as he can from the terms and the precision of Linnaeus, not yet being able to envision everything, or to be certain of everything, within the empire of his laws, and is often obliged to waver. This is perhaps the most brilliant work that has appeared to date on this subject. But the beginners' progress will be very slow if, deprived of any artificial method, they have no other teacher than this book, otherwise incomparable.[714]

As for the confiscation of church property,[715] I say that the Levites, in the Mosaic law, had no share in the distribution of the promised land; that Jesus Christ, author of the evangelical law, having only wanted a place to lay his head, deprived those who fell in behind him to become his ministers of their property. Showing the banner of poverty to ecclesiastical beneficiaries, under which they had to enlist when having themselves entered into the clergy, reminds them of their origin. If you want to take the trouble of retrieving from Monsieur [Henri] Gagnon my memoir on the elimination of mendicity, with the title **Beatus qui intelligit super egenum et pauperem**, you will find in it the wishes I was expressing two years ago regarding the reform of abuses, and about which we are now witnesses. *But having food and covering, we shall be therewith content.* I Timothy 6. 8.

I have let Monsieur Serre know your opinion regarding his dried plants, along with your interest in the state of the pregnant woman who you visited in La Roche. There is likelihood that our surgeon will establish himself at Veyne, where he has been residing for

~

[713] Chaix had told Villars earlier in the year that he would not send his books to be bound until he had received the second part of Villars' Vol. 3 on the cryptogams. As he does not thank Villars here for the reception of the terminal fragment, it would appear that there is at least one letter missing between 6 October and 22 December 1789, the period in which that fragment had to have been published.

[714] Chaix was in particular prepared to accept Jussieu's natural system as he had already opted for Linnaeus's rudimentary natural method when compiling his catalogue for the plants of Gap.

[715] Decree Confiscating Church Property, 2 November 1789, a financial expedient by the National Assembly to provide revenue for the bankrupt state: The National Assembly decrees that,

    1. All ecclesiastical property is at the disposal of the nation, upon condition of providing in a suitable manner for the expenses of worship, the maintenance of its ministers, and the relief of the poor, under the supervision and according to the instructions of the provinces;

    2. That, in the provision to be made for the maintenance of ministers of religion, *not less than 1,200 livres per annum* be assured for the endowment of each and every living, exclusive of lodgings and gardens pertaining thereto. See J. H. Stewart, **A Documentary Survey of the French Revolution** (New York: Macmillan, 1951), pp. 158-159.

a few days already with his wife, quite dividing the assets of his fortune, whether to forestall a [potential] rival to his position or because of the troubles he has experienced in La Roche. I doubt whether this course will be more advantageous for him. His ecclesiastical brother, who had been admitted in philosophy at Gap, defrocked himself. After a disappearance of about two weeks on a trip to Provence, with the pretext of enlisting in the military service, he has come back with the expressed intent of directing the tiny school in my parish. What a choice! [*Each man hath his own gift from God,*] *one after this manner, and another after that.* **I Corinthians** 7. 7.

If Monsieur Serre was unable to fill entirely the orders of those who want alpine plants, I could supply him with many of the species that he certainly does not have; because he has botanized very little in the Alps. By no means do I want to take his place. I merely offer to make up for deficiencies. Monsieur Pourret had promised me some plants he owes me, but I am not banking on getting them.

The widespread alarm within the kingdom during the later days of July was disastrous for many pregnant women.[716] One of my married nieces in Rabou, who was at my home on the day of your visit, and whose husband came to consult you the following day about his unsettled health, died on November 2nd during an unsuccessful childbirth, having experienced great disturbance in her pregnancy following those alarming threats. She has left four very young children without any other woman in a house that is otherwise quite suitable. This loss has been very saddening to all her relatives and to me. May God be blessed! Another woman from Corréo is in the same danger from the same cause, and so on.

You must urge Monsieur Duvall, your admirer for good reason, to have copies of your work sold in England, Switzerland, and other countries where he has correspondence. It is surely appropriate for you to defer an appendix for *errata* and additions. Time will contribute to a greater perfection of the edition. Take good care of your health.[717]

P.S. If I can come to Grenoble next spring, as you have had the kindness to invite me, I would certainly take great pleasure in leafing through your herbarium with you; and you would have many things to give me. Whatever may happen, I shall draw up a list of plants for you that are reported in your work and that I do not have, once I can go through my bound copies again. In the meantime, if any mosses or lichens come into your hands, have the goodness to send them to me. I enclose here six species of *Lichen* that I beg to you determine for me when you do me the honor of writing me.

Rursus P.S. I have just learned that the woman from La Roche, about whom you wanted news, is now cured. But if you know what has become of Dominique Bonnardel, son of my neighbor Antoine Bonnardel, give us a word about him. His relatives, and in particular his very infirm poor father, are very anxious for news of him, having heard nothing for more than two years. It would be a charity on your part.

~

[716] A reference to the peculiar panic that swept over parts of France after the fall of the Bastille, now called by historians the Great Fear of 1789. It is briefly and well described in J. M. Thompson, **The French Revolution** (New York: Oxford University Press, 1966), pp. 88-89.

[717] We have indicated earlier that the projected supplement would never be published. This is the first letter that links that deferral to the need to sell more copies of the flora: to financial reasons.

Monsieur Serre, after the advice I passed on to him from you, . . has confirmed the complete recovery of the woman in question; and he promises to send you his observation of this illness from here by Twelfth-Day.

I reread your paper ["Observations sur une fièvre épidémique qui a régné dans le Champsaur et le Val Gaudemar" (1781)] with much pleasure. Could you not send me the paper you presented to the Société royale de médecine de Paris that won you the title correspondent?

Note: When you send back my bound books to me, please take care to give them to Monsieur Clavel, one of the messengers from Gap, so that the same person will be the carrier each way. He will also bring you the money that I shall owe you.

138

Les Baux, 10 February 1790

I must acknowledge reception of your last two letters, one from the end of December, the other from the end of January. I thank you for the demonstration of your kindness and also for your verdict about my uncertain *Lichens*. If the binder's illness is the cause for the delay in the shipment, I should be unreasonable to seek prompter service. I hope he will devote himself to the work I gave whenever it will be possible for him. Meanwhile, I have borrowed your third volume from Monsieur Serre for diversion during my leisure hours. I declare to you, and without any flattery, that the more I read your work, the more I admire the learning, the judgment, and the correctness in it. It is entirely proper that you will defer publishing a supplement or *emendenda*. The passage of time will provide us with new material.

I am sending you my opinion of your *Genista humifusa* [Vill.][718] and beg you to tell me your feeling about it. Monsieur [Jacques] Meyer, [now] curé de Lachau in the Baronnies, had sent me some plants to determine for him. The said *Genista* was among them, as well as *Chrysanthemum monspeliense* [L.], *Briza uncialis* Chaix,[719] *Melampyrum cristatum* [L.], *Jasione montana* [L.], *Achillea ptarmica* [L.], *Centaurea pectinata* [L.], and *Ruta chalepensis* [L.]. My eagerness to see this good friend, and to botanize in his country, may lead me to go down there with your brother.[720]

Dom Grangier has written to me from the [Chartreuse de] la Verne [in the Massif des Maures], sending a box of pretty plants with roots for replanting. But a delay of two and a half months [in transit] has left me a resource beneficial to my herbarium alone. Those that would have given me the most pleasure were *Styrax officinalis* [L.], *Arbutus unedo* [L.], *Cytisus montspessulanus* [L.] [= *Teline montspessulana* (L.) C. Koch (1869)], and *Cistus monspeliensis* [L.]. This good monk and warm friend speaks to me about the uncertainty of his situation. I have responded to him about present circumstances. In asking me for news of you, he charged me to assure you of his continuing friendship. If he should reenter

⌒

[718] Villars had listed it in both **Prosp. Pl. Dauph.** 41 (1779) and **Hist. Pl. Dauph.** 3: 421 (1788). [=*Genista pulchella* Vis. (1830)].

[719] An unpublished species, later found by Verlot, **Cat. Pl. Dauph.**, p. 370, in the notes for Villars' unpublished supplement. Verlot believed it to be *Eragrostis minor* Host. (1809).

[720] Jacques Meyer earlier had been vicar in La Bâtie-Neuve, a neighbor of Jean-François Villar, the curé de La Bâtie-Vieille.

the world, I should see him in Les Baux, and you in Grenoble.[721]

My brain is too aged and too limited to have Monsieur de Jussieu's method put into it. I have to be content simply to admire his prodigious work encompassing about 2000 genera. Is one head alone capable of such a task?[722] His style is very correct, laconic, and Linnaean. He has reproached you for frequent exceptions required to use your method; but the introductions to his classes and orders are filled with *vel, aut, sive, seu, nune, modo.* These are labyrinths, therefore, where Ariadne's threat would be quite necessary. His genera are hardly different from those of Linnaeus. I would set more value on his secondary characters. This great man has shed excellent light upon the field of botany; but it will require time to expand upon the principles he has set down and to test the insights.

You previously told me in a letter that some day you would expose the shameful plagiarism by Monsieur de Lamarck, recorded in his **Encyclopédie** [**méthodique: Botanique**]. He merits it, but will your modesty permit you to do it? Yes, if the public interest requires it.

As Monsieur Serre must go to Grenoble, I am sending him your Jussieu to carry it along with 24 livres to you. Thus, you will have the total amount for the two herbarium folders and for the binding of my four books in-octavo. kindly acknowledge receipt of the money and settle my doubts about the items under question herein. . . .

139

Les Baux, undated
Between 21 February and 17 March 1790

As you had the kindness to correct my bundle of algaes last year, I beg you to do as much for my mosses that I am sending you, so that I do not put the species in this family at risk when I put them into my herbarium. You need not send back the specimens to me, as I have them all in my collection, but simply correct them by letter. It appears to me that Gérard found many of his plants in his books on his desk, especially the cryptogams. I do not want to do as he did. . . .

It is said that an avalanche has covered, and possibly destroyed, eight or nine houses in the Valgaudemar. Many people, no doubt, will have perished. I know neither the circumstances nor the place that has had this unfortunate occurrence. On Sunday, February 20th, the eldest son of my nephew from La Grangette, whose mother I told you had died [in November], was returning from holy mass in Rabou to the home of his maternal grandmother, where he is staying for school, when he and a companion were covered by an avalanche a short distance from the church. They were only pulled out an hour after the snowslide thanks to the work of twenty men. They were buried three feet under the snow, where they would have suffocated if they had remained there another half hour. Fortunately, the father of my grand-nephew's companion, who was following them at some distance, was not enveloped in the snow and could call for help. It is greatly to be feared that many such mishaps will occur in the mountains. May god have mercy on his poor people! There are many places where the people have not been able to grind in order

∾

[721] Although the National Assembly, when confiscating all ecclesiastical property, promised to provide financial support for public worship, they held out no future for the monastic communities. All religious houses would not be declared sequestered until 18 August 1792.

to make flour, where forage is lacking for livestock, and without the possibility of bringing in supplies from elsewhere. The city of Gap has had a shortage of wood and no means of transport for more than three months. No one remembers a winter as difficult as this one. The will of God be done.

[The undated fragment was perhaps mailed with letter no. 140.]

140

Les Baux, 18 March 1790

As our friend Serre had to go to Grenoble because of his brother, I had given him 24 livres to take to you along with your copy of Monsieur Jussieu's book. But matters turned out otherwise, and I got my parcel back. I put off writing to you until receiving the books that you had the kindness to have bound. All has worked out for the best; I am very pleased with the results, for which a thousand thanks. If, in order to settle the 24 livres 6 sols it comes to, the plants that you mention to me will suffice, I shall be under a double obligation to you. If not, I will send you said sum when advised by you. Perhaps you alone will take some of the excess in my collections. Monsieur Pourret, who is never short on promises, has not responded following the shipment I made to him last autumn, which passed through your hands. Can you give me any news of him? Could he be sick? Or could he have gone to join his cardinal?[723]

When reading from my other letter herein enclosed, have the kindness to excuse me for not having taken the trouble to redraft it in this one. You will have to ignore the passage in it regarding the parcel.[724] When your duties permit you to think about my note regarding your *Genista humifusa* [L.], please have the kindness to tell me your opinion of it, as well as of my *Briza uncialis* [Chaix], and to determine the *Lichen* that I am sending you, which is very common and appears to be to be quite different from *Lichen stellaris* [L.].

I turn next to the investigation in the Linnaean herbarium that was communicated to you by Monsieur Duval:

1. Whatever the *Sonchus canadensis* [L.] and the *Sonchus alpinus* [L.] from Lapland may be, our sonchus is the same as that of the earlier authors, none of whom had been either to Canada or to Lapland. If the description of Linnaeus is misplaced, your description and your synonymy are precise.[725]

2. If you accept the change on some *Arenaria*, the inconvenience is not great; and your hesitant adoption attests further to the sagacity of your discernment. But, if your *Arenaria*

〜

[722] In fact, as readily acknowledged, Antoine-Laurent de Jussieu had built his remarkable system upon a foundation worked out by his uncle, Bernard de Jussieu.

[723] The reference is to Archbishop Loménie de Brienne who had fallen from power in August of 1788 following the failure of his financial reforms. In December he was made a cardinal and departed for Italy. Abbé Pourret, in charge of Brienne's botanical collections, did not follow him to Italy. In 1790, however, he deemed it prudent to emigrate. As a native of Narbonne and a specialist in the flora of the Pyrenees, Spain was his logical choice for exile, first in Barcelona, then in Madrid, and finally in Orense. There, at the time of Napoleon's invasion of Spain, angry patriots would burn Pourret's books and collections. He fled to Santiago where he died in 1818.

[724] This would be letter no. 138, evidently not mailed when written in February. It seems probable that no. 139 was an annex to it.

[725] This obscure passage seemingly referred to one of several disagreements with Linnaeus that Villars published in his treatment of *Sonchus*, Hist. Pl. Dauph. 3: 157-162 (1788).

juniperina [Vill.] is *Arenaria grandiflora* L., what then is *Arenaria juniperina* L.?[726]

3. If your *Salix helvetica* [Vill.] is *Salix arenaria* L., what would be the *Salix arenaria* and the *Salix lapponum* in your flora at the Linnaean herbarium? for we do not have *Salix lapponum* L.[727] I should like to be instructed about this matter. Have you any further notions about *Salix amygdalina* L. and *S. fragilis?*

Even though Linnaeus is the prince of botanists, he is not the God of botany; and we must not, like a Gouan, kiss even the tracks of his aberrations. The synonyms reveal more than an inappropriate practice. The justice rendered you by the English gentlemen [Duvall] is merited. You have indeed sorted out many difficulties that have kept botanists in trouble. I say this without meaning to flatter you, and I can say it because of the knowledge I have of certain plants. Even though eagles have very penetrating eyes, their lofty flight prevents them from fixing upon many objects that the tiny kinglets observe perfectly.

The announcements we receive from Grenoble keep us informed about the principal affairs of state. Although anticipating the fate of the regular clergy, I could not read about the general suppression of all the orders without trembling.[728] I should have much preferred that they had all been reduced to work, to withdrawal, and to poverty in accord with their original institution such as the orders of Sept-Fons and La Trappe, and that the door should not have been closed to those who should want to embrace that life. Agriculture and the arts could have gained from that [solution], and society would have found in them asylums for a sometimes inconvenient population.[729] If all the simple benefices, which offer the state nothing but the vain glory of a title, had been suppressed, that would be the reform of a great abuse. We are frightened by rumors of the bankruptcy of [state] finances. I very much fear some violent attempt upon Alsace.

It is said, and the matter seems very probable, that parishes with fewer than forty active inhabitants will be abolished. As a consequence, mine would be among that number, in which case I know not what my fate would be. But I hope they will leave me bread for the remainder of my days, either by employing me otherwise, or by granting me a pension for necessities. The vicissitudes of things human increasingly warn me to cast my lot with the permanent state of the other life.

[Note enclosed from Dr. Serre to Villars.]

Please have the kindness to send the parcel from my brother at the first opportunity. My wife's advanced pregnancy is the reason I have postponed the trip about which I spoke

~

[726] *Arenaria juniperina* Vill., **Hist. Pl. Dauph.** 3: 624 (1788), = *Arenaria grandiflora* L. But it appears from Villars' description of the lower leaves being *fasciculatis* that he was describing what is now var. *incrassata* (Lange) Cosson. As for *Arenaria juniperina* L. [= *Minuartia juniperina* (L.) Maire & Petitmengin (1908)], it is limited to SE Europe.

[727] *Salix helvetica* Vill., still held to be a good species, is treated within the *Salix lapponum* group.

[728] The Decree Prohibiting Monastic Vows in France had been adopted on 13 February 1790. It had come to be commonly believed in the 18th century that the regulars had no utility for society, but that the secular clergy exercised a useful, necessary function.

[729] Even though many of the religious houses had become decadent by the 18th century, Chaix's statement reflects a knowledge that monasteries had often managed their lands and forests conservatively, as at the Durbon he knew so well. He was also aware that monastic institutions had traditionally accepted and sheltered social misfits and outcasts.

to you in one of my earlier letters. Perhaps I shall go to the Federative Assembly.[730] But, meanwhile, let me have your news when you send the parcel, and please include an account of my brother's expenses in Grenoble. For I believe that I am obligated to you, and it would grieve me to have you be unrecompensed.

141

Les Baux, 8 April 1790

As the price for your work is marked for each volume separately, it is pure knavery on the part of the bookstores to have been unwilling to add up the three sums to get the total for each set. Either they must give you an exact accounting for what you claim, or they recover your six sets for you. Their procedures are dreadful. For a long time, I have regretted that you have been distributing so many copies gratuitously, mostly to people who will only read the preface. I am truly sorry about this liberality from your excessive kindliness. You frightened me by telling me that "you have paid 8,000 livres to cover the costs of printing and engraving, and that you still have 4,000 livres and more to pay." That comes to more than 12,000 livres! When will your sales bring you such a sum? If the government does not recompense you, you yourself will have paid for your labors and your night hours. That troubles me greatly. I had come to believe that the total would not surpass 5,000 livres, and I think that you had so informed me. Can you not tell me something that will ease my mind?

Observation on *Genista humifusa* Vill. (*verrucosa* Chaix).[731] I possess three specimens of this plant. 1) The one I have from you is more pubescent, its leaves more abundant, and slightly larger. 2) The one from Monsieur Danthoine, collected in Provence, is nearly glabrous and nearly nude, the shoots terminating in more noticeable spines. 3) The one from Monsieur Meyer, collected in the Baronnies, the climate midway between yours and Monsieur Danthoine's, and, as a consequence, being a habitat midway between one and the other. I see in the three only accidental differences caused by a warmer, drier, or more unfruitful climate, and *vice versa*. But their shoots always terminate in a true spike, more or less pronounced depending upon time and place; their verrucae or tubercles are constant; their flowers and fruits are the same. Monsieur Danthoine's specimen is perfectly represented by the minor figure of *Genista pungens* in Chabrey [**Stirpium icones**], p. 86. The figure of your *Genista humifusa* [Vill.] depicts the spine clearly, at least on a shoot; but the tubercles were forgotten. As for the major figure in Chabrey, *loc. cit.*, I think it is *Spartium purgans* [L.], described in Gérard, which I have, having received it from Monsieur Pourret, and which gives me the true character of *Spartium*.[732] Does the one you have

～

[730] The federations were local patriotic leagues formed in many cities to assume responsibilities left unattended as royal officials abandoned, or were driven from, their posts in 1789. Chaix had indicated in earlier letters that Serre was keeping abreast of political events. This is the first indication of his active participation that would culminate in his election to the Convention in 1792.

[731] See letter no. 138. Chaix had sent a separate list of questionable plants to Villars with that letter. Chaix gave his own (unpublished) name to Villars' *Genista humifusa*, namely, *Genista verrucosa* Chaix, probably to emphasize the absence of verrucae on Villars' figure. The list was seen by Grenier, **Flore de France** 1: 351 (1848).

[732] *Spartium* is unarmed.

possess the calyx of a *Spartium*? So, the difficulty that Gérard's description[733] gave me is overcome, and I therefore adhere to my opinion on the subject of the *Genista* in question.[734]

The Académie Delphinale has honored me with the title of *Associé libre* (Honorary Member). I am embarrassed to see my name among those of so many illustrious and learned people of distinction; but, it having been done, I must report to you the honor that has been done me. If I can contribute in some manner to the work of the Académie, I will not be negligeant. Consequently, I have written to Monsieur [Henri] Gagnon, who did me the honor of writing to me when sending the formal letters of affiliation.

Write to me as often as you can. I experience no greater pleasure than hearing you converse through your letters and to respond subsequently in turn. It is the only compensation for absent friends and lessens the cruel necessity for their remoteness.

142
Les Baux, 24 June 1790

The share that I have in everything that concerns you, and the closeness of our ties for the past twenty-five years of our acquaintance, do not absolve me from expressing my condolances to you. The crushing blow that has struck you had already grieved me even before you, as the fatal news reached me on the day of the sad event. The dear curé confrere wrote to me immediately, bathed in his tears.[735] What is there to do? That is the command of providence. For a longtime, I have been subject to similar deprivations. Our relatives precede us, and we shall soon follow them. *Thus*, the apostle says to us, *let us not grieve as those who have no other hope.*

I enclose still one more commentary here, perhaps the final one I shall make on your botanical books, which you may add to those I have transmitted to you other times on the same subject to be of use to you if you should publish a *mantissa*. In that event, and if I am still alive, will you try to have me read your draft before sending it to the printer? My revision should not injure the work. One eye alone, when focused on so many subjects as yours, cannot absolutely encompass all points of view.

Enlighten me on the subject I ask you about at the end of my above said memoir, and write to me when your duties will permit.

P.S. Next Sunday we are going to name our electors for the Administrative Assembly.[736]

---

[733] *Spartium foliis simplicibus sessilibus ramis inermibus.* Gérard, **Galloprov.**, p. 480, no. 3. [= *Cytisus purgans* (L.) Boiss. (1839)].

[734] *Genista humifusa* Vill. = *Genista pulchella* Visiani (1830).

[735] This refers to the death of Villars' mother in Le Noyer. The news would have reached Les Baux more quickly than Grenoble.

[736] The reorganization of local government, to make it more rational and decentralized, and to replace traditional names of historical use with names from Nature, became one of the enduring legacies of the French Revolution. The new principles that divided the kingdom into departments, and each department into districts, may be found in the Decree of 22 December 1789. (The actual decree, carrying out the division of France into departments, was that of 26 February 1790.) The reorganization provided for both electoral and administrative assemblies at the district and departmental levels. In effect, old Dauphiné was divided into three departments: Drôme, Hautes-Alpes, and Isère; each with a chief town: Chabeuil, Chorges, and Moirans respectively. In subsequent reorganizations, the chief towns would become Valence, Gap, and Grenoble respectively.

~

Les Baux, 14 July 1790
Supplement to Declaration of Patriotic Donation[737]
The annual income from my parish in Les Baux, which I cited in my carefully framed declaration, comes to 434 livres. As it is the sole source of my subsistence, I did not believe myself obligated to yield to the exact wording of the decree for the total figure of my patriotic contribution. Obligations I had contracted before the promulgation of the law put me in a position of enjoying a *net* income lower than the 434 livres and, thus, powerless to make a larger contribution than the 30 livres to which I obligated myself.

I had forgotten to add to my pledge statement, and I add it here today, that in the event the nation provides me with a larger fortune, I would obligate myself to pay a quarter of one year's income as a patriotic donation in conformity with said decree.[738]

143

Les Baux, 22 July 1790
The last letter I received from you was dated April 16th. I hope it will not seem importunate if I no longer postpone asking for your news. Perhaps you have written to me since that date, or perhaps you did not receive what I wrote to you thereafter. In my last letter, written towards the end of June [the 24th], entrusted to Monsieur Serre, I sent you some new notes on your botanical work. No response. I add the following:

In 3: 835, no. 3, you have confused two very distinct species: *Equisetum palustre* [L.] and *E. fluviatile* [L.].[739] The synonym that you cite from Haller [**Hist.**, no. 1675], and the description you give of it, is for *Equisetum fluviatile* that you failed to report in this entry. Therefore, you must revise the entry on *Equisetum palustre* and add one for *Equisetum fluviatile*.

I am nearly convinced that your *Hieracium piloselloides* [Vill.] is really *Hieracium dubium* L., Haller [**Hist.**], no. 52, although I have not seen any stolons.[740] And that *Hieracium dubium* L. is really *Hieracium auricula* L., Haller [**Hist.**], no. 53.[741] From there I think that *Hieracium auricula* [L.] is only a variety of *Hieracium cymosum* [L.], and that it is about [the latter] that what Linnaeus says in **Flora svecica**, p. 272, must be understood: *rarissime occurrit, hieracio dubio major, forte hybrida, seu alia qualis cunque varietas, omnibus partibus major nec repens.* Haller, **Hist.**, no. 54, *calycibus glabris*, which you cite [as a

~

---

[737] As the Estates-General had finally been summoned in 1789 to cope with the bankruptcy of the state, the succeeding National Assembly was necessarily seeking new sources of income, the seizure of ecclesiastical property being one expedient. In the autumn of 1789, some prominent citizens began to make voluntary contributions of cash and valuables, which encouraged the Assembly, through the decree of 6 October 1789, to request "patriotic donations" from all citizens of means. Chaix had evidently pledged 30 livres.

[738] A.H.-A. Z Guillemin 6005.

[739] Villars' entry in Hist. Pl. **Dauph.** 3: 835 (1789) was for *Equisetum palustre* L. alone. [= Haller, Hist., no. 1677.]

[740] *Hieracium piloselloides* Vill., **Prosp. Pl. Dauph.** 34 (1779), remains a good species. Haller, Hist., no. 52 = *Hieracium auricula* L. = *Hieracium x floribundum* Wimmer & Grab. (1829) [*H. caespitosum / lactucella*].

[741] *Hieracium x dubium* L. also remains good and is the same as Haller, Hist., no. 53. [*H. caespitosum / cymosum / lactucella*].

synonym] for your *Hieracium piloselloides* [Vill.], whose calices are hirsute, cannot fit it. Allioni refers it to his *Hieracium florentinum* [All.].[742]

You already know that I owe you 24 livres 6 sols. Please let me know whether I should pay this sum you are out of pocket for me, or whether I should still wait in the hope that you can have me earn it with a shipment of plants about which you wrote me.

I had the pleasure of seeing and dining with the curé de La Bâtie-Vieille [Villar] in Chorges at the home of the local curé during the meeting of the electoral assembly of our department, where I had the honor to be the deputy from our canton of La Roche.[743] Monsieur Serre was conspicuous there and has been named a member of the administration.

P.S. I am on the verge of being obliged to seek a new governess from among my relatives, because a marital opportunity has been presented to my niece, the last daughter of my late brother, who has been serving me. So many changes are saddening for a sensitive and affectionate heart like my own; but it is my duty to approve them.

144

Les Baux, 29 October 1790

You have had the kindness to have a copy of your memoir on the school of surgery sent to me through Monsieur D'heralde.[744] I thank you very kindly for it. Your plan appears to be to be excellent. I certainly hope it will be supported by the administration.

In our Département des Hautes-Alpes, we are frightened by the expenses entailed in having four districts in such a poor region.[745] On this subject, we issued a protest in the electoral assembly of the district. I do not know how the protest will be received in the other districts or in the National Assembly where our project is to be sent. We are also outraged by the directory of our district for having worked to abolish nearly all our rural cantons. Under the pretext of economy, the city [Gap] and the three towns of St.-Bonnet, Tallard, and Veynes should become dominant to the detriment of the countrymen.

The curé de La Bâtie-Vieille [Villar], who came to Gap for this occasion, gave me your news and that of Madame Villars who was then still in Le Noyer. He took the responsibility of transmitting 8 livres to you, the price of your third volume, from Monsieur Martin, curé du Saix, who had given the money to me. Please acknowledge receipt to me, and tell me when you want the 24 livres 6 sols I owe you.

~

[742] *Hieracium florentinum* All. = *Hieracium piloselloides* Vill. ssp. *piloselloides*, which Villars already knew; and the calyx may have few to numerous hairs.

[743] The district of Gap (see letter no. 144) was divided into 13 cantons. The 3rd canton, of which La Roche-des-Arnauds was the *chef-lieu*, included the communities of Manteyer, Montmaur, Furmeyer, Châteauneuf-d'Oze, Châtillon-le-Désert, Rabou, and Pelleautier. See Paul Guillaume, **Inventaire sommaire des Archives départementales postérieures à 1790. Hautes Alpes**. Série L, Période révolutionnaire, 1790-1800 (Gap: Impr. & Librairie alpines, 1911-1939, 1: xix.

[744] Villars, "Sur l'école de chirurgie, le jardin de botanique et les pépinières établies à Grenoble." (1790).

[745] In fact, the four districts were soon reduced to three, administered from Gap, Briançon, and Embrun. The intention of the National Assembly had been to provide rational uniformity and considerable local autonomy, with all officials to be elected frequently for limited terms. The system soon proved to be complex and costly, as Chaix reflects here. Worse, the population was too inexperienced politically to exercise so much self-government effectively: zeal unsupported by knowledge and experience.

Meaning to complete my herbarium, which I had been neglecting since last spring because of my difficulty in determining the mosses, where I am a beginner, I have been occupied with them into this month of October. My collection being small, determinations are made more difficult by lack of comparative material. Moreover, although I have arranged my specimens the best I can, I have not yet dared to paste them in my herbarium or to tag them, fearing error and another slap from you. You sense correctly that I have profitted from information you have provided me about certain species, but my uncertainty has not withered away. I beg you, therefore, not only to determine the plants enclosed, referring back to me the numbers I have given them, but to steal a few moments from your important duties to make up a small shipment of your mosses in order to provide me with some comparative material for plants about which I am still doubtful. You can entrust your shipment to one of our messengers from Gap, addressed to me in care of Monsieur Armand, master wigmaker in Gap, who will pay the shipping costs.

The changing circumstances could deliver a great blow to the progress of botany. Your work in this regard is completed. I should like to know very much whether you have sold many of your copies and whether you are beginning to be recompensed for such large advances. If that be so, I should hope that you will again take courage to give us your supplement and you *emendanda*.[746] As for me, whether on account of age, or for want of emulation, this study that heretofore has been so spirited and so beloved is becoming lifeless and unresponsive for me. It seems that created beings were not made to bring joy to the heart of man.[747]

Note: Vill. 2: 555, line 4, *with three capsules;* read *with one capsule, trivalve polyspermus,* in order not to be in contradiction with the following line 10, where only one capsule is stated.

Do you have *Saxifraga pensylvanica* [L.] and *Mentha auriculata* L., **Mantissa** 81, in the garden at Grenoble? Monsieur Liottard sent two plants to Monsieur Serre that are analogous to them, and which I have not yet determined. Try to add *Saxifraga umbrosa* [L.], ready for planting, to your shipment of mosses. I should like very much to cultivate that pretty species, as I have nothing left in my pots.

⌒

[746] Georges de Manteyer, in **Les Origines de Dominique Villars le botaniste** (Gap: Imprimerie L. Jean & Peyrot, 1922), p. 192, stated on the basis of his intimate knowledge of Villars' remaining papers that the first volume was entirely subscribed and sold, but mostly to people who became émigrés. As a consequence, the second and third volumes went largely unsold.

[747] Chaix's new pessimism indicates he had come to recognize that the changes being worked at Versailles and Paris, however he may have welcomed reform, were having the effect of removing the traditional patrons of botanical learning from the scene. He had examples close to home. His Carthusian friends, who had welcomed botanists in the field, were being dispersed. The Flotte family in nearby La Roche-des-Arnauds, which had inhabited the seigneurie since 1060, had emigrated in 1789. Could he have known that the Flottes, after such a history, would never return, his depression could have been more immediately deeper. See J. Roman, **Tableau historique du département des Hautes-Alpes** (Paris: A. Picard, 1887-1890), 1: 105; G. de Rivoire de La Batie, **Armorial de Dauphiné, contenant les armoires figurées de toutes les familles nobles et notables de cette Province** (Lyon: Auguste Brun, 1867): 236-237. The royal intendants in Grenoble had financed Villars; the subdelegate in Gap had recognized the usefulness of Chaix's work. Was there any reason to expect similar patronage from local officials elected for short terms?

145
Les Baux, 28 November 1790

More than a month ago I had the honor to write you by mail. I have not yet received any response from you. This long silence troubles me, not knowing its cause. . . . In anticipation of your response and your shipment of cryptogams, I put off the completion of my herbarium. My nephew Rambaud, husband of my niece Marguerite, my first governess, going to see two brothers, one in Vienne and the other in Lyon, will post this epistle. On his return, in ten or twelve days, he will come by your home and will very gladly bring whatever you want to give him for me.

Monsieur [Jacques-Nicolas] Belin de Ballu, a member of the Académie des belles lettres in Queue in Beauce, has written to me. In replying to him, I have urged him to take a portion of my dried specimens, or even to find me a buyer for my herbarium of five volumes in folio enclosing about 3,000 plants. The abbé Pourret, having finally gone back to Narbonne, has also just written to me. In responding to him, I shall not fail to remind him of past promises in my favor.[748]

If you have made up your mind to publish a supplement and an *emendanda* for your botanical history of Dauphiné, your former printer, Monsieur Allier, now established in Gap, could serve you well. And I would gladly take the responsibility to oversee the printing to make the edition correct.[749]

Three people from this small village, the only three Protestants remaining here, all in the same family, were poisoned on the 22nd of this month. If we had not given them help promptly, by having them drink much warm water, etc., they would have perished within an hour. The pulse was gone; their sight had failed; the tendons in the arms and legs were oddly stricken; the intestines were afflicted by a dreadful tingling; there were fainting spells. Monsieur Serre, who was called, and I, who was the first to reach them, we believed that it was the effect of arsenic. The next day, however, our patients being recovered, we are given to suspect--based on information they gave us--that roots of *Veratrum album* [L.] had been put in their cabbage soup. The malicious or imprudent hand that mixed this poison is unknown. I would never have believed that white hellebore could be so violent a poison. What do you think about this?[750] Please send me news of your family, of your present situation, and of your expectations.

◆

---

[748] It would appear that Pourret gave no hint of his imminent emigration, but there would be no more letters or shipments of specimens from him.

[749] Joseph Allier became designated the official printer for the Département des Hautes-Alpes in 1791. See Guillaume, **Inventaire sommaire** 1: 489. Série L. 418.

[750] *Veratrum album* had long been known to be a violent poison, but it was also used in diluted forms in various medications. See Mrs. Grieve, **A Modern Herbal** (New York: Dover Publications, 1982), 1: 391. It was readily available in the high-mountain meadows of Champsaur.

146

Les Baux, 18 January 1791

[This is the first example of a new and simplified address: To Monsieur Villars, physician at the Military Hospital of Grenoble, rue Pertuisière, across from the college church.][751]

Allow me to request the mediation I have been asked to solicit from you. The enclosed letter will give you information about the issues. The actual murder and the condemnation of Sieur Thomé, as the principal author, were only known to me through public hearsay. I do not know the true circumstances of the deed. But if, according to the solicited depositions, an innocent, or the least guilty, is paying for the real criminals, the administrators of justice would have to be critical of themselves, either for being no respecters of persons or for haste in reaching judgment. Consequently, I beg you to recommend the importance of this affair to the public defender, who is named in the letter, so that fuller information can be gathered before a definitive judgment is rendered. A father who entreats for his son, whom he believes has little guilt, can touch any human heart. My heart has been deeply touched; and yours, so well formed, will be even more so. Thus, lend your hand to a matter that is required of one human being to his fellow-man.

I owe you the acknowledgment of two letters, one from November 30th, the other between eight and ten days later. When I mentioned to you the [possible] sale of my herbarium, without having made any decision about it, I only indicated my plan; and I am persisting in it. For what use will my herbarium be to me if kept another ten years in my room (if I live that long)? My plants are consigned to my memory and in your work. Until my death, I want for nothing more than to extend my taste for botany. I shall not, of course, get rid of what I have in stock. I will provide a catalogue that will certify for buyers the number and the species of plants that I have arranged according to your method in my five volumes in folio. Monsieur Belin, an academician from Paris and an administrator in the Département de Seine-et-Oise at Versailles, has promised to support my goal. I am counting even more on the efforts of a close and old friend like you.[752]

In your last letter, you were so free in praise and regard for me that I would have believed the comments to be out of place had I not taken into account that an outburst of your affection was the cause. We know each other too well to need a diet of reciprocal compliments. . . .

I have finished arranging my few cryptogams in my last volume. Information from you having reassured me about certain specimens, I remain doubtful, ambiguous, about certain others: the problem of not having enough comparative material. In the end, I have done what I could. . . .

$\sim$

[751] The elimination of the title, Professor of Botany, suggests that Chaix had become aware of titular changes in Paris in 1790, made for political reasons. The Jardin du roi was renamed Jardin des plantes, recasting a royal institution into a more national mold; and its professional staff stopped using academic titles, simply calling themselves officers of the garden, an effort to avoid the charge of elitism in a moment of egalitarian upheaval. The surest proof of patriotism was a title that made one's usefulness to society obvious. Anything suggesting esoteric preoccupations was to be avoided, as the fate of the cloistered clergy had already demonstrated.

[752] In subsequent letters, Chaix will offer several reasons for offering his herbarium for sale. The reader should remember that his decision initially was made in the fear that the state might deprive him of his regular income. He had, to be sure, a substantial supply of duplicates for comparative study in the future. Given the botanomania of the later 18th century, there was a market, as Chaix knew, for ready-made herbaria.

In the epigraph of your first volume, instead of *exposcendum*, it should read *exposenda*. How poorly you were served by your censors in their report! Either ill-temper or prejudices blinded them. Your dictionary is complete; your method, well expressed, is excellent; the particulars on the virtues of plants are sufficient.

The primary and electoral assemblies, where I was present, have not yet given me any grief. Several envious opponents wanted to quibble about my nomination to be an elector, but I have not since been bothered about it. I shall have no trouble about taking my civic oath, as I have already taken it in our assemblies. Similarly, I shall take the oath to uphold the [Civil] Constitution of the Clergy.[753] The changes in discipline and in ecclesiastical police do not alarm me. If it should come to dogma, by the grace of God I will carry my head to the scaffold in support of its inviolability. But the faith has suffered nothing in all that the National Assembly has done. By abolishing the abuses, that majestic diet has restored the old form and the original procedure in the Church. . . .[754]

147

Les Baux, 3 May 1791

By what fate did it happen that my letters did not reach you nor your letters reach me? Having written to you twice, first by mail toward mid-March, a few days after our electoral assembly, to ask your advice regarding the proposal made to me by our elected bishop;[755] secondly by way of your son to whom I remitted the 24 livres 6 sols that I owed you, repeating my request for your advice, I had reason to complain about your silence, especially about a matter that much concerned me. Yesterday, I was in Gap for the fair, intending to inform your brother about my trouble. I dined with him, and he, too, complained about tardiness on your part.

Your letter of April 15th, which came in the mail, now informs me that you are in the same situation in regard to me. That available people, to whom one may entrust letters, may be unreliable is not surprising; but I do not understand why sending [letters] through the mail is not dependable. When you do me the honor of writing me, always prefer the latter course to the former, as will I.

My decision is made: I shall not go to Embrun. Monsieur Caseneuve repeatedly

∽

[753] The Civil Constitution of the Clergy was an integral part of what came to be the new Constitution of 1791. It provided for a reorganization of the ecclesiastical structure of the French Church, a widely anticipated reform; and clergy, if desiring to retain their positions or livings, were required to take an oath to uphold the Civil Constitution. Religious doctrine was untouched, and there was no separation of Church and State. But the pope, if still acknowledged as the spiritual head of the Church, lost his previous administrative powers in France. A second potentially troublesome provision was the *reintroduction* of the principle of election for both bishops and priests: bishops by the same departmental electors who nominated the members of the departmental councils; the parish priests by the district councils, amounting to a great encroachment of lay authority upon the ecclesiastical.

[754] Chaix did, indeed, take the oath to uphold the Civil Constitution of the Clergy on 20 January 1791 in La Roche-des-Arnauds. A.H.-A., Série L. 242.

[755] Ignace de Caseneuve of Gap (1747-1806) was elected the constitutional bishop of Hautes-Alpes on 8 March 1791. He had been a cathedral canon; but he had gained notoriety as a member of the municipal council of Gap after 1781, was known to favor reform, and had been elected mayor of Gap in July of 1790. See Félix-A. Allemand, **Dictionnaire biographique des Hautes-Alpes** (Gap: Alpines, 1911), pp. 129-130. According to Villars, Chaix received a few votes for bishop in the electoral assembly. See Villars, "Notice historique sur Dominique Chaix, botaniste." **Bull. Soc. Etudes H.-A.** 3, no. 1 (1884): 309.

pressed me to accept the post of superior in the diocesan seminary there. I replied to him respectfully, but excused myself citing incapacity and inability, derived from a solitude of 32 years and having reached the age of 60. If, at this stage of my life, I were already exercising the duties such as are proposed to me, I would be insisting on withdrawing in order to enjoy a retirement. No suggestion of advancement has ever pleased me, much less at the age where I find myself.[756] Only death will remove me from my flock. If things should be arranged to combine it with that of La Roche, I will be what I can for one and the other for the rest of my days. My bones will rest with theirs, as my heart has always been fond of them. Moreover, my temporal lot is quite better than it has ever been.[757] With my income, I shall enjoy my meadow and garden: the former has been granted to me as my half-acre, and the latter I am allowed to retain on account of my plants, which are of public interest.[758]

I am quite annoyed that you have not received my letter in which I reported to you the high opinion Monsieur Belin, the administrator in Versailles, has of your botanical work and of your method, thus indicting quite sensibly the report by your censors. In good time, I shall put his letter into your hands. . . . I have not yet drawn up the catalogue for my herbarium, but the present circumstances are hardly favorable for its sale.

148

Les Baux, 27 July 1791

Although the botany of our former province of Dauphiné would seem to have run its course under your auspices and ought to rest at the end of its run, a more limited, minor botanist just may still glean if following in your steps. Here is proof of it:

1. *Campanula bononiensis* L., **Mantissa** 337 [**Sp. Pl.** 165 (1753)], cultivated by me from seed sent from Paris, has recently been found around Gap by the Curé Jacques, an eager amateur.[759]

2. *Anthemis montana* L. *Chamaemelum montanum* All., **Fl Ped.** 1: 187 [= *Anthemis cretica* L.]: found by Curé Meyer at Lachau in the Baronnies, somewhat immature.

3. *Chaerophyllum alpinum* Vill. [= *Anthricus sylvestris* (L.) Hoffm. (1814)],[760] . . . recently abundant, but not previously collected, in the rubble on the north open slope before entering our upper Loubet woods. Similar to, but distinct from, *Chaerophyllum*

⁓

[756] Sixty would have seemed older to a man of Chaix's generation than to us. Baron Bonnaire, the enlightened prefect, calculated that the average life of males in Hautes-Alpes at the end of the 18th century was 28 years, 2 months, 19 days; of females 29 years, 11 months, 28 days. Despite the purity of the air, living conditions for the majority were detrimental to health, notably a diet often limited to salted meats and coarse bread. During the long, cold winters, rural people habitually slept in stables in the proximity of manure, where there was little circulation of air. Woollen blankets for sleeping were never washed, if occasionally aired to eliminate insects. Filthy and impregnated with sweat, they encourage skin diseases, notably scabies. See Félix Bonnaire, **Mémoire au ministre de l'Intérieur sur la situation du département des Hautes-Alpes** (Paris: Impr. des Sourds-muets, Year IX): pp. 10, 17-19.

[757] The National Assembly had granted parish priests a minimum salary of 1,200 livres a year.

[758] On 30 March 1791, the departmental directory, in response to Chaix's petition, allowed him to retain his small garden as it served as a botanical garden. The action exempted his tiny plot from being confiscated as ecclesiastical property to be put up for auction by the government. Guillaume, **Inventaire sommaire** 1: 89. Série L. 54.

[759] The specimen was found in the Villars herbarium by Verlot, **Cat. Pl. Dauph.**, p. 227.

[760] Grenier, **Flore de France** 1: 742, recognized a variety, *tenuifolia* DC., as Villars' plant.

sylvestris [L.] produced here. But where am I to find your similar *Chaerophyllum cicutarium* [Vill.] [= *Chaerophyllum hirsutum* L.]? . . .

~

Villars to the Member of the Directoire du Département des Hautes-Alpes.
Grenoble, 18 October 1791[761]

Monsieur Chaix, curé des Baux, whose learning and name are known to you, having already appeared favorably in the history of plants of our former province, and whose virtues are too well known to suspect me of flattery, possesses a rich collection of dried plants in a herbarium, well-prepared, classified, and all with names. The table shows more than 3,000 plants. It is a unique collection in that it includes nearly all the plants in my [published] work, in addition to some others whose affinity or analogy provides necessary comparative material. This worthy pastor, whose aim is to benefit his less fortunate relatives, has nearly made up his mind to sell the collection for their profit, the fruit of more than twenty-five years of his most delightful and virtuous leisure time. Moreover, moved by the purest and most disinterested patriotism, he will hope to make this collection useful to his country. I had approached Monsieur Duvall, the learned English naturalist living in the canton of Bern, about its purchase; but Monsieur Chaix told me to defer that offer. In addition, he has a small collection of well selected botany books. It appears to me that these items merit becoming part of the equipment to be used for public instruction within the department. . . .

Natural history ought to become one of the principal foundations of the national education. For it is the field that accustoms the young to examine things in detail and to see the harmony–the admirable aggregate of all the parts–in species and individuals; that develops the sharp and experienced eye, that solid judgment, that love for the beautiful and the true, which will protect the young equally from error and from boredom. In fact, Messieurs, what distinguishes the ignorant from the learned man? The first has seen nothing, examined nothing, and hardly reflected. The second, whose eyes have beheld the beautiful productions–the beautiful phenomena–of Nature, has adorned his mind and memory with the finest pictures and the richest thoughts. . . .

According to Monsieur Talleyrand's report,[762] the departments are the natural heirs to the monastery libraries and the other national monuments within their borders. It may be that the Department of Hautes-Alpes has little hope of such inheritances; but the rich and generous nation, for whom the department is the frontier and, so to speak, the guardian, will come to its aid. It is a matter, therefore, of laying down a preliminary base for public education, and what is truly of importance is to know the productions of its soil. Monsieur Chaix's collection is unique for the [local] plants. . . . It was a fortunate set of circumstances that, when I was charged by the administration of the former province to

~

[761] Extract from "Mémoire du Citoyen Villard sur l'agriculture," A.H.-A., Série L. 725. Théodore Gautier, in La Période révolutionnaire, le consulat, l'empire, la restauration, dans les Hautes-Alpes (Gap: Guillaume, 1895), p. 14, dates this document 25 October 1791, the day Villars had had it presented and read, citing Série L. 47.
[762] On 10 September 1791, the Legislative Assembly heard Talleyrand read a report on public instruction. The proposed bill was not enacted, but later replaced by the famous plan that Condorcet would read to the Assembly on 20-21 April 1792. See Georges Lefebvre, La Révolution française (Paris: Presses Universitaires de France, 1957), pp. 567-569.

compile a history of its vegetal productions, I found an equal, a friend, in Monsieur Chaix, whose talents, in anyone less generous than he, would have made him a dangerous or discouraging rival. Such sentiments never arose. Between us, going beyond the natural history of plants, there is much affection and little jealousy. . . .[763]

149
Les Baux, 13 November 1791

Our worthy and respectable curé de La Roche [Jean-Pierre Reynaud] died on the 4th of this month. His devotion and his conduct during 42 years in that parish provide his eulogy, heard from the lips of all who knew him.

My parishioner, Dominique Ricard, who consulted you in Gap, died on the 11th instant. His intemperate conduct, drinking white wine with friends for an entire day and then going out on a subsequent day, a quite cold one, to guard his flock, hastened his death. His humors [morbid fluids] had so invaded his brain during the week that followed his first loss of consciousness that he had only brief intervals to think about his salvation. The rest of the time was occupied by dreams and mental derangement. Therefore, you will not need to send the hemlock pills that you had promised him.[764]

Monsieur Meyer, curé de Lachau in the Baronnies, devoted student of your botanical method, who came to see me at the time of the fair in Gap, has found *Briza uncialis* Chaix[765] in abundance in the olive groves at Nyons, whose seed I am sending you for Monsieur Liottard's attention; *Valeriana rubra latifolia* [Bauhin, **Hist.** 3: 211][766] at La Roche-sur-le-Buis; and *Chenopodium botrys* [L.] at Venterol near Nyons. He brought me a bit of the seed from *Verbascum meyeri* Chaix, with the odor of scrophularia, which I forgot to show you in my herbarium. In time, we will make it known, God willing.[767] The finest plant that he has yet found is *Staehelina dubia* [L.] in warm, rocky places, along the way between La Magdeleine and Valserres in the region of Jarjayes, some distance from the Durance River. He has given me a very fine specimen of it. This new enthusiast, as you can see, is enriching the flora of our former Dauphiné more and more every year.

Monsieur Allier had me look at the tables prepared for the publication of your memoir to the administrative body of our department [dated 18 October 1791]. You will

∼

[763] On 29 November 1791, the departmental council of Hautes-Alpes authorized the directory to acquire Chaix's herbarium and library. Guillaume, **Inventaire sommaire** 1: 7. Série L. 46.

[764] These would have been pills containing the poisonous *Conium maculatum* L., the usual dose being one decigram per pill, given as a narcotic sedation.

As for Dominique Ricord's habits and fate, Delafont, the last royal subdelegate in Gap, wrote in 1784 that the local peasantry possessed considerable humanity and compassion for the unfortunate, but was not truly industrious or competitive. He meant that the peasant only knew how to do what his father had done and was, thus, simply active or hard-working but not ambitious; that he was usually poorly nourished and consumed too much wine, drunkenness being widespread; and that, finally, he was superstitious and not imbued with "all the good faith" that might be desired. See J. Roman, "Mémoire sur l'état de la subdélégation de Gap en 1784," pp. 167-168.

[765] See letter no. 138.

[766] *Valeriana rubra* L. = *Centranthus ruber* (L.) DC. in Lam. & DC. (1805).

[767] *Verbascum meyeri* Chaix, an unpublished species, was found in the Chaix herbarium by Edouard Timbal-Lagrave, **Mém. Acad. Sci. Inscript. Belles-Lettres Toulouse** 6, série 4 (1956): 115, who called the plant *Verbascum* x *nigro-chaixii* Timb.-Lagr.

understand with what pleasure I learned this news! I begged him to keep aside at least two copies for me. Your brother, the curé, tells me that you will possibly come to fetch your young ladies about the 20th of this month. If I can know your arrival time in Gap, I shall hasten there to greet you.

150
Les Baux, 3 March 1792

Monsieur Serre has just now sent one of his soldiers to let me know that he is leaving for Grenoble tomorrow. I am taking advantage of his courtesy to send you the books you had the kindness to lend me:

1) Necker, whose unintelligible language frightened me to a point of reading very little of him.[768]

2) **Le Voyage en Barbarie**, quite interesting and instructive, but perhaps a bit romantic.[769]

3) [Richard] Pulteney, translated by [L.-A. Millin de] Grandmaison, interested me very much, especially on the subject of Linnaeus's academic graciousness. Those works are included here.[770]

As for Monsieur Talleyrand's **Instruction publique**,[771] it is now in the hands of your brother the curé, who will take the responsibility to return it to you. This work is perhaps too learned, too metaphysical, too speculative, and insufficiently practical. It could be edited to make it more moving.

Your brother has also had me read **Les Accords** by the constitutional bishops, and [Pierre] Durand de Maillane, **Histoire apologétique**,[772] works that have indeed informed me about the religious and constitutional principles I take pride in professing. . . .

I have prepared a short discourse on the subject of botany that I shall go to read before the directory of our department when Monsieur Moynier, the public prosecutor, has arrived from Paris. I do not believe this step to be improper, since, following your memoire, the departmental council authorized the directory to acquire my herbarium and my botanical books. After I have given my reading, I shall have a copy of it sent to you.

Let me have your news, as I have not had any for more than three months, a deprivation very hard and very long for someone who lives only for you.

~

[768] In retirement at Coppet in 1791, Jacques Necker wrote Sur l'administration de M. Necker, an apology for his ministry.

[769] This was probably Père Jean-Baptiste de La Faye, Relation en forme de journal de voyage [en Barbarie] (Paris: L. Sevestre, 1726).

[770] Revue générale des écrits de Linné, 2 vols. (London & Paris: Buisson, 1789), originally published in English in 1781.

[771] Rapport sur l'Instruction publique, fait an nom du Comité de constitution, à l'Assemblée nationale, les 10, 11 et 19 septembre 1791 (Paris: Imprimerie nationale, 1791).

[772] Histoire apologétique du Comité ecclésiastique de l'Assemblée nationale (Paris: Buisson, 1791). An attorney, Durand de Maillane specialized in canon law and Gallicanism. He was one of the deputies who constructed the Civil Constitution of the Clergy, claiming its premises were consistent with the Gallican liberties dating from the 16th century. The argument, if only partly true, was reassuring to clergy caught in the confusion of revolutionary events.

151

Les Baux, 23 March 1792

Your letter of February 27th, included with one to Monsieur Serre, was delivered to me only the day after I had sent you your books by way of said Monsieur Serre. Thus, I was unable to respond to it. After you sent me those books, I had received neither letter nor news from you, from which you will know whether anything went astray. As for me, enclosed in my den during the rain, snow, and cold of winter, I had put off writing to you since about mid-November. But now my fingers have lost their numbness.

Monsieur Serre has reassured me a bit about the inflammation of your eyes about which you had greatly alarmed me. Monsieur Allier, who had also seen you, said much the same to me about it. I hope that by taking the same precautions you would give to others--more moderate work and never at night, a quietude of mind and body, food and drink that revives--will remedy these troublesome attacks and prevent such accidents in the future.

Your son is giving you troubles. I do not know of a well-born son whose one, two, or three misbehaviors are not pardoned by paternal indulgence accompanied by paternal correction. But these reiterated lapses are foreign to your character. But you tell me, "He has a good heart." I should like to believe that, but it must be otherwise to grieve you so. "He would like to be upright, but penchants carry him away." Penchants? A reasonable person, educated, who takes pride in having feeling, can he become the lax and shameful toy of some passion? Let reason prevail! Let unchecked passion give way! May religion triumph and the soul become dominant! That is what makes a man.[773]

You have agreed with my way of thinking about Monsieur de Talleyrand's **l'Instruction publique**. In actual fact, what does the public expect? Out-of-sight dissertations? Abstract, metaphysical speculative rationalizations? *No*: a very simple, practical, scientific, moral and religious plan.

[There follows a rodent-damaged paragraph in which Chaix reiterated his intention to read a paper on botany before the directory.]

If, during your leisure time, you could make a small collection for me drawn from your herbarium including plants that I do not have, and a list of which I enclose, you would give me great pleasure. One of our messengers from Gap could carry it, the package being addressed to me. But this must not be at the prejudice of your health nor do any injury to your botanical treasury. Oh! If, as you suggest to me, a repository for natural history is ever made in the departments, my small works would have some place there, and the sale of your botanical history would become very gratifying. God gives us the peace, the morality, and the religion that combine for the happiness of the peoples.

A thousand compliments to Madame Villars, to the dear son who will someday be a faithful copy of such a worthy father, and to the two virtuous young ladies, the consolation of their adoring parents.[774]

∼

[773] Of Villars' two surviving sons, Pierre, the eldest and an incompetent, had been provided an income by Villars with the proviso that he remain in Le Noyer, where he would die in 1795 at the age of 28. The troublesome son, therefore, had to be Dominique, born in 1774 and 18 at the time of this letter. See Manteyer, "Les Origines de Dominique Villars, le botaniste (1555-1814)." **Bull. Soc. Etude. Hautes-Alpes**, sér. 4, 40 (1921): 129-141.

[774] Villars had two daughters: Marguerite, born in 1777; and Marie-Anne, born in 1780.

152

Les Baux, 6 July 1792

Your last letter was dated March 27th. That is quite a long time for two hearts that have been closely united for nearly thirty years, a tie that nothing can alter. For my part, I can no longer remain in silence, hoping that you will also emerge from yours. Recently, your brother did give me news of you.

Monsieur Serre, billeted with his battalion at Brignais near Lyon, has informed me that he had heard in Valence that your young soldier had been home on a visit.[775] Martial discipline, along with your wise counsel, will do for him what perhaps an overly soft existence seemed to have arrested. Time, which forms the body of the young, rectifies the sensibility when preceded by a good education.

Your great medical responsibilities during this time when troops are being assembled around your city, the change affecting your military hospital, along with the lessons you give to students, do not allow you to fall back on our former pleasures for relaxation. As for me, the solitude of my desert allows me more leisure. You must, therefore, excuse me for this digression that I put before you for a moment.

Vill., **Hist.** 2: 177, l. 6: *nor perpetuate themselves by stem suckers*, an assertion contradicted by experience.[776] *Carex myosuroides* [Vill.] [=*Kobresia myosuroides* (Vill.) Fiori (1896)]. According to me instead of *spica hermaphrodia*, it should be *floribus hermaphroditis*, since many other sedges have hermaphroditic spikes.

I have trouble believing that *Triticum unilaterale* L. [= *Vulpia unilateralis* (L.) Stace (1978)] in Vill., **Hist.** 2: 165, differs from *Triticum biunciale* [Vill.] on p. 167. The difference in your descriptions of the two plants seems to depend upon the age when you described them.[777] Both *Hieracium dubium* [L.] and [*H.*] *auricula* L. have given you difficulty,[778] and they trouble me every time I read Linnaeus and Haller. *Folia ovato, oblonga* (*H. dubium* L.) does not well fit our *H. dubium*, whose leaves are more *oblongo-spathulata*. Haller [**Hist.**], no. 52, *floribus umbellatis*, does not fit it at all [= *H. auricula* L.]; but his no. 53, *paucifloro* [= *H. dubium* L.], suits it very well, but not at all *H. piloselloides* Vill. Consequently, I should be tempted to believe that I have never seen *H. dubium* L., and that I only have *H. auricula* L. As for *H. auricula* in your work, it is, as you say very

---

~

[775] The Wars of the French Revolution had begun with the French declaration of war on Austria, 27 April 1792.

[776] In the citation for *Lolium temulentum* L., a noxious grass that invaded cereal crops in the 18th century (darnel), Villars had written: It would be desirable not to have this plant in Dauphiné where it is only too common. Everyone knows that darnel invades the wheat-grasses, and every effort should be made to get rid of it. The task would not be impossible if everyone would abandon a belief as old as it is without foundation. Some people, otherwise enlightened, still believe today that defective wheat seed transforms itself into darnel. I know of nothing with so little foundation, nor anything more impossible. Species may destroy each other, but they never alter themselves except when producing monsters [hybrids], infertile beings incapable of regeneration, *nor even of perpetuating themselves by stem suckers.*

[777] Chaix, **Pl. Vap.** 10 (1785); Chaix in Vill., **Hist. Pl. Dauph.** 1: 314 (1786), had published *Triticum gracile* Chaix that he recognized to be close to *Triticum unilaterale* L. Villars, ibid. 2: 167 (1787), then published *Triticum biunciale*, citing *T. gracile* Chaix as a synonym. Chaix is here recognizing both, correctly, as synonyms of *T. unilaterale* L.

[778] See letter no. 143.

well, a hybrid variety of *H. pilosella* [L.] and *H. cymosum* [L.]. . . .[779]

You have promised me a shipment of plants I do not have from the list I gave you. I hope you will accomplish this when you have the time.

153

Les Baux, 19 October 1792

Monsieur [Ernest] Curten [of Grenoble] asked me for some shrubs from our mountains that I shall send through you. I have endeavored to fulfill my promise and beyond. On notice from him, I could send him additional species in the spring. The bundle packed with straw is meant for him.

The other package is for the garden confided to the care of Monsieur Liottard. If any species be superfluous to the said Liottard, Monsieur Curten could put them to use.[780]

Monsieur Curten gave me some things from his garden. It would give me great pleasure if he could send me any *Amaryllis* except for *lutea* [L.] [= *Sternbergia lutea* (L.) Ker-Gawler ex Sprengel (1825)], which I have; any *Galanthus; Leucojum* except for *vernum* [L.]; some *Narcissus multiflorus* [Lam.];[781] his geranium with white-streaked leaves or any other fruticose ones from Linnaeus's first division,[782] except for *zonale, vitifolium*, and *capitatum* that I have; *Aster cordifolius* [L.]; cuttings from *Populus heterophylla* [L.] and from *Salix babylonica* [L.].

I would ask Monsieur Liottard for *Betonica alopecuros* [L.]; his *Dianthus carthusianorum* [L.] *flore simplici; Lobelia syphilitica* [L.]; his *Tripsacum dactyloides* [L.]; or at least some seeds that he knows I do not have.

Your brother the curé must find me a coach driver in order to send you the Ray. If you find it appropriate to keep it at your own expense, you know what I paid for it, and you may apply that sum toward the purchase of other botany books for me, Buchoz's dictionary, for example, in which it appeared to me that the virtues of useful plants are quite well indicated. . . .[783]

P.S. Please point out to Monsieur Curten that my two labels for *Rhamnus saxatilis* Vill. ought to read *Rhamnus rupestris* Vill. [= *Rhamnus pumila* Turr. (1764), the form with rounded, entire leaves.] I made the error thoughtlessly.

~

[779] See *Hieracium* x *hybridum* Chaix ex Vill., Hist. Pl. Dauph. 3: 100 (1788). The current treatment gives this as a hybrid of *H. cymosum* L. and *H. peleteranum* Mérat. But also see *Hieracium cymosum* L. var. C, Chaix ex Vill., ibid. 3: 102 (1788): *Hieracium spurium foliis ovato-oblongis pilosis, caule subnude, pedunculis sparsis inaequalibus,* from Les Baux. [= *Hieracium* x *spurium* Chaix ex Froelich in DC. (1838)]. This is currently considered a hybrid of *H. cymosum* and *H. pilosella.*

[780] Ernest Curten was a nurseryman in Grenoble, an active member of the local agricultural society, a promoter of enlightened forestry, and the author of numerous reformist pamphlets.

[781] Probably a form of *Narcissus tazetta* L.

[782] Geraniums with seven fertile stamens, all of African origin, subsequently in the genus *Pelargonium* L'Hérit. See letter no. 161 for more on *Geranium.*

[783] He probably meant P.-J. Buchoz, **Dictionnaire raisonné universel des plantes, arbres et arbustes de France** (1770-1771).

~

The effective overthrow of the monarchy in August of 1792 produced a decree providing for the election of a National Convention that would govern France while producing a new republican constitution.[784] *An electoral assembly for Hautes-Alpes was summoned for 2 September 1792, and Chaix was chosen to be one of the 6 deputies to represent the canton of La Roche-des-Arnauds.*[785] *The assembly deliberated from 2-6 September. On the 3rd, Chaix rose to speak on the importance of the national elections, urging the electors not to send to the Convention any man without "the firmness of character to die in the defense of liberty and equality."*[786]

As an elector from the district of Gap, he continued his local political participation. The minutes of the district assembly show that, as one of its eldest members, he was elected to preside on 18 November 1792; and on 21 November 1792 he was chosen to be one of four directors of the district postal service.[787] *On the same date, the district electors, by a vote of 60 out of 62 voting, chose him to fill the recently vacated parish of La Roche-des-Arnauds. The following day, they elected his replacement for Les Baux, Jean-Pompon Gérard, who had previously been in charge of the chapel [desservant] in nearby La Freissinouse.*[788]

The previous March, Chaix had told Villars he had prepared a short discourse on the subject of botany to be read before the departmental directory. That moment came in December when Chaix had reached his greatest political and ecclesiastical authority, but when the authorized purchase of his herbarium and library had yet to be accomplished. The published text of his short speech ran to nineteen pages.

Récit historique et moral sur la botanique, lu devant le Conseil d'administration du Département des Hautes-Alpes, séant à Gap, le 18 Décembre 1792, l'an Ier de la République Françoise, par le Citoyen Dominique Chaix, Curé de la Roche-des-Arnauds.[789]

Citizen administrators, the spectacle of marvellous things, strange events, or unexpected revolutions strike every reasonable man; but it is left to the true *philosophe*, who reflects more profoundly, to take pleasure in them. All men are incapable of the same degree of admiration, but they can all, more or less, read from the book of nature. . . .

We can say that the knowledge of plants followed closely upon the creation of man and was propagated among all civilized and primitive peoples. [That knowledge] was principally cultivated by the Egyptians, who had received their knowledge from the Chaldeans; then by the Greeks (Aristotle, Theophrastus, and Dioscorides); and the Latins

~

[784] It also produced the decree of 18 August 1792 suppressing faculties and titles, formally costing Villars his rank and titles. The practice of medicine was no longer constrained, virtually authorizing charlatanism. Military, and then civilian, authorities were soon driven to renominate the qualified to their military and civil medical titles. Villars, having been outspoken about charlatanism, was permanently affected, becoming increasingly anxious and embittered. See Joyeux and Dejarnac, "Le Médecin Dominique Villars," p. 131.

[785] A.H.-A. Série L. 118.

[786] A.H.-A. Série L. 132.

[787] A.H.-A. Série L. 117.

[788] Paul Guillaume, *Clergé ancien et moderne du diocèse de Gap. Abbés, prieurs, curés, vicaires, chapelains . . . de toutes les paroisses du diocèse actuel* (Gap: Jean et Peyrot, 1909), pp. 34, 96, 184.

[789] Published in Gap: Allier, 1793. See A.H.-A., 8 A pièce 3839.

(Virgil and Pliny). The Romans brought that knowledge to the Gauls with a degree of perfection, as the latter had only a primitive knowledge of acorns, uncultivated fruits, and pastures. . . .

Botany entered its adolescence through the sagacity of Gesner and through the immense works of the Bauhins. It reached its adult age through artificial methods of classification: some based upon the fruit (Cesalpini, Morison, and Ray); others on the corolla or on the parts comprising the exterior ornament of the flower (Rivinus, Tournefort, and Pontedera); still other on the calyx, the coriaceous or herbaceous part of the flower (Magnol and Linné); and, finally, still others on the various parts of the fructification (Royen, Haller, and Wachendorff).

A natural and universal method that would classify all plants by order, based upon the analogy of all their parts, would certainly be preferable to any artificial method. But all the greatest efforts by our botanists to date have been limited to the arrangement of certain [plant] families, numerous to be sure: the grasses, the lilies, the labiates, the crucifers, the umbels, and the chicoraceae. . . . After such incomplete trials to date, it remains to use artificial methods and to adopt the one that, like the thread of Ariadne, can most easily and quickly lead the beginner through the paths of the vast labyrinth.[790]

Tournefort's method, based upon the corolla or the petals, was quite generally used until the publication of Linnaeus's method based upon the sexes, that is to say, the stamens and the pistils, the essential parts of fructification. This method, which is not irreproachable in its universality, is infallible in its application. Leading more promptly to the goal, it is preferable.

Some botanists, . . in order to discredit Linnaeus's system, have denied the existence of sexual parts in the cryptogams; but the reality of these organs, supported by Linnaeus himself, and before him by the discerning Dillen, . . has been demonstrated by Dr. Hedwig of Leipzig in 1782. . . .

Citizen Villars, in his history of the plants of Dauphiné, fashioned his method upon Linnaeus's; but he simplified it, abridging it by half and rendering it more natural, having to accommodate only the plants of one province. [He described at length the organization of the work.] Because of Citizen Villars, many unknown species indigenous to Dauphiné have been brought to light, and *Berardia* now figures gloriously as a distinct genus. The family of Umbelliferae owes a sounder distribution to him. His censors did not render him sufficient justice in their first report, but they recognized his merit in their subsequent reports. Thus, whoever desires to undertake the career of botany, especially in the departments of Isère, Drôme, and Hautes-Alpes, cannot choose a more competent guide.

[*Under the heading* Progress, *Chaix cited many botanists who had chosen to specialize in particular plants families or genera, or who had undertaken extensive and sometimes dangerous explorations in Europe and overseas.*] In nearly all the sovereign states of Europe, the government has established chairs, gardens, and public demonstrations in its principal cities. In France, in particular, at Paris under the direction of the learned Thouin; at Montpellier, Lyon, Aix, and Dijon; in our vicinity, at Grenoble in the hands of the

---

[790] As Chaix had used Linnaeus's incomplete natural system for his own catalogue of the plants around Gap, and as he had studied and admired the natural system of Jussieu, the argument he presented here seems to be a loyal attempt to justify Villars' choice of a largely artificial method.

indefatigable cultivator Liottard, a zealous botanist. And Citizen Curten maintains, outside the porte de France, a garden enriched by the most valuable plants, exotics in particular, a garden truly worthy of attention by the interested. . . .

No expense has been spared by wealthy individuals to procure all possible plants for themselves, indigenous and foreign, alive to be planted, or dried to be conserved in their museums: the Monniers, the Jussieux, French; Clifford, Dutch; Jacquin, Austrian; Tessin, Swedish. Mr. Smith, English, has just bought the herbaria and the manuscripts of Linnaeus at great expense. He counts himself very fortunate to possess one of the finest monuments of this century. This learned man is preparing a new edition of the **Genera** and the **Species**, authenticated with the dried specimens in his acquisition. . . .

In recounting the exploits of the heroic explorers and the examples of private generosity that redound to the glory of botany, I should like to inspire most citizens with that same passion. Each one of us has his own duties to fill; but it would be strengthening for religion, beneficial for morality, advantageous for society if [we all emerged from our daily occupations] to learn to appreciate the thousand objects available to our eyes, both for recreation and instruction: a flower, a moss, a bird, an insect; a crag, a stone, a cloud, a flash of lightening. . . . After moments given to laudable distraction, each one would resume the exercise of his profession fresher, calmer, and happier. [He then devoted many paragraphs to the more conventional utility of plants: as food, fuel, a source for cloth and dyes, for medications; and to botany in particular as the foundation for agricultural progress, reforestation, and reclamation of the degraded alpine slopes.]

Conclusion: I have merely claimed to come here, learned administrators, in support of the memoir that Citizen Villars presented last year to the departmental directory, and which, published by order of that administrative body, was made available to the departmental council during its session. The departmental council greeted it favorably as it authorized the directory to acquire my botany books and my herbarium: about 3,000 plants within five large volumes, for the most part indigenous to our former province of Dauphiné. Entirely concerned for the public weal of this department, you will, I think, embrace the goals of your predecessors; considering, in particular, that it is the aim of the National Convention, among its plans for public instruction, to put a repository for natural history in each department. . . . I am truely embarassed to be unable, for the reasons explained by Citizen Villars, to sacrifice without charge the monument dearest to my heart to the country. . . .

Le bonheur le plus grand, le plus digne d'envie,
Est celui d'être utile et cher à sa Patrie.[791]

~

[791] The official response to Chaix's narrative was published with the memoir. François-Antoine Moynier du Bourg, the elected public prosecutor in Gap, proposed that Chaix be designated the naturalist of the Department of Hautes-Alpes. He required the discourse to be published and recommended reconsideration of the council's decree of 1791, relative to the purchase of Chaix's herbarium and library, which had not been pursued for want of funds. The council then conferred upon Chaix the title Botanical Demonstrator of the Department. They meant to distribute copies of the discourse not only within Hautes-Alpes, but to all 84 departments as evidence that the Revolution would make this most useful science available to the citizens of Hautes-Alpes, a science whose study was essential in agricultural nations.

154

La Roche-des-Arnauds, 28 January 1793
Year II of the French Republic

Honorable citizen, my truest friend,

Although separated in locality, our hearts are united by the knot of the closest friendship and will be until death. I am herein enclosing my discourse on botany. . . . I have read your project on medicine with much interest and satisfaction. I hope that the powers that be will give it the value it merits, and that they will benefit the public by putting the plan it contains into action. Your petition concerning it is extremely well done.[792] In your two citations from Hippocrates in Latin, you have allowed two grammatical errors: *summam* for *summa, alvus* for *alvi*. What did you mean on p. 52 in saying "the English earlier discovered the secrets of theology for the people?" In the usual sense, that expression would seem to shock our Catholic faith. The errors of England in theology are known. Explain your meaning to me.

When going to the electoral assembly meeting in Briançon, a close codisciple, Citizen [Paul Roux], curé de St.-Crépin, stopped me on the road to renew an acquaintance interrupted for the past 37 years. He is so grateful to you for advice you gave him about his health, that, not knowing how to repay you, he gave me on my return trip a pair of stockings for you, a package enclosing a letter. This was subsequently given by me to your brother, the curé, who took the responsibility for sending it to you. The said Roux, troubled about his package, having received no news about it, has just written to me about his concern. Have the kindness to make him happy with a word of response; or indicate to me what it would be, and I will calm him myself.

I cannot, dear friend, talk to you about matters concerning myself without violating the humility that has a claim upon my silence. But if I ignore that, it is because your friendship causes me to speak. Our canton named me its first elector; the electoral assembly at Briançon chose me to be its provisional president; I was elected president of the assembly of the district at Gap; the latter assembly named me unanimously to the parish of La Roche. That parish had asked for me and welcomed me with distinction at my installation. Since then, I have exercised my duties there, circumstances not yet having permitted me to establish my permanent residence there. I am also still serving in Les Baux as my successor, Gérard, continues to reside in La Freissinouse.[793] Our bishop [Ignace de Caseneuve] and his episcopal council have extended many invitations to me to accept a place as episcopal vicar. I have been unable to persuade myself to accept and have endeavored to excuse myself to both one and the other.[794]

~

[792] Villars, Projet d'un plan d'institution élémentaire de l'art de guérir à établir dans les départements, présenté à l'assemblée administrative du département de l'Isère le 22 décembre 1792 (Grenoble: J. Allier, 1793).

[793] It remains unclear whether Jean-Pompon Gérard ever actively served in Les Baux during 1793. He was elected to be vicar in Montmaur on 1 January 1794, renouncing his clerical functions entirely on 16 March 1794. In effect, Les Baux would have no priest in residence from 1793 until 1803. Guillaume, Clergé ancien et moderne du diocèse de Gap, pp. 34, 96.

[794] It seems probable that Chaix had only agreed to move the 4 kilometers from Les Baux to La Roche because of his expressed fear that the Convention would no longer fund parishes as tiny as Les Baux. Beyond that, he would not go. Both Bishop Caseneuve and the surgeon Jean-*Joseph* Serre had been elected to the Convention from Hautes-Alpes and were identified as Girondist sympathizers.

My discourse, read before the administration [of the department], where several other citizens from the city were present, was applauded as you can see, as they conferred the title Botanical Demonstrator of the Department upon me. I am going before the same council, when it resumes its meetings at the beginning of February, to pledge that I will go to Gap to give botanical lessons, at whatever place is indicated, every Thursday during the months of May, June, and July. If I obtain a certain number of attendants, and if the enterprise should be successful, I would hope it would provide the justification for a botanical garden in that city, which would augment the sales of your botanical history. Given that goal, I am still keeping my herbarium and the few books that I have. And I shall do nothing without having consulted you.

N.B. Citizen Paganon, when turning over the papers of the commune of La Roche to me, of which he had been the custodian, did not tell me that the principal documents of legal cases were in the office of the clerk of the court. It is urgent for our commune to obtain them. Citizen [Joseph-François] Chaix, notary and mayor, and the municipal officers have asked me to ask you to kindly go yourself to said clerk of court to have a look at said documents. Based upon the information you will give me as soon as possible about your inquiry, the municipality will send you a power of attorney in order to obtain a release, if you are willing to oblige us by accepting it; or, otherwise, it will send a deputy. We hope that you will render us this service. Citizen Chaix had written to Citizen Paganon about this matter, from whom no response was received. The matter, however, is pressing for the interests of the commune. Pardon the importunity.[795]

155

La Roche-des-Arnauds, 23 March 1793
Year II of the French Republic

I am sending you the three additional copies of my discourse on botany you asked for. I presume you intend them for people not content with a superficial reading. I should like one of them to be sent to Citizen La Tourrette on behalf of both of us. If you had instructed me earlier about the desire of that excellent naturalist for plants that I have, I would have long ago responded with what I have. I am now only waiting for a list of the alpine plants he wants.

Citizen Curten says he has prepared a bundle for me. Will you please have it sent on to me immediately, when you get it, by the first messenger from Gap, addressed to me in care of Citizen Armand, wigmaker in Gap. Please have the three roots of shrubs, which I am sending to you with our good friend Serre, carried to the said Curten.

You say nothing about the present that I had sent on to you on behalf of my former codisciple, Citizen Roux, curé de St.-Crépin. Has it become lost? I am as concerned about

~

[795] This reference to municipal legal papers is imprecise. Let us only note, therefore, that the representatives of La Roche-des-Arnauds had brought legal action against their seigneur much earlier in the 18th century before the Parlement de Dauphiné. The case had lasted for 60 years before reaching a friendly settlement sometime before 1789. As noted earlier, the Flotte family had subsequently emigrated, and it may be that the municipal officers were engaged in recovering papers that had been sent to Grenoble. See Jean Egret, **Le Parlement de Dauphiné et les Affaires Publiques dans la deuxième moitié du XVIIIe siècle** (Grenoble & Paris: B. Arthaud, 1942), 2: 52.

it as is that worthy curé, who is full of gratitude and veneration for you. You also do not tell me anything about this commune's papers, although I asked to have them sought in the office of the clerk of court. Consumed by affairs, you have been unable to give the matter any attention. At least honor me from time to time with your cherished news.

As for me, I am content to be in La Roche. I expect to continue to be so, because I want everything that it pleases Providence to dispense, whether consoling or mortifying. I try to do my duty, after which I have an easy conscience.

P.S. Citizen Thouin sent me a box of 113 packets of seeds that have given me great pleasure, even more so his very interesting letter. I have sent him a copy of my discourse as well as one to Citizen [Pierre] Broussonnet. I promised the administrative council to give a botanical lesson in Gap every Thursday in May, June, and July if I have any interested candidates.

My good confrere, the curé de La Bâtie-Vieille [Villar], was here recently to see our Representative Serre and to greet me at the same time.

156

La Roche-des-Arnauds, 13 August 1793
Year II of the French Republic

Your letter, dear and estimable citizen, written on August 4th, a day that the religion renders us especially memorable to one another, was delivered to me on the following Sunday, the 11th. I was all the more eager to get it, as I had not had one from you for an infinite time. . . . At this confusing moment, I believe in our young surgeon [Serre]. He will betray neither his talents nor his virtue he got from his father; his work and his conduct are firm guarantees of that! *A wise son maketh a glad father.* **Proverbs** 10. 1. *A wise son heareth his father's instruction.* Ibid. 13. 1.[796]

Our dear curé [Villar] came to spend two days with me accompanied by Citizen [Jacques] Bernard, curé de St. André-lès-Gap. They have both attended my botany lessons regularly and made progress. If the said Bernard maintains his taste for this science, he could soon take my place because of his acuteness and the prodigious memory he possesses. Your brother showed me the letter about the new post that has been proposed to you. We have nothing to say to you about it. You are oriented; you know the terrain; your prudence guides you. I think, nevertheless, that you sail with more assurance in a ship that you have steered for a very long time.

~

[796] Word of Jean-Joseph Serre's arrest in Paris had obviously reached La Roche-des-Arnauds. Although the disposition of the king by the young Republic was only one of the issues dividing Girondists from Jacobins, it became an inflammatory issue easily exploited for party advantage. In November of 1792, Serre had publicly argued that the trial of the king should be held in an ordinary court rather than in the Convention, and he had reported the frightful excesses of the September massacres to the electors of Hautes-Alpes. In the Convention in January, he voted for the king's guilt, but against the death sentence and for reprieve. On 2 June 1793, he was among the Girondist sympathizers who signed a protest against the expulsion of the Girondist leaders from the Convention. His own arrest occurred exactly one month later. See Félix Allemand, **Notice sur Jean-Joseph Serre, ancien député, ancien sous-préfet (1762-1831)** (Gap: Jean and Peyrot, 1907). Alison Patrick, **The Men of the First French Republic** (Baltimore & London: The Johns Hopkins University Press, 1972), pp. 51, 218-19, 333. Patrick was mistaken about Dr. Serre's profession.

That our friend Serre is under arrest affects us keenly. I, myself, am broken-hearted about it. He may have committed some imprudence, but they will never convict him for a national crime. His patriotism is known throughout his department; he has demonstrated it in the Convention. If there be a zealous republican, it is surely Serre. He may have erred in deed, but his intentions have always been upright. I am as assured of it as if I had read his soul. He is not outside Paris, as you suggest to me,[797] as he writes every week, or even more frequently, to his wife from Paris under the frank of the Convention; and responses are always sent to his usual address. He merely tells his family that he is well, nothing more. His wife, his mother-in-law, and his mother are not fully informed, I think, of his disgrace. If they can be ignorant of what everyone else knows here, I at least have always kept the news from them.

The debasement of the assignats here is pushing commodities to a frightening price: 20 to 22 livres a *hémine*[798] of wheat; 80 livres for a *charge*[799] of wine; 100 écus for a hundredweight of wool; 20 sols for a pound of meat, etc. How will pensioners be able to live if payed only in paper? The 1200 livres that is given to curés is not worth 400 livres in an earlier day when they also enjoyed the *casuel* [surplice fees]. God nourishes the birds in the sky, provides for the lilies in the fields, and He will take care of those who put hope in His providence.

When you see Citizen Curten, assure him of my regard. Tell him that I received his shipment of live plants in good state despite the great heat; and that most of them have taken root, the geraniums in particular, but that the *Heliotropium peruvianum* is still dead. I have no luck with that plant. I fear that I irrigated it too much, for it had given some sign of life. I had sown seeds of it, sent by Citizen Thouin, but despite my care they have not come up. Ask Citizen Curten if they only come up the second year, and if he will try to preserve some seeds from this valuable plant for me.

Do not put off writing to me as you have recently. It is the only consolation for absent friends. Take care of your health.

～

Despite Chaix's appeal, their correspondence ceased for two years, the most turbulent period of the French Revolution. The expansion of the Wars of the French Revolution, the threats of invasion, and the open sympathy of the refractory clergy (those who had not taken the oath of uphold the Civil Constitution of the Clergy) for the enemies of the Revolution, combined to convince the Jacobins then in political ascendency that Christianity was an enemy that should no longer be tolerated. As a substitute faith, one dedicated to the Jacobin cause, the regime sponsored the Cult of Reason derived from eighteenth-century natural philosophy. The advances of the Cult of Reason in old Dauphiné by the spring of 1794 drove most of the clergy to resign their functions, Chaix submitting his resignation on 18 April 1794.[800]

～

[797] Many of the Girondists were known to have escaped from the city. They were henceforth called federalists and were subject to arrest if caught.

[798] A measurement for grain used in the Midi, before the advent of the decimal system, and about the equivalent of a half hectoliter.

[799] Another measurement in the Midi that varied considerably depending upon locale.

Villars, meanwhile, as a military physician in a frontier region, had little time for correspondence. He did make an effort, in 1793, to produce a 48-page pamphlet wherein he recommended the use of many alpine plants as food to increase the food supply during that time of troubles and to lessen the need for bread as the sole vegetal nourishment.[801] On the surface, the work seemed to be a patriotic contribution. Yet, the original introduction to the pamphlet expressed political views that his publisher induced him to omit as unsuitable for that moment, leaving room for doubt that this project had been encouraged by members of the Isère departmental directory as he claimed. He knew in advance that there might be some public resistence to experiment with new foods, and he took care in his preface to remark that what constitutes food for one person may be a medication for another. Some foods nourish and refresh--while purging others.

Among the nearly 3,000 plants growing naturally in old Dauphiné, Villars claimed that hardly 100 of them had been used for food, whereas most all of them offered nourishment. Anyone with a sophisticated knowledge of plants (a category that could not have included many sans-culottes in Grenoble) could have found among the plants to which he attributed culinary value no small number of species long known for their purgative properties by the rural population. It is hard to resist the suspicion that Villars' pamphlet was a malicious botanical proposal for the elimination of Jacobins. At its end, he observed that mosses could be regarded as species of vegetal wool suitable for use on republican beds--as was the practice in Lapland.

The popular response in Grenoble may have surprised Villars. He was understood to have slandered bread and, thereby, to have expressed counterrevolutionary ideas. In particular, the anger of local women was aroused, something to be avoided at any time, but especially in revolutionary times. They quite rejected the prospect of alien vegetation in their pots. Had Villars lived in Paris, the reaction might have cost him his liberty and position, even his life. In Grenoble, he was merely lampooned. Toward the end of the 19th century, Alfred Chabert, a surgeon and botanist from Chambéry, discovered a fragment of a song in Grenoble dating from the Year II, which was highly ribald and defies translation, attributing dubious adventures on beds of moss to Villars. How threatened he felt by the popular response cannot be determined.[802]

In the Year II, during a convalescence from a hospital fever, Villars went to La Roche to spend a few days with Chaix. Although Villars did not date his visit precisely, circumstantial evidence suggests it to have been after Chaix's resignation from his living in April of 1794 and before the fall of Robespierre in late July. Villars had not seen Chaix for six years and found him to be much altered: wan, gloomy, quite dismayed by events, and very much aged. It had never occurred to Chaix that the Convention would go to the point of preventing him from saying mass. In his despair, he told Villars to take his herbaria, his manuscripts, and his books. He would no longer need anything! Villars urged Chaix to recognize that the revolutionary turmoil

⁓

[800] See Timothy Tackett, **Priest and Parish in Eighteenth-Century France** (Princeton: Princeton University Press, 1977), p. 299. Paul Guillaume, **Clergé ancien et moderne du diocèse de Gap**, pp. 34, 462, gives a somewhat earlier date: 19 December 1793.

[801] Villar [he had temporarily dropped the *s* without giving any explanation], **Catalogue des substances végétales qui peuvent servir à la nourriture de l'homme et se trouvent dans les départements de l'Isère, de la Drôme et des Hautes-Alpes** (Grenoble: Alexandre Giroud, Year II-1793): 2, 4, 23.

[802] Alfred Chabert, "Villars sous la terreur." **Bull. Herb. Boiss.** ser. 1, 5, no. 10 (October 1897): 821-825. C.-E.-B. Bonner, "Villars sous la terreur." **Les Musées de Genève** 16, no. 10 (1959): 2.

could not last; that everyone was obliged for the moment to live under the despotism of an infidel; that they all should take consolation in God and in friends while waiting for a better order of things. Chaix did seem to revive somewhat under the influence of Villars' presence and arguments, but there was no doubt that his disenchantment had been devouring him and undermining his health.[803]

It is evident that the two discussed the disposal of Chaix's herbaria and library. When their sale was first broached in 1791, Chaix feared he might lose his living. He now, indeed, had to provide for himself. The following document shows that Villars attempted to revive the stalled negotiations for the sale in Gap:

~

The Directory of the District of Gap
to the Committees for Domains and for Public Instruction
Gap, 16 Thermidor Year II [3 August 1794][804]

We do not have, Citizens, any public botanical garden within the confines of our district. Citizen Chaix, former curé of a commune located near [Gap], who has applied himself to research on plants for a long time, possesses a herbarium in six volumes, which contain more than 3,000 plants, secured and classified, . . . and about 30 books on botany or natural history. This citizen had brought these works to the attention of the department in 1791 and had offered them to the department in order to provide the nucleus for a library.

Citizen Villars, botanical demonstrator in Grenoble and a former student of Chaix, has just reiterated the offer and asked that Citizen Chaix be granted the post of district librarian and an acre of land in order to establish an agricultural and botanical garden, which this citizen would direct, would cultivate, or have cultivated. We are interested, Citizens, as is the department, in finding the means to support the aims of this honest citizen in order to conserve and propagate those plants that are believed to be the most useful and that are the fruit of a labor of more than 40 years. We shall obtain their nomenclature and have it passed on to you.

The Commission of Agriculture and the Arts
to the Administrators of the Department of Hautes-Alpes
Paris, 8 Fructidor Year II [25 August 1794][805]

The Committee of Public Safety has provisionally entrusted the supervision of botanical gardens and all rare plants that exist within the confines of the Republic to us. But, in making us responsible for their conservation, the Committe has put at our disposition no funds whatsoever.

After that observation, you will sense, Citizens, that we cannot buy Citizen Chaix's

~

[803] Villars, "Notice historique sur Dominique Chaix, botaniste." **Bull. Soc. Etudes Hautes-Alpes** 3, no. 1 (1884): 308-309. [Actually published by Paul Guillaume.] The manuscript for this piece was dated 1800, thus written shortly after Chaix's death.
[804] This letter from A.H.A. Série L. 133 was reproduced by Félix Allemand, **Notice sur Jean-Joseph Serre**, in 1906; but I was unable to locate the original in Gap.
[805] A.H.-A., Série L. 193.

herbarium or assign to him any emoluments as a teacher of botany. These arrangements concern the communication of Public Instruction, and it is to that *[agency]* that you should turn for satisfaction.

Dominique Chaix, former curé de La Roche,
to Administrators of the Department of Hautes-Alpes
La Roche, 11 Frimaire Year III [1 December 1794][806]

The departmental administrative council voted in 1791 (old style) to make the acquisition of an herbarium in five large volumes, . . . the entirety of my collection and of my botanical work; and, moreover, to acquire the books on botany that I own. This same plan was reiterated in 1792 following a botanical discourse I read before the council. And, Citizen Villars, health officer at the military hospital in Grenoble, having supported this same matter in a petition presented to the directory last spring, the directory again received the idea favorably.

The administration has only been constrained by a want of funds--I am more than certain of that. But in the course of these delays, I have already lost the opportunity to sell said herbarium and books to an Englishman named Duvall, a resident of the canton of Bern. At the moment, a young man on his way from Montpellier to Grenoble has appeared, who has spoken very strongly for them. I shall accept offers from foreigners only when I give up hope of putting the fruit of my labors in botany for over nearly 30 years to good use in my own department. Thus, I ask that the administration give me a prompt and positive response. *[He appended a rough and incomplete list of his botany books.]*

Administrators of the Department of Hautes-Alpes
to Citizen Chaix, former curé des Baux and La Roche
Gap, 28 Frimaire Year III [18 December 1794][807]

We inform you, Citizen, that the administration is taking advantage of the offer you made of your herbarium and of the books that accompany it, in consideration of an indemnity of 2,400 livres granted to you for the value of the work and the care you have expended on this subject. We enclose here a copy of the decree that authorizes acceptance of your offers. We invite you to come as immediately as possible to withdraw the purchase price.

~

That the sale of the herbarium had finally been consumated was confirmed in a letter from the departmental Agriculture and Arts Committee to Villars, dated 11 Pluviose Year III [30 January 1795], a letter asking Villars to recommend a gardener for the nursery the department would be establishing. Villars' response, 25 Pluviose Year III [13 February 1795], reflected his immense satisfaction over the founding of both a nursery and a botanic garden, and he enclosed an index for Citizen Chaix's herbarium.[808] Consequently, we shall be surprised to learn, a year hence, that the sale had aborted.

~

[806] A.H.-A. Série L. 1007.
[807] A.H.-A. Série L. 85 [Departmental Register], no. 22, p. 69.
[808] Ibid. no. 134, p. 45.

157

La Roche, 18 Thermidor Year III [5 August 1795]

For a very long time, very dear friend and estimable citizen, I had received no news from you. I have just received some. My joy is complete, my heart holding no one dearer than you on this earth. Give it this sweet satisfaction as often as you can. . . .

Citizen [Jacques] Roux of Geneva spent a week botanizing here. He has been on the [Pic de] Bure, but his research is too hasty to observe well. He will be going into the Oisans before going on to Grenoble. He will see you if he has not already done so. Give me news of him. He is a young man of great merit who will acquire great knowledge. He has promised to write to me.

Our department has designated me to be the professor of botany, but I would like someone younger than I am to get ready to fill the position more suitably. I had talked to your brother about taking the matter into consideration and talking to you about it. In place of him, Citizen Martin, the former curé du Saix, could have taken over if his health had permitted. Beyond those two, Citizen Bernard, former vicar de Saint-Bonnet and recently curé de St. André-lès-Gap, should be quite ready to fill the position. My age, 65, does not permit me to pursue a career; and, besides, I sense my dullness in expressing myself in public on a subject, having had so little practice. I will not refuse my help in the determination of plants that are known to me.

What is the fated fortune of France? Having been delivered from a bloody tyranny,[809] she is now racked, at least in these mountains, by irrational superstition. The unfortunate victims who fall into this odious trap are prey to hypocrites or to ignoramuses who are carrying the day, common sense to the contrary. The number [of victims] in this commune, however, is small. It has not increased since I resumed my functions. These false apostles have tempted and appalled me by proposing a retraction of my oath. I have known what to say to them. I am unmoved by solicitations: I must have reasons. St. Paul teaches me to be reasonable in my obedience *rationabile obsequiem vestrum*, **Romans** 13. Some despicable deserters, rebels against their own country, now come, with the pretext of religion, to define their country's laws! One cannot betray one's country by declaring oneself to be antichristian; one betrays it by rejecting its constitutive laws.

P.S. Best wishes to Citizeness Villars, to your son and daughters. Citizen Gaude, having been outrageously treated in Furmeyer by his parishioners following the general removal of priests, has come at my invitation into my country to resume his ministry in the parish of Rabou, which was without a pastor. He sends his greetings to you.[810]

~

[809] The terrorist dictatorship of Robespierre and his Jacobin faction had been overthrown at the end of July, 1794; and Chaix's friend and neighbor, Serre, who had been lucky to survive his imprisonment, had been released. The subsequent reaction included a decree by the Convention, 31 January 1795, that allowed public religious practice. Many refractory priests returned from exile or came out of hiding; some of the clergy who had taken the oath now renounced it; some, like Chaix, adhered to their oaths or would agree to take oaths later when required. It meant a divided and quarrelsome clergy.

[810] While the Department of Hautes-Alpes did see instances of intemperate behavior during the French Revolution, those who knew what was transpiring elsewhere in the Midi were of the opinion that no department in France had a gentler, more humane, or more affable population that Hautes-Alpes. No bloody discord! no loss of life! See Pierre-Antoine Farnaud, **Description abrégée du département des Hautes-Alpes** (Paris: Imp. de la République, an VII) p. 7.

158

La Roche, 5 Brumaire Year IV [27 October 1795]

Our assembly of constitutional priests in Gap was quite numerous and very peaceful.[811] A number of citizens also attended. The points contested by the refractories were discussed there and resolved according to conscience; for the evangelical law commands obedience to the powers when they do not run counter to the religious cult. No refractory spoke or appeared there. What do you think will be done about them? Can one ally to the Republican government while formally opposing the laws of the Republic? I cannot distinguish nonsubmission from opposition. To my mind, a refractory is a rebel, an insurgent, an enemy of the country, a traitor, a rogue, an aristocrat, or a royalist. Tell me something about this topic.

The storm has again rumbled in Paris on the 13th Vendémiaire [5 October 1795].[812] The Republican ship of state, assailed previously on several occasions, has again escaped shipwreck. Can the French constitution consolidate itself once and for all on the debris of the throne and triumph over its evil sappers? Let the mad tigers feed their open mouths nothing but futile hopes!

It is rumored here that La Crotte and La Grangette will not be put up for sale, whether because of the forest they hold, or because of the enormous devaluation of the assignats, which would bring very little to the nation.[813]

If you write to the worthy citizen of Geneva [Jacques Roux], ask him for some seeds of *Linnaea [borealis* L.] that he cultivates, of some African geraniums, and of *Heliotropium europaeum* [L.].

My expulsion, brought on by terrorism, has deprived me of everything. I am going to confine myself again to my former world, and there I hope to nourish my decided taste for plants again.[814]

Your reform of *Cheiranthus erysimum* L., which you reduce to *Erysimum hieracifolium* L., is very sound. . . . I am enclosing some seeds here from *Erysimum cheiranthoides* [L.], with siliques, and more seeds from the corrected *Erysimum hieracifolium*.

I have always suspected that *Hieracium dubium* [L.] and *H. auricula* L. are only

~

[811] The decree of 31 January 1795, allowing the resumption of public worship, had been supplemented by a decree on the Exercise of Worship, 29 September 1795, that provided for state surveillance of the exercise of worship. The law was meant to provide for public security, that is to say, to assure the loyalty of the clergy to the Republic by requiring them to take a new oath if they wished to perform the ministry of any religion: "*I recognize that the universality of the French citizens is the sovereign, and I promise submission and obedience to the laws of the Republic.*"

[812] The royalist insurrection protesting the Two-Thirds Decree passed by the retiring Convention to guarantee the reelection of two-thirds of its members to the new assembly of the Directory.

[813] Chaix had been born at La Crotte, and La Grangette was nearby. Both were within the Berthaud domain, a monastic property under the control of the monks of Durbon in the 18th century. The monastery at Durbon had been evacuated in July of 1790, the municipality of St.-Julien-en-Beauchêne taking the statements of the members of the community as to preferences about future status. The ecclesiastical property was nationalized and subject to local auction for the benefit of the state; but such purchases had to be paid for in assignats, the revolutionary paper issued ostensibly in equal value to the property confiscated by the state. For Durbon, see Charles Charronnet, **Monastères de Durbon et de Berthaud (Diocèse de Gap). Documents historiques** (Grenoble: Alphonse Merle, 1863), pp. 47, 70.

[814] As he did not return to Les Baux, he must have merely meant that he would return to his seclusion of yesteryear and pursue his botany.

varieties.[815] I would say nearly the same for *Valeriana celtica* [L.] and *V. saxatilis* L.[816] A word from you about this, please.

If your *Galium anysophyllum* [Vill.] should fall into your hands, send me a specimen of it in a letter. I have never seen it.

P.S. Gaude, my friend and neighbor, begs you to get him some glasses that magnify, as suitable to his age, 73. Citizen Serre sent him a pair from Paris that he cannot use. He will reimburse you for the expense either through Citizen Villar of La Bâtie-Vieille or whoever you designate.[817]

~

Chaix to Citizen [Pierre-Antoine] Farnaud
Secretary to the Administration of the Department in Gap
La Roche, 20 Pluviose Year IV [9 February 1796]

You will remember, I think, the arrangement made last year between the departmental administration and me in regard to my herbarium and my botany books. Circumstances (the suspension of the national schools and the devaluation of the assignats) upset carrying it out. I want to know what the present plan of the administration is in this regard. I have just had a request from Grenoble for most of my botany books. I shall respond to this request after you give me the decision, yes or no, on behalf of the administration. I await your response.[818]

~

Farnaud's response, ten days later, was less definitive than Chaix had sought. But the administration expressed hope that a work done by Chaix, and comprising the plants native to Hautes-Alpes, would not become the personal property of a foreigner when it could become so useful at home, especially at a moment when the Ecoles centrales were close to establishment, that is, the projected new public secondary schools, one for each department. After reviewing the history of prior negotiations for the sale, Farnaud mentioned in particular the disrepute of the assignats to account for the deadlock. The administration hoped that the project could be renewed when circumstances should become more favorable, adding that the establishment of the Ecoles centrales would provide resources for a professor of botany. In sum, if it is possible for you, please do not give up your herbarium and botany books to anyone requesting them, but rest assured that the agreement would be honored once the means should be sufficient.[819]

The negotiations ended here, as the promised moment never came. Chaix would still possess his herbarium and books until the moment of his death.

~

[815] Both are considered to be hybrids.
[816] Both are good, but closely related, species. The cauline leaves are present in *celtica*, usually absent in *saxatilis*.
[817] Citizen Serre had just been elected to the Legislative Body of the Directory from Hautes-Alpes.
[818] A.H.-A. Série L. 1007 [Also in Série L. 85, no. 74, p. 115]. The letter was signed D. Chaix, former curé, further evidence that he had been forced to suspend his ministry late in 1795.
[819] Farnaud to Chaix, 30 Pluviose Year IV [19 February 1796]. A.H.-A. Série L. no. 74, p. 115. The file in Série L. 1007 included a rough catalogue of Chaix's books, not in his handwriting, showing a total value of 406 livres, not as inclusive as a later list dated 1799.

159
La Roche, 29 Prairial Year IV
[17 June 1796]

For a month, dear friend, I have been waiting for your response. Did not my last letter dated in Floréal reach you?[820] I apprised you in it of the death of the venerable Gaude, my old friend. My poor niece, Margueritte, preceded him by only about forty days. Such is the precariousness of the human condition.

Doctor Botta came to see me twice and promised me the **Auctuarium** by Allioni and Bellardi. If he keeps his word, he will give me great pleasure.[821] Although rusty in botany and sluggish in scholarship, I have not totally lost the taste for it. My added notes prove that to you.

1. A gynandrous plant having the habit of *Orchis militaris* [L.], even the root, appeared to me recently in the Manteyer preserve. But it certainly had only four petals, including the nectary. The three upper petals, united from their middle down (the two lateral petals longer than the middle one), are upright, connate, and a smoky color. The nectary is a small tongue, quite entire, raised, that covers the sexual parts, white with some red spots, terminating behind in a serratiform prominence. I found only one complete example, the others having been picked by children. Close by, I saw the very handsome *Orchis militaris*. What do you think of my discovery? Is it a trick of Nature? Is it a new species? I know that *Orchis* or *Satyrium repens* [L.] [= *Goodyera repens* (L.) R. Br.] is tetrapetalous, although I have never seen it. But this is not it. See these two flowers, glued here.[822]

2. *Astragalus glaux* [L.] in Vill. [3: 459 (1788)]. *Totus parum hirsutus, gracilior, stipulae breves, pedunculi recti, foliis duplo longiores, bracteae albidae, calice triplo breviores, exfunae, glabre; calices pilia nigrescens, apice nigriores; vexillum violaceum, emarginatum, carinam linea superans.* Close to my *Astragalus exilis*. Specimen included. . . .[823]

If you know *Rottboellia* [L. fil.], tell me its character. I do not have this plant in my books, and, having sown it, I neglected the label on it.

From the appearance of this fleshy leaf, specimen enclosed, from a plant sent by Curten and Liottard, but which has not yet flowered, it is fructescent. You know it. Please give me its name.

~

[820] A letter that is still missing: May of 1796.

[821] Allioni, **Aauctuarium ad floram Pedemontanum cum notis et emendationibus**, had been published in 1789. Bellardi, **Appendix ad Floram pedemontanam**, appeared in 1792. Neither title appeared on the final list of Chaix's books.

[822] Despite the confusing characterization of *Orchis* here, this appears to be a hybrid that was found in the Chaix herbarium and published as *Orchis purpurea-militaris* Timbal in Gren. & Godr., Fl. Fr. 3: 290 (1856). See Timbal-Lagrave, "Observations critiques et synonymiques sur l'Herbier Chaix." **Mém. Acad. Sci. Toulouse**, sér. 4, 6 (1856): 95.

[823] Verlot, **Cat. Pl. Dauph.**, pp. 92-93, says that Villars' *Astragalus glaux* L. ought to have been *Astragalus purpureus* Lam.; and that the unpublished *Astragalus exilis* Chaix was actually *Astragalus hypoglottis* L. Chaix also called *Astragalus onobrychis* L. into question here, suggesting that the local plant might well be *Astragalus arenarius* L.; but Villars was surely correct.

160

La Roche, 27 Thermidor Year V
[14 August 1797]

Your shipment, my very dear friend, and the letter enclosed therein, brought me the greatest pleasure. I learn about the favorable state of your health that has held up despite the fatigue inherent in two consecutive [field] trips when making the discovery of new plants, along with physical, natural, mathematical, and mineralogical observations. You think that the highest summit of our [Pic de] Bure is 1,300 *toises* above sea level.[824] I shall remember that. Tree trunks buried in the marshes, nearly 400 *toises* above the present woods and near the perpetual snow, prove that an extraordinary disturbance occurred in that ground. All circumstances considered, that disturbance can only be related to an upheaval that shook the globe. That upheaval was called deluge. Thus, there was a deluge, irrefutable proof for unbelievers, just as are the elephant remains found on the high mountains of Siberia, those wild animals now living only in the torrid zone; or the deposits of shells one finds on the most elevated plateaus. Philosophes! stop your reasoning when experience contradicts you.

If you do not have a specimen of *Ranunculus, foliis graminis parnassi, palmaris, glaber, inherbidis*, in Berard's theater,[825] to compare with his *Ranunculus, foliis damasonii minoris, triuncialis, villosus, in caulibus sterilibus*, I would presume that the first one is only a variety of the latter one, which may have disappeared from the herbaceous places [of the high slopes] as it has not been found after so much searching by you and other precise investigators.[826] The pubescence of plants is more readily seen in high and open locations; for example, *Myosotis nana* [L.] [= *Eritrichium nanum* (L.) Schrader ex Gaudin (1828)], *Achillea nana* [L.], etc., and such plants grow much less high than in the grasslands. See the dwarf variety of *Carex atrata* [L.] that we have raised to a species [Vill., **Hist. Pl. Dauph.**] 2: 205.[827] If one cultivated this *Ranunculus parnassifolia* [L.] that you have sent me, it would slowly shed its hairs. These are conjectures. Otherwise, the leaves of the *Parnassia* and of *Alisma damasonium* [L.] [= *Damasonium alisma* Miller (1768)] are quite similar.

Your specimens of *Veronica tournefortii* [Vill.][828] and *Potentilla frigida* [Vill.] will have a place in my herbarium alongside those same species that I already had from you. I thank you for them.

Your project to compose an abridged volume of all your plants from Dauphiné, whether already published or subsequently discovered, is superabundantly in accord with what I was proposing. I adopt it with great joy and approval. But it will have to be concise,

~

[824] 1 toise = 1.949m., which would be 2,534 meters. The elevation today is given at 2,709 meters.

[825] Petro Berard, **Theatrum botanicum (pharmaceopaeo Gratianopolitano)**, a manuscript in the public library of Grenoble dating from 1653. See Villars, **Hist. Pl. Dauph.** 2: 676 (1787).

[826] Villars, **Hist. Pl. Dauph.** 3: 734 (1788), cited the first of these as a synonym of *Ranunculus parnassifolius* L., a plant known to Tournefort. Villars indicated that Berard had found the plant on Mont de Lans in Oisans, but that he and other botanists had been unable to find it there. He did not cite the second *Ranunculus* mentioned by Chaix here. *R. parnassifolius* L. has since been found elsewhere in the Alps. Note that he described one as glabrous, the other as villous.

[827] *Carex humilis* Chaix, non Leyss., **Pl. Vap.** 8 (1785); Chaix in Vill., **Hist. Pl. Dauph.** l: 312 (1786). [Probably = *Carex humilis* Leyss. (1761)].

[828] This seems to be an oval-leafed form of *Veronica officinalis* L. that Villars had tried to separate from *Veronica allionii* Vill., **Prosp. Pl. Dauph.** 20 (1779); Vill., **Hist. Pl. Dauph.** 2: 9 (1787).

limiting you to the trivial name for each species, the author's diagnosis, its native place, and a description in two lines of small type expressing its most essential qualities, in imitation of Linnaean precision. You can note the plants that are yours with the letter V., and use some sign to indicate those that had been omitted, and another sign for those that are totally new. I shall be quite anxious to see this new product of your genius appear, the complement to the Dauphinois flora. But I wish that the sale of your large history had been completed for fear that this species of **Nomenclator**, or **Mantissa**, or **Emendanta** will injure further sales. I offer you my feeble help if you think it worth anything: for instance, the description of *Scabiosa gramuntia* [L.] [= *Scabiosa triandra* L.]. I should like it to be as follows: *Corolla quinquefidis, stellaris, brevissimis; foliis hirsutis, pinnatis, superioribus minutim pinnatifidis.*

The *Astragalus* in question: I named it *exilis*,[829] because it is really slender, thin, upright, and small compared to most of the rest of this genus, particularly growing in herbaceous places on our Montagne de Noissière and near Chaudun. It cannot be called *A. purpureus* [Lam.], because, as you say, that is what Lamarck has named our *A. glaux* [L.]. It does not have a purple corolla as does Lamarck's, but is a clear violet.

I have cultivated a geranium from Africa and still have it in a pot, born from a seed sent, among others, from Paris, and which has some afinity to *Geranium zonale* [L.], but differs from it.[830] It is taller, more upright; its leaves larger, their lobes less deep, its bracts also larger; its corolla red (*rubra*), not poppy-red (*rubicunda*), as in *G. zonale*; it is less woody with less resistance to our winters. I would have believed it to be *G. tabulare* L., but I was sent seeds from Paris under that name, quite different, as their flowers are very small and whitish, having no resemblance to *G. zonale* that Linnaeus compares to *tabulare*, and to which he ascribes purplish flowers. Would you please look in the last editions of Linnaeus by Murray, Reichard, etc., at those species closest to *G. zonale*? Perhaps the one that concerns me will be designated there; for I believe it to be a species different from *G. zonale*, and that the Parisians strayed from Linnaeus in labeling their *G. tabulare*. All of this at your leisure and when you are ready to respond to my letter.

What troubles between the Legislative Body, the Directory, the troops, the Jacobins!

P.S. Not sent by post until 2 Fructidor [19 August 1797].

~

Chaix had been told in the summer of 1795 that he would be the professor of botany in the local Ecole centrale. Said schools were authorized in Paris by a decree of 25 October 1795, but the school in Gap did not open until December of 1796. An official inspection the following October exposed a total enrollment of ten students during 1797, served by a staff of six professors, one librarian, and one concièrge on state salaries. Two of the faculty members were found not to be scholars whose residence was elsewhere than Gap: an insupportable situation.[831]

In 1796, Chaix had resumed addressing Villars as Professor of Botany, an indication he was

~

---

[829] See letter no. 159.

[830] These African geraniums now belong in the genus *Pelargonium* L'Hér.

[831] Guillaume, **Inventaire sommaire** l: 177, 179; Théodore Gautier, **La Période révolutionnaire, le consulat, l'empire, la restauration, dans les Hautes-Alpes** (Gap: Guillaume, 1895), pp. 51-52, 131.

aware that Villars had been appointed to the faculty of the Ecole centrale in Grenoble in addition to his military medical duties. The school occupied the quarters of the former Jesuit collège, a sketch of which was made by Stendhal for his **Life of Henry Brulard***. Villars would teach natural history there for the seven years the school survived, but with mixed success if Stendhal can be believed. "The appointment of teachers to the Ecole central," he remarked, "MM. Gattel, Dubois-Fontanelle, Trousset, Villars (a peasant from the Hautes-Alpes), Jay, Durand, Dupuy, Chalvet, to name them roughly in order of their usefulness to children, the first three having some merit, cost little and was soon done, but there were big repairs to be done to the buildings."[832]*

A substantial amount of Villars' teaching material is preserved in the municipal library in Grenoble to suggest the character of his courses. In general, the point of view he had expounded before the Académie de Lyon on 1 April 1788 reemerged in his teaching, as it would again in a paper he would read before the Société des sciences de Grenoble on 4 January 1799. One can find four dominant themes: (1) That mankind constitutes the final goal of Nature's achievement, the final rung in the ladder; and that it is incumbent upon man to insure that supremacy by cultivating his moral and intellectual potentialities through instruction. (2) That species are fixed, namely, as products of the laws of Nature. (3) That science employs names as useful tools to sort out natural objects, and that the order of the science is a precise substitute for the order in Nature. And (4) that there exists a chain of being in Nature that, by a progression from the simple to the complex, leads us from inorganic matter ultimately to man, the most complex of beings.

Such views, to be sure, were characteristic of the 18th century and consistent with natural theology.[833] He also stressed the usefulness of the natural sciences and a faith that science would assure social reform. He asserted that the young Republic would become the vehicle for that progress, not only by providing general public instruction, but by providing the liberal climate that encourages research.[834] Whether that assertion honestly reflected his opinion remains obscure. For someone in Villars' position, it was politic to speak well of the Republic. But his early emancipation owed to the patronage of the monarchy, and he could not have ignored the constraints upon the liberties of his two intimates, Chaix and Serre, worked by the Republic.

~

[832] Stendhal, **The Life of Henry Brulard** (New York: Minerva Press, 1968), pp. 168-169. The autobiographical novel was written between 1835 and 1836.

[833] Despite these orthodox expressions, both Linnaean and Christian, there remain letters from Villars to Lapeyrouse between 1793 and 1796, when Lapeyrouse was engaged in a special study of the *Saxifraga*, that reveal Villars' perplexity about those "accidental plants" we call hybrids. Three alpine species, *Saxifraga rotundifolia* L., *S. cuneifolia* L., and *S. hirsuta* L., Villars reported to Lapeyrouse, had been planted by Liottard along an east-west wall in their garden. For six years they had grown normally and undisturbed. But in the seventh year, 1793, an intermediate plant, a hybrid species, had appeared: "One can describe it but not explain it. But Nature always enjoys casting a veil over these operations. She likes to conceal herself from our eyes, the better to exercise our imagination, and also perhaps in order to vary and improve her productions. I have lately come to presume the genus *Saxifraga* to be augmented, in this region, by several species or varieties of new creation. I mean to say hybrid species, born of two other neighbors." Villars has to be counted among the first to risk breaking with the strict Linnaeans on the matter of speciation, and he was critical of Gouan and Cusson in particular for being determined to reduce everything to established Linnaean species. See Timbal-Lagrave, **Opinion de Villars sur les plantes hybrides, d'après sa correspondance avec Lapeyrouse** (Toulouse: Douladoure, 1858), pp. 3-12.

[834] Emile Callot, "Un discourse de Dominique Villars sur l'histoire naturelle." **Revue d'histoire des sciences et de leurs applications** (Paris) 20, no. 3 (1967): 281-184. See "Leçons de botanique, par M. Villard, professeur d'histoire naturelle, à Grenoble." Bibliothèque Municipale de Grenoble, R. 1608.

The latter, elected to the Council of Five Hundred on 26 October 1795, where he sat until May of 1798, had not forgotten his near fatal encounter with Jacobinism in the Convention and had become an ardent royalist.[835]

Chaix, in addressing Villars in this period, also recognized the election of Villars to be an Associé non résidant of the Institut national des Sciences et Arts, thus a correspondent. In response to that election, dated 22 March 1796,[836] *Villars drew up a report on his projected activities that occupied him from March 26th until April 19th. If politely addressed to the members of the Institut in general, the report obviously courted the attention of A.-L. de Jussieu and René Desfontaines in particular.*

The main thrust of his paper was an appeal for order that derived from his long research on the flora of Dauphiné and his correspondence with Chaix for a quarter of a century. Linnaeus, with the talents of a regenerator, had given botany not only a new language, but his works had spawned a crowd of botanists and botanical works. Vast numbers of new species had been described in the course of the century, and immense progress had been made toward their classification thanks to the originality of Bernard de Jussieu. But the vigor in discovery had not been matched by energetic cataloguing, and Villars meant to warn that the evident progress could be undermined by inattention to conservation and cataloguing.

It was commendable, he noted, that private herbaria, like those of Tournefort and Vaillant, had been acquired for the collection in Paris. Yet, a catalogue for the Jardin des plantes had not been made since 1626. As herbaria may perish thanks to insects and weather, there was desperate need for a cumulative catalogue, a new **Pinax**. *"In order to know what we lack, we must know what we have." One should begin by making catalogues for every collection, from which the* **Pinax** *could be produced: an alphabetical index, a general table of genera and species known in Europe, with simple descriptions of their distinctive characters, and a reliable citation for each one.*

He did not expect the professors in Paris, so variously occupied, to undertake this work; but he appealed for their direction and supervision of it, suggesting that the correspondents of the Institut could be fruitfully employed, and evidently sought encouragement to devote his future botanical energies to the remedy of the current confusion. His paper, finally, made clear that, whatever compromises he had made with natural classification in 1779 or 1786, he had become a convinced follower of the Jussiaean system. No doubt the gesture was sincere, but it was also politic. Even so, although his paper was read in Paris in the Year V, it was not recommended for publication. Believing that his proposal merited the attention of all botanists, Villars published it independently four years later.[837]

~

[835] Félix Allemand, "Notice sur Jean-Joseph Serres, ancien député," pp. 260-263; Allemand, **Dictionnaire biographique des Hautes-Alpes** (Gap: Alpine, 1911): 428.

[836] Sieyès to Villars, 2 Germinal Year IV [22 March 1796], cited in Manteyer, **Les Origines de Dominique Villars**, pp. 182-183.

[837] Villars, **Mémoire sur les moyens d'accelérer les progrès de la botanique** (Grenoble: Chez Villier, Year IX), pp. 14-19, 25-30.

161
La Roche, 26 Fructidor Year V
[12 September 1797]

I was quite surprised yesterday, my very dear friend, upon seeing our Bonnardel in La Roche, completely cured, whose illness had led me to fear unfortunate consequences. He is solely indebted to your care and to your skill at healing. He held you in regard previously; but after such a benefit, he will bear you close to his heart for life, as will his family. I am no less touched, and it becomes one more reason for gratitude.

After your clarifications, the *Geranium tabulare* from the Parisians is indeed Linnaeus's plant that he places among the herbaceous [species]. But the synonyms he has adopted for it have led him to give it purplish flowers, whereas they are consistently yellowish-white without any closeness to *Geranium zonale* [L.]. It is a variety of the latter that has no belted leaves at all; while the variety that Linnaeus has described is marked with a purplish zone near the umbilicus. [A.-J.] Cavanilles has well named it as *Geranium elongatum*, as its peduncles are nearly a foot in length. Murray did not know it when merging it with *Geranium zonale*, giving it a shrubby stalk. He meant, no doubt, to speak about *Geranium tabulare*, which Cavanilles described very well, and which has the greatest affinity with *G. zonale*; but its stalk is erect; its branches upright; its leaves longer, thicker, nearly without any belt; its flowers red, not poppy-red. Therefore, I will call it the *Geranium tabulare* of Cavanilles. *Geranium elongatum* Cav. has leaves somewhat sticky above and a very noticeable odor of pine-resin or turpentine. In our climate, the seeds of *G. tabulare* Cav. will mature, but those of *G. zonale* [L.] abort.

Geranium inquinans [L.] has often been sent to me in seed from Paris; but I have been unable to detect the stain that Linnaeus says its crushed leaves will make on one's fingers. If Citizen Curten should have seeds from any rare African species of geranium, ask him for some for me. He will give me some of them, I do hope. This [plant] family pleases me in the narrow edges of the garden that I cultivate in precarious tenure.

Would you kindly inform me whether Citizen [Henri] Reymond, bishop, is still in Grenoble, and whether he is pursuing his functions? or whether he has gone to Paris to participate in the national synod that we hear has been convened? whether the pope authorizes this assembly? and what results can be expected from it?[838] The memorial by [Jean-François de] La Harpe, which Monsieur Serre got to me, says not a word about it.[839] But then it was not his subject. He devotes himself to censuring the Legislative Body and the members of the Directory. It appears that Monsieur Serre is listening quite eagerly to this journalist. Using the pretext of wanting to maintain the Catholic religion in France and to reunite all the priests, he has adopted the view of that outrageous minority that

$\diagdown\!\!\!\sim$

[838] Henri Grégoire, constitutional bishop of Blois from 1791-1801, a convinced republican and a member of the Council of Five Hundred, had assumed the leadership of the Gallican Church and sought to give it vitality by convoking a synod in August of 1797. Although the Civil Constitution of the Clergy had become a dead letter after the displacement of the Constitution of 1791, Grégoire opposed the restoration of Roman jurisdiction in France, thus obstructing the desired reconciliation of the clerical factions. He would resign his diocese in 1801 after Bonaparte's concordat was signed.

[839] La Harpe, critic and poet, had been a vociferous Jacobin during the Convention, yet became a suspect and was arrested in 1794. During his incarceration, he experienced a religious conversion and emerged an ardent and reactionary Catholic, in no way sympathetic to clergy who had taken the oath.

exempts priests from submitting to the laws of the Republic. If that opinion had prevailed, the refractory priests, who are openly intolerant, would have destroyed those who took the oath by styling them as apostates, schismatics, degenerates, and as damned. Their persecution would have been murderous. We have examples of that in our South. From that have come the disobedience within the armies and the uprisings within the country. Just how could the Legislative Body guarantee the observance of its laws by issuing the requested formal declaration of nonsubmission [by priests]! Should it be said of any foreign clan coming into France that it is to be excused from any obligation to submit to the laws of the Republic? It is asserted that priests are not public functionaries, or that what is not required of the ordinary citizen ought not be required of them. They may not be functionaries in the eyes of the law; but in fact they are very much so in the name of the divinity, an authority of quite greater importance. Moreover, the ordinary citizen, far from allowing himself to be deported or jailed, has taken an oath in his primary and electoral assemblies. Citizen Serre, who had the speeches of the two parties sent to me, must know what I would think of them. But do not say anything to him about this. That could antagonize him in regard to me. Let the water flow freely in the stream.

If the great minds, about whom you tell me, want to wait until *all* the plants are known before publishing them, they will have disappeared, both they and their heirs, before that moment arrives. It would be pleasant to study Nature, especially the botanical part, if, by shedding the so-called philosophical mind, one could remember the author of Nature. Otherwise, one takes pleasure as a brute, and that amounts to nothing.

P.S. We learn today that the Directory is at loggerheads with the Legislative Body.[840] Only time will tell where all this will end. God gives us His peace! Amen! Fiat! Fiat!

162

La Roche, 8 Vendemiaire Year VI
[29 September 1797]

The anxiety, very dear friend, inseparable from the present state of things, the disturbing uncertainty about the future, should be the constant preoccupation of any thinking man. But I am emerging from this gloomy silence, and for a few moments I will reach a truce with these black thoughts. Amiable botany will always be my gentle consoler during crises of dejection, and my firm hope in God my support during the instability of revolutions. Thus, I bring you back again to the African geraniums.

Geranium alchimilloides [L.] has stems, peduncles, and leaves shaggy with white hairs; the leaves lightly marked with a brown stain, or often without the stain, five-lobed, dentate or sublobed; flowers medium, yellowish-white; anthers red; perennial root.

G. elongatum Cav. resembles the above in flower and duration; but its stems are more elongate, less hirsute; its peduncles are much longer and quite glabrous; its leaves are also glabrous, green, stained or not stained depending upon the variety. The description of *G.*

---

[840] A reference to the Coup d'état of the 18th Fructidor [4 September 1797]: a purge of royalist deputies through the nullification of 200 recent elections.

tabulare L. fits it except for the purplish flowers, perhaps reported by following Ray's synonyms and accepting the subherbaceous stem in L., fruticose in Murr., as they are truly herbaceous, as is *G. alchimilloides* [L.].

In contrast, *G. tabulare* Cav. is woody, upright, tufted; leaves alternate (those above have opposite leaves); and has the greatest affinity with *G. zonale* [L.]; but its flowers, instead of being poppy-red, are red. I believe it to be *Geranium folis . . . dentalis, glabris, caule fruticoso* Burman [**Plantae Africae**] tab. 1, fig. 44. Linnaeus seems to have confused two plants that Cavanilles segregates for us.

I have an *Inula* in the garden for which I have not retained the label, and which I can only relate to *Inula undulata* [L.], **Mantissa** 115. Nevertheless, I am left with some doubt. I have found *Inula hirta* [L.] here on a dry slope toward the vineyards of La Roche, which is identical to the plant you attribute to me in your flora, 3: 219, under *Inula germanica* [L.]. Because of my limited knowledge, I have been misled by your confidence more than once. . . . I do not have the latter species. [J. G.] Gmelin [**Flora sibirica**] has illustrated it, tab. 78, fig. 1.⁸⁴¹

If you send me your Gmelin, which you offered to do in your last letter, you will give me great pleasure. Having a passable knowledge of most of the plants of our region, I should devote myself to determining the exotics that others send to me, if they happen to be in this author, along with the small number I already possess: A *Helianthus laevis* that barely relates to the one in **Species Plantarum** [from Virginia]; a *Helenium species nova* from North America; an *Alcina prefoliata* Cav. [= *Melampodium* L. from Mexico]; a *Rudbeckia amplexicaulus species nova*, American etc., all intrigue me with reason. Your author, combined with the clarifications you will give me, will soon dissipate my doubts and distract me from my ominous meditations. You can get this shipment to me through the Gap messenger, who also serves as drayman, addressed to me in care of Citizen Garnier, whom I will gladly reimbourse. Do not fail to tell me about how Madame Villars is, whose illness you made known to me.

If the Legislative Body has adjourned, we shall soon see our friend Serre. He has grown very disgusted with his post. France is in a very shaky condition.

P.S. Citizen [Jean-Joseph] Jacques, former curé, is here with me. He currently lives in Manteyer. For two or three days, we have had your brother here with Martin, now professor at the Ecole centrale. The Mondet and Chaix families are in good health.

163
La Roche, 29 Vendemiaire Year VI
[20 October 1797]

I have received your [J. F.] Gmelin, two volumes of [Caroli àLinné] Systema naturae.⁸⁴² It is quite a shame that it is only a *nomenclator* modeled on that by Haller; for

⁸⁴¹ The plant given as *Inula germanica* L. by Villars, a good species, was in reality *Inula spiraeifolia* L., the former not being found in France. Chaix was complaining about the fact that he had listed *Inula hirta* L. in his catalogue, Chaix in Vill., Hist. Pl. Dauph. 1: 370 (1786), as collected at Reynier near Tallard; and Villars had arbitrarily called it *Inula germanica* L.., ibid. 3: 219 (1788).

⁸⁴² This was the 13th edition of Linnaeus's work that Gmelin had begun publishing in Leipsig in 1788. The Gmelins were a distinguished scientific family of Tübingen: Johann Georg (1709-1755); Johann Friedrich (1748-1804).

he includes a great number of species, but with a single sentence. Nevertheless, I can ruminate at leisure. **La Bibliothèque du Dauphiné** is a new book to me. I noted the sketch about you in the preface with pleasure; I do not know whether the author wanted to describe me as *the citizen of the Alpes, little known*. You ask me for a sketch of my life. I respond to your request with reluctance; but as they must make it known only after my death, I shall not blush for my bold presumption. I have merely set down the plain facts here. Add to it and cut out anything you judge to be appropriate. You will find it herein.
. . .[843]

Citizen Serre can no longer obtain the records and speeches in the legislature under his signature in the Council of Five Hundred for me, this device being prohibited by law. I have read the first twelve journals of the Council in Paris. I can readily dispense with the remainder if the above are all I may have. If the government continues to protect this distinguished assembly, the benefit that will be obtained for both religion and the state will be very great. The goal is to pacify religious discord, strengthen the constitution, bring back morality, and reform the clergy and the faithful. *But God is faithful, who will not suffer you to be tempted above that ye are able; but will with the temptation make also the way of escape that ye may be able to endure it.* I. Cor. 10. 13. If you know things of interest [about affairs], such as matters concerning the peace, please bring them to my attention.

Our former bishops, Vareilles of Gap, Leyssen of Embrun, have no desire to return to administer their dioceses, not wanting to disown their conduct; and Cazeneuve, [Constitutional] Bishop of Hautes-Alpes, is nothing more than a passive personage, indifferent and uncaring. I shall urge my confreres to send a representative from this diocese to Paris, or to confer power on a member of the synod. We ask that we be given a pastor. For too long already we have been a dispersed flock, a headless body, without any credit.

Take care of your health and that, too, of your worthy companion, Madame Villars, whom I greet.

P.S. If I had been more timid, I would have taken flight long ago; for a sinister raven does not stop warning me of evil days ahead. If the sky falls, not being able to prevent it, I shall perish under its collapse. If it holds firm, I shall walk on, so long as God wishes, under its vault.

164

La Roche, 12 Pluviose Year VI
[31 January 1798]

I am very eager, oh best of my friends, to get news of you at the earliest, having been deprived of it for a very long time, with the exception of what I get indirectly about you

~

[843] In 1797, Villars prepared notes on various Dauphinois botanists for Guy Allard, publisher of **La Bibliothèque du Dauphiné**, which went unpublished at the time: "Chaix was by nature solemn and serious, but sensitive and affectionate. Inviolably attached to the Catholic faith, he was unshakeable despite the variations in [fashionable] opinion during the Revolution. His mind was slow and tentative, but also judicious and precise, as he loved the sciences and was passionate for botany." Antonin Macé, **Notes inédites de Villars sur quelques botanistes dauphinois, lecture faite à l'Académie delphinale dans la séance du 24 mai 1861** (Paris: J. Techener, 1862), p. 8. But the information provided by Chaix later served Villars when he would write a eulogy for Chaix in 1800.

from Citizen Mondet because of his trial. This important case is to be tried tomorrow in court before which he is to appear as his own defender. I quite fear he will lose.

Be kind enough to tell me what regimen and remedies must be used by a young man in this parish, age about 22. He has been tormented by a dry chest cold, with diarrhea, for two and a half months without knowing the cause. Citizen Pieron, a surgeon in this arrondissement, who saw him, prescribed the use of milk, some *tussilage*,[844] and a little good wine. (He does not tolerate the wine well.) He told him that he did not have any fever, but I fear that he has a very slight fever that is draining him imperceptibly. His father, who lost his elder son two years ago following hardship and exhaustion suffered in the Army of Nice, is disposed to do everything to prolong the life of this younger son, the only son left to him; and, moreover, he is quite well off. Your recommendation will be scrupulously followed, and I ask for it immediately from you for him.

I have a young niece six years old who, from her first years, has had a soft wen about the size of a nut on the upper part of her forehead. Is it possible to get it to disappear? . . . When giving me news, please say a word about the two above infirmities.

The sad events around Rome worry us a great deal. Is the second tragedy of Verona fatal to the Venetian Republic? From whence came the coup? It is a mystery that history will clarify in time.[845] Is Citizen Reymond, the bishop, currently in Grenoble? Does he enjoy any credit there?[846] Is the dissident party still dominant there?

165

La Roche, 18 Ventose Year VI
[8 March 1798]

I wrote to you more than a month ago, dear and very dear friend, to consult with you about a sick man in my parish. Having had no response without knowing the cause, I am obliged to write you again [about him]. . . . I fear he is declining into consumption. Dr. D'heralde [of Gap] and Pieron of Veynes, who have seen him, have not prescribed much for him, saying that his symptoms are not serious. These gentlemen treat the sick who are unknown to them with indifference. I have recommended some simple remedies to him that cannot hurt him even if they are not curative. Please prescribe a treatment for him. I have told his parents that, having had no response from you, I would write to you again. Respond at the earliest by mail. The ill man, having previously been in good health, does not know the cause of his sickness.

I am sending a specimen of *Galanthus nivalis* [L.] in my letter, found in the parish garden in La Roche.[847] I do not know from where it had been brought. Perhaps it is not foreign to our region. The three or even four exterior petals are connivent and convex; the three interior, called nectaries, are shorter by a half, emarginate at the end, streaked with

---

[844] Coltsfoot, from *Tussilago farfara* L., a traditional remedy. *Tussilago* means cough dispeller.

[845] Chaix evidently refers to recent events in Italy: Bonaparte's creation of the Cisalpine Republic in the north, and the revolutionary activity in December in Rome at the end of 1797 that provided a pretext for French intervention. Chaix's apprehensions were justified. The Roman Republic would soon be established, and Pius VI would be taken by the French to Valence as a prisoner.

[846] Henri Reymond, the second constitutional Bishop of Grenoble. He had been imprisoned in 1793 but was saved from execution by the fall of Robespierre.

[847] A cultivar in Amaryllidaceae, widely naturalized and quite variable.

green. Note that the nectaries fill the center of the flower in this specimen, making the flower quite full. Do not conclude that this is *Leucojum vernum* [L.].[848] Do you have this plant?

Citizen Forbin writes me often from Marseille to ask for plants.[849] He forgets that we inhabit a different climate from his, and that the cold puts our plants to sleep for six months of the year. Citizen [Dr. Joseph] Guérin [of Serres] sent me an announcement of his medical journal. My bare instruction in that excellent branch of natural history, and my advanced age, only permit me to admire his learning and his zeal. Already knowing him for a character quite rare in these times, I have been touched by this mark of his generosity.
. . .

How is your Ecole centrale going? In Gap, there are more professors than active classes. That is how the great promises for public instruction are turning out! I fear that it will all slip away, as the primary schools are not being organized. Instruction will be available, therefore, only for the rich; and, as a consequence, only the rich will be capable of governing, and our government will become aristocratic along the lines of the republics in Italy and the large cantons of Switzerland, imperceptibly degenerating into a new revolution. I shall not see these changes, but I foresee them.

166
La Roche, 16 Thermidor Year VI
[3 August 1798]

I have been expecting daily, my very dear friend, the fatal news that you finally gave me on the 27th of last month.[850] I have fulfilled all my duties of piety for that beautiful soul. I would have done so even without your reminder; and I shall not stop there, for the memory of a husband and wife so worthy of respect will always be in my mind. I am expecting to suffer a similar grief in the case of my poor brother, the only one remaining in my family. He contracted a cruel rheumatism in the kidneys, accompanied by sharp pains in the chest, which forced him into bed for more than a month. He can now move about only with difficulty, aided by two crutches. He has begun to suffer from swelling in his legs, is very drawn, urinating with difficulty, afflicted for a long time with a hernia. He is younger than I, but more worn from the labors required by his occupation. Citizens Guérin and Cler of Serres have visited him, but I have little hope for his recovery. None of us must lose sight of our final moments.

I have read the merited tributes you pay Citizens Mouton and Guérin in the latter's journal. Guérin, having spent two weeks here, has gone on to the region around Barcelonnette. I hope to see him again, as he has promised to return in order to visit the high mountains of Champsaur, Valgaudemar, and Oisans, finally returning by way of Grenoble, Lyon, etc. He has been on the Montagne d'Aurouze and de Céüse.

Your *Aira* from Manteyer,[851] which I have also found at Les Baux along the Buëch, is a very early annual that I would have named *verna* if I had not found it referable to *Aira*

~

848Another occasionally naturalized cultivar in Amaryllidaceae.
849Forbin was a noble family from Marseille.
850Jeanne Disdier, wife of Dominique Villars, died in Grenoble on 13 July 1798. Manteyer, **Les Origines de Dominique Villars**, p. 187.
851Villars did not publish any *Aira* specifically collected at Manteyer.

praecox L. It grows amongst *Aira juncea* [Vill.] [= *Deschampsia media* (Gouan) Roemer & Schultes (1817)], but is much smaller with tighter panicles, *floribus paniculato spicatis*, L. [**Sp. Pl.** 65]; Gér. [**Fl. Galloprov.** 87, no. 8], etc.

Are you not making the common *Cheiranthus* of our mountains simply a variety of *Cheiranthus alpinus* [L.]? They have the greatest affinity. I believe that you are giving the name *Cheiranthus erysimoides* [L.] to your plant from Tréminis, from Oisans, which I have earlier grown in Les Baux, and from Molines-en-Queyras and the islands in the Drac at Pont-du-Fossé, much taller, the flowers small. If that is so, where is *Erysimum hieracifolium* [L.]?[852]

Respond to me when you can, and take consolation in your son, your daughter-in-law, and daughter, who now comprise your lovable family. Your brother, the curé, made a brief appearance here with friend Martin, but they have promised a second trip.

P.S. Do you know this plant, the seed for which was sent to me from Paris with the name *Hyssopus bracteatus*? Although planted in a large pot, it has only grown to be an inch or two tall. You are seeing the complete plant. It is not in Gmelin's **Systema**.

Tomorrow is our festival. *Dominicus* means *from the Lord;* not that I am *from the Lord,* only the name. But if I be only the name in this time, that will provide for me in a time as yet only a dream, only a phantom in the darkness.

167
La Roche, 19 Fructidor Year VI
[5 September 1798]

Chemists, physicists, and philosophes, dear friend (and you think as I do), have made marvelous discoveries in recent times about the secrets of Nature. But with all the digging and probing, what will it all come to if they do not finally recognize the first cause? I, at least, do not know any of them who dare to name it. Despite himself, Robespierre dredged up the name of the *Supreme Being*. Our contemporaries will not deign to pronounce the name and only use the name of *Nature*, which they do not know adequately. It is beneath them to use the name of *God.* They would debase themselves if He intruded into their writings. Oh atoms! Oh nonentities! Where will your pride take you? The learned [Noël-Antoine] Pluche, without having the benefit of all the recent discoveries, has reasoned as well as any of them, always reaching better conclusions than they do. I admire the depth of their learning, but I abhor their shameful, or their prideful, obsession. How I congratulate myself on my deficiency and on my simplicity! How pleased I am in the obscurity of my solitude! I would count myself wiser than they if only I knew my beginning and my end. *Noverim te, Deus, noverim me.*

I am finally satisfied by your clarifications on our *Cheiranthus.* I was truly surprised some time ago to notice a vesicular stipule on our *Coronilla* from La Grangette that I had never seen on those from Les Baux, which tempted me to make a distinct species of it. I do not know whether that bladder forms the double stipule you speak about. I shall pay

~

[852]This passage is not clear. It is *probable* that the plant Villars had called *Cheiranthus erysimoides* L., and which he said was difficult to distinguish from *Erysimum hieracifolium* L., was actually what would be published as *Erysimum virgatum* Roth. (1797).

attention in the future to the individuals I run across.[853]

I still have remorse about *Hieracium auricula* L., Haller's, and ours. If the critical stolons of *H. auricula* did not make an obstacle, I would be glad to merge it with your *Hieracium piloselloides* [Vill.]. I dare to speak out frankly: either *H. auricula* L. (Haller) is only a variety of *H. dubium* [L.], or they are wrong by presuming *H. piloselloides* to have stolons, a [plant] very common in our marshes. You appear to be close to my opinion in your work. . . .[854]

Geranium zonale [L.] never produces any seeds here, either because of the castration of its anthers, or because they are always situated below the stigma. In contrast, *Geranium tabulare* Cav., whose anthers surround the stigma, seeds very well. Beyond which, *Geranium tabulare* from the Parisians does appear to be Linnaeus's; but I only have it with very small white flowers approaching those of *G. alchimilloides* [L.]. If Citizens Curten or Liottard have seeds from any African geraniums, ask for some for me. Our ex-Representative Serre brought me some cuttings from different species from Paris, but found them rotting by the time he reached Lyon. I shall be very grateful to them.

Do you believe that you must still defer your supplement and appendix? I quite desire to see it appear.

Our friend Guérin, who has visited Turin and its learned men (and Milan, too), must be coming to Grenoble via the Montagne du Chaillol, Valgaudemar, La Bérarde, and Oisans. He does less with botany than with mineralogy and meteorology. This young natural philosopher, blessed with wealth, talent, and taste, will elevate himself to the apogee of the sciences and make himself capable of fulfilling the tasks that will be confided to him with distinction. Greet him on my behalf and encourage his efforts on the matter of my herbarium. Otherwise, I fear it will become the pasture for worms or be carted off to the green-grocers. Do not save on the costs of mailing, which I pay very gladly when it comes to your letters. Guérin should also write to me when he can.

P.S. The more I reread your works, the more I admire your botanical genius and your astonishing discernment. Allioni, and Bellardi in particular, render you appropriate justice. . . .

~

In late March or early April of 1799, while walking some distance to visit a sick parishioner, Chaix fell to the ground, unconscious, and was perhaps revived (or so Villars surmised) by the snow he found in his hands upon regaining consciousness. He was alone. It is not clear whether he returned home immediately or completed his errand. Thereafter, he spoke of the incident to Serre and also reported it to Villars in a letter that has not been preserved.

While Villars supposed that the seizure had been slight, as there had been no subsequent weakness in the extremities nor any irregularity in the facial muscles, he described it to Chaix as a warning. He wanted Chaix to take some precautions, prescribing soothing beverages, dry rub-downs, moderate exercise, the use of thermal waters, and a few grains of ipecac to settle his

~

[853]It would appear that he refers here to *Coronilla minima* L. as characterized by Villars, **Hist. Pl. Dauph.** 3: 396 (1788). And it may be that Chaix had run across *Coronilla vaginalis* Lam. (1786) that they did not recognize. [854]Villars had noted, ibid. 3: 100 (1788), that he had never seen stolons on *Hieracium piloselloides* Vill., but that, otherwise, his plant had great affinity with *H. auricula* L.

stomach. *Chaix was also advised to stop his intellectual work, to avoid traveling alone, and to eat little at night.*

Dr. Serre would later tell Villars that it proved to be difficult to get Chaix to follow any regimen. But it was also true that Chaix was temperate by nature. He was not noticeably overweight nor did he have a flushed appearance. His letters in recent years had clearly revealed growing sadness and exasperation with the course of the Revolution, and Villars suspected that his courage and strength had been slowly ebbing. In such a state of mind, he was inclined to trust his destiny to Providence, no matter his faith in Villars.[855]

168

La Roche, 8 Prairial Year VII
[27 May 1799]

I return a second time to appeal to your humanity, this time in favor of Jacques Thomé of La Roche, wounded in the arm near the Adige [River] during an unfortunate day for France.[856] I do not doubt that you will provide care for this unfortunate conscript. He is poor, but he comes from a very respectable family of this village.

I feel quite recovered from my indisposition, which was a kind of nervous attack brought on by the cold of this winter in the church. But, my memory has been noticeably altered, attributable also to my advanced age. The instrument you had sent to me is quite useful. I thank you for the gift you made of it to me.

If you have any news from Citizen Guérin, please let me have it. He no longer writes to me, perhaps believing I am not in a condition to respond.

169

La Roche, 15 Messidor, Year VII
[3 July 1799]

Last year you had the kindness to promise me to take charge of my botany books after my death, to keep them for your own use or to dispose of them in any manner you should judge appropriate. I owe their acquisition to you, and you accept the obligation for their sale. At the moment they are useless to me. I no longer read any of them; I am not in a state to get any benefit from them. Thus, you could do, while I am still alive, what you are prepared to do after my demise. The proceeds would be useful to meet my needs in my old age or to see me to my grave. I would retain only your history of the plants of Dauphiné, which you could reclaim after my death. I have made some notes in it that could be of use to you. You could also lay claim to my manuscript in which I pay tribute to you. Better than anyone else, you can have these books purchased by your students, or, taking advantage of the great reputation you enjoy, make your correspondents aware of their availability. Please indicate to me when I may send them to you, and just when you will

∽

[855]Villars, "Notice historique sur Dominique Chaix, botaniste." **Bull. Soc. Etudes Hautes-Alpes** 3, no. 1 (1884): 309-310. The manuscript for this article, dating from 1800, was lent for publication by Eugène Chaper, a former deputy from Isère, enabling Paul Guillaume to edit and publish it. The manuscript was identical to one published by Jean-Michel Rolland, "Notice sur Chaix." **Mémoires de la Société d'émulation des Hautes-Alpes** (Gap: J. Allier, 1807). Also see A.H.-A., Z Guillemin 6005.
[856]The Battle of Adige, 28 March 1799, fought against the Austrians between the Adige and Lago di Garda: one of a series of French defeats that cost the French control of Italy.

be in Grenoble to receive them. I shall consult with Citizen Serre about how to make the shipment. You alone are to set the prices and can give me an accounting for what you receive for them. I shall add a list of the books within the shipment.

I am thereby giving you a burden, but remember that it is for a person whom you have always loved, and whose straits at the end of his days you feel.

I am quite well at the moment, but I dread the winter cold that was the source of my indisposition. Alas! I cannot hold back the course of events. I began my seventieth year this past June 8th. May God deign to have pity on this unfortunate sinner!

Take care of your health, and send me news of your family and of current affairs.

170
La Roche, undated.
[Between 3-21 July 1799]

I wrote you recently to beg you to accept my botany books, which you promised me here last year to make use of. Fearing that my letter did not reach you, and taking advantage of a trip to Lyon by Citizen Corié, a former Feuillant and my friend, I reiterate my request. The books are currently useless to me. Thus, it is far better that I rid myself of them while I am still living, not knowing into whose hands they might fall after my death. You, in particular, will be able to dispose of them, to sell them to your students or to bookstores familiar to you. I will benefit in my old age from whatever money they will bring. I say the same thing about my herbarium, since our department is unconcerned about it, and as the public advertisement by Citizen Guérin has not produced any response. Get out of it whatever you can. I leave you in charge of the matter, and you may take whatever time you judge to be appropriate. I shall make the shipment whenever you give the word. You may deduct the shipping expenses from whatever money the books bring in. You are the only one who can render me this service, and you will greatly oblige a person whom you have always loved, and who has limited expectations for this life. At the moment, I have completely recovered; but I am not blind enough to count much on prolonging my days.

~

Dr. Serre to Dr. Villars
6 Thermidor Year VII [24 July 1799]

You will not learn without grief of the distressing event that deprives us of the best of friends forever. Citizen Chaix, that venerable old man, suffered an attack of apoplexy on the 3rd instant [21 July 1799] while he was celebrating mass. He expired the following night at two o'clock in the morning, mourned by his family and quite generally by all the citizens of the commune. No symptom had preceded this unfortunate accident, so that there was no reason to suspect his impending demise. The day before, he seemed better than on any other day since his initial accident, spending a part of that day writing the catalogue of his books in order to send it to you.[857]

~

[857]Villars, "Notice historique sur Dominique Chaix," pp. 310-311.

POSTSCRIPT

Villars assumed that Chaix's legal heirs were unaware of arrangements to dispose of the books and the herbarium. Having every confidence in Dr. Serre, he charged Serre with the responsibility for shipping those materials to Grenoble. The heirs refused to release anything to Serre, demanding payment. The impasse was the more irritating to Villars as he had recently lent some of his own volumes to Chaix and could not recover them. It became necessary to acquire the legacy for a sum of money not specified. Villars then approached René Louiche Desfontaines, Charles-Louis L'Héritier, and Pierre Ventenat in Paris, urging them to acquire the Chaix material. While interested in a few individual specimens, they were not interested in a herbarium of 3,000 plants, most of which they already possessed.[858]

In the meantime, Villars notified Philippe Picot de La Peyrouse, a professor of natural history at Toulouse and a member of the Institut national, that the Chaix herbarium would be available if an interested party wanted to take it: "The herbarium has most of my species except for the mosses and lichens, as my respectable friend was not as strong on those two groups. But few herbaria are as well arranged. If it should come to you, I would review the labels; but they have already been verified. He was candid, reliable, unpretentious, but a good observer."[859] Having corresponded with La Peyrouse for a number of years, Villars knew his interest in the comparative study of Alpine and Pyrenean species. Thus, the Chaix herbarium was shipped to Toulouse in 1800, but only after Villars had corrected a few labels, drawing a line through names Chaix had given. La Peyrouse later acknowledged his use of "the fine herbarium of Villars' worthy friend, Chaix," attributing his possession of it to his friendship with Villars.[860] It was always understood in Toulouse that Villars had been, in fact, an intermediary, and that the money paid for the herbarium went to Chaix's heirs.

An undated official document detailing the division of the estate of Dominique Chaix reveals that, although he had been a man of unquestionable piety and faith, he left no provision for charity. For over thirty years, he had lent substantial sums from his small income, mostly to members of his family or relatives by marriage; and he had contributed to six of his nieces' dowries. Most of the debts were not repaid *until* the estate was settled. Chaix had evidently owned no real estate. His furnishings were valued at 440 livres; and the settlement listed the sale of a property for 358 livres, which may refer to the sale of his herbarium. There was no windfall for any of the numerous heirs. In effect, the assets of the estate had been lent over many years to the very relatives who would repay the loans in order to be eligible for the redistribution of funds.[861]

∾

[858] Villars, "Notice historique sur Dominique Chaix,", pp. 311-312.
[859] Edouard Timbal-Lagrave, "Observations critiques et synonymiques sur l'Herbier Chaix." Mém. Acad. Sci. Toulouse, ser. 4, 6 (1856): 86.
[860] Philippe Picot de La Peyrouse, Histoire abrégée des plantes des Pyrénées (Toulouse: Bellegarrigue, 1813), p. xiv. The name is more frequently spelled Lapeyrouse.
[861] A.H.-A. Série F. 676. État de la succession de feu Dominique Chaix prieur curé des Baux.

The fate of the Chaix herbarium is known if imprecisely. La Peyrouse died in 1818, his principal works on the flora of the Pyrenees having been published toward the end of his life. The Chaix herbarium was subsequently acquired by a Colonel Dupuy of Toulouse, an avid student of the natural sciences, then left by Dupuy to his nephew, a Dr. Judan of Toulouse. As Judan's main intellectual interests lay elsewhere, he never exploited the herbarium but gave Edouard Timbal-Lagrave free access to it.

When Timbal-Lagrave began his critical study of the collection, he soon recognized with great regret that the herbarium was in an advanced state of decay. In the half-century since Chaix's death, little more than ten percent of the original 3,000 plants were found suitable for examination. Even to reach that percentage, he had to seek opinions and information from abbé Armand David, who was laboring to revise Mutel's earlier **Flore du Dauphiné**; and from J.-B. Verlot who had access to Villars' material in Grenoble, and who would publish his own flora of the region in 1872. Consequently, when Timbal-Lagrave published his observations of the Chaix herbarium in 1856, he was painfully aware that his conclusions were less than satisfactory.

As for the ninety percent of the original collection that was beyond recall, he had found obvious gaps that suggested certain species of particular interest had simply been removed and never returned by prior owners. A substantial number of surviving species were incomplete specimens, the material too fragmentary for reliable identification. One has to suspect that those snippets, frequently exchanged within letters or books by Chaix and Villars, became herbarium specimens. But the greatest damage by far had been done by insects, the very peril anticipated by Chaix should the herbarium fall into careless hands.[862]

The debris was eventually brought home to Gap for a proper burial in the departmental archives. Responsibility for the arrangements remain obscure. Dr. Jean-Joseph Serre, who was serving as subprefect in Embrun at the time of his death in 1831, had had a son, an army officer, who was garrisoned at that time in Toulouse. He, too, was an avid student of botany. Captain Serres, as he spelled his name, had a personal herbarium and published several titles on the flora of the region around Toulouse. He venerated Villars and must have known that his colonel possessed the Chaix herbarium.[863] While it is improbable that the transfer to Gap occurred before Captain Serres' death in 1858,[864] he must have brought its existence to the attention of those who effected the transfer. Word of its presence in the Musée de Gap was published in 1874,[865] and it was thereafter removed to the departmental archives, a sad and useless remnant of an ardent interest in botany.

Meanwhile, in his later years, the then Colonel Serres had frequently botanized in his native region, alert for species that had been known to Chaix but not satisfactorily

∾

[862] Timbal-Lagrave, "Observations critiques," pp. 87-90. Also see Dr. Victor-François Bally, **Notice historique sur la vie et les travaux du docteur Villars, naturaliste, correspondant de l'Institut** (Grenoble: Imprimerie de Maisonville, 1858), pp. 23-24.

[863] Captain Jean-Joseph Serres, **Flore abrégée de Toulouse, ou Catalogue méthodique des végétaux phanérogames qui croissent naturellement aux environs de cette ville** (Toulouse: J.-M. Corne, 1836), p. vi.

[864] Adolphe Rochas, **Biographie du Dauphiné** (Paris: Charavay, 1856-1860), 2: 404; Dr. Miquel-Dalton, "Les Médecins à la Convention." **La Chronique médicale** 10, no. 6 (15 March 1903): 178-179; A.H.A., Z Guillemin 9024.

[865] Bull. Soc. bot. Fr. 1874, sess., extr., p. cxiv.

determined in his time. Villars, in **Hist. Pl. Dauph.** 3: 106 (1788), had listed *Hieracium villosum* Jacq., citing a variety of it (B) that Chaix had found at La Grangette in the Clos de Tiniets: *Hieracium villosum alpinum magne flore alterum.* Serres found var. B again in the same location and sought to raise it to specific rank to honor Chaix: *Hieracium chaixii* Serres.[866] Although noted by Verlot in 1872, the segregation did not stand.

Villars anticipated, when placing Chaix's herbarium in Toulouse, that Chaix's contribution to science would not be lost. When Villars wrote his historical notice on Chaix in 1800, he still possessed Chaix's books, his annotated copy of Villars' flora, and his manuscripts. While Chaix had accumulated approximately fifty volumes, he had also borrowed books from Villars over the years, taking notes from them and making extracts for later use. The voluminous works of Tournefort, Haller, Scheuchzer, Vaillant, Linnaeus, and Gmelin's **Flora sibirica** had all been extracted in his hand: slow, painful work done for no reason other than his use; but, incidentally, a superior way to exercise and fortify the memory, to learn the technical terms of the science, and to learn the merits and defects of a particular author.

In Villars' opinion, the best of Chaix's manuscripts had been entitled **Dominici Chaix, Bauciensis parrochus, notae botanicae cum propriae, tum ex celeberrimis scriptoribus compendiosae excerptae,** 1771. The manuscript of nearly 600 in-4 pages contained descriptions and observations of plants that Chaix was cultivating in his garden. Villars decided against its publication on three grounds: (1) That the text was frequently interrupted with extracts from other authors; (2) that Chaix had often repeated observations of a plant made at different periods; (3) and because, very frequently, the descriptions were incomplete. He had usually noted the native place of plants, the names of correspondents who had sent him the plants, and descriptions of vegetation, factors quite characteristic of his letters to Villars.

Beyond which, Chaix's age (he was fifteen years older than Villars), his habits, and his great isolation left him unable to profit from the most recent work in a rapidly changing field, the work of Gaertner, Hedwig, and Jussieu in particular, work that would have taught him to analyze the organs of fructification more than he did. In contrast, he had mastered Dillenius, Tournefort, Haller, Boerhaave, and Linnaeus, all important but dated when compared, for instance, to Jussieu. The manuscript, consequently, if of great sentimental value to Villars, was not deemed a document of sufficient current scientific interest to merit publication.

As for Chaix's annotated copy of Villars' flora, Villars meant to use the notes on possible new species, perhaps in a later edition of the work. Their botanical work had never been separate during Chaix's lifetime. It would continue to be united after his death in the **Histoire des plantes du Dauphiné.**[867]

There is a curious and sobering parallel in the final years that each man lived. In the decade of revolution after 1789, Chaix lived to see his vocation abolished, his patriotic hopes smashed; and his morale collapsed as a consequence. Villars' tenure at the hospital in Grenoble had never been serene, and even the great moment of the publication of his

⌒

[866] **Bull. Soc. Bot. Fr.** 4 (1857): 437-438.
[867] Villars, "Notice historique sur Dominique Chaix," pp. 313-318.

flora of Dauphiné had been cursed by the dismal sale of the second and third volumes. The loss of Chaix in 1799, as a friend and collaborator, coincided with a new attack upon Villars in Grenoble. The retreat of the French forces from Italy that year amounted to an avalanche of misfortune for the military hospital in Grenoble. Beyond the care of troops legitimately injured in combat, the local physicians were besieged by citizens seeking medical exemptions from military service as the Republic desperately scrounged for recruits, or by those seeking medical discharges to avoid further service. The physicians had to cope with a variety of feigned illnesses and self-inflicted mutilations by those seeking discharges. The region around Grenoble, moreover, experienced an epidemic of typhus, spread both by Austrian prisoners taken earlier by Bonaparte and by the retreating French troops in lamentable condition.[868]

Some families of means unquestionably bribed medical officers to sign such exemptions and discharges. On the basis of local gossip, Villars was charged with having signed, with the compliance of a second physician, Antoine Cabanne, a certificate of exemption from service in the Army of Italy. The alleged beneficiary was Michel Boudillon, age 26, whose mother supposedly paid Villars 72 francs. Villars, long the subject of malicious jealousies in Grenoble, joined Cabanne in demanding a court-martial to examine the charge. During the trial, the indictment crumbled like a château of cards. Most of the witnesses retracted their prior assertions, and the remaining few admitted to having only heard rumors. Both doctors were acquitted. But subsequently, perhaps weakened by the ordeal, both men contracted typhus. Cabanne died shortly after; but Villars, though gravely ill, survived. It is said that he never entirely recovered.[879]

Indeed, the accumulation of anxieties, outrages, and illnesses by the end of the century remolded Villars into an embittered and unsociable man. His financial future, already undermined by the losses suffered in the publication of his flora, was soon in serious jeopardy. The Consulate, recognizing the unsatisfactory record of the Ecoles centrales, closed them down in 1802, costing him his teaching position. Plans for a reorganized secondary school in Grenoble, a lycée, did not call for a professor of natural history. The government next closed the military hospital in 1803, transfering its military patients to a civil hospital. As his career seemed to be broken, Villars for a time contemplated returning to Le Noyer to a faithful clientele and to his beloved plants.[870]

Meanwhile, a new medical school was being organized in Strasbourg. In January of 1805, Villars wrote to Laperouse that he had agreed to accept a staff position in Strasbourg: "The city of Grenoble has left me, after 30 years of residence, without hospital and without quarters. I created the surgical and botanical schools, and I had paid for my quarters by

⌒

[868] Manteyer, Les Origines de Dominique Villars, p. 215.
[869] Yves Mazars, "Maladies simulées, certificats de complaisance ou les tribulations de Dominique Villars." Les Cahiers de l'Alpe (La Tronche-Monfleury) 14, no. 66 (1975): 4-6. (Based upon research done by Dr. Alexandre Bordier in 1897.) Also see Jean-Charles-François, baron de Ladoucette, "Notice biographique sur M. Villars . . . lu à la séance publique de la Société royale et centrale d'agriculture, le 29 mars 1818." Annales encyclopédique (Paris: Imp. Mme Herissant Le Doux, 1818), 3: 13; and Dr. Etienne Trousset, Histoire de la fièvre qui a régné épidémiquement à Grenoble pendant les mois vendémiaire, brumaire, frimaire et nivose de la présente année (Grenoble: Giroud, Year VIII), pp. 64, 67.
[870] Joyeux and Dejarnac, "Le Médecin Dominique Villars," pp. 131-132.

having them refurbished. To accomplish that, I removed myself from familiar country and ways, from my friends and from my beloved plants."[871] It had reached a point that he was too short to pay the carrying costs for letters and packages of plants. He would move, in sum, for financial reasons and not, as Antoine Gouan surmised, out of inordinate ambition to achieve greater fame than he had already acquired in Dauphiné.[872]

On 16 February 1805, Nompère de Champagny, the minister of the interior, notified him that the emperor had appointed him to be a professor of botany at Strasbourg.[873] He left behind a gracious farewell letter, dated 11 March 1805, which was published several weeks later in the **Annales du département de l'Isère**. He mentioned the library of rare books he had collected at considerable personal expense, books once owned by Nicolas Chorier, by Pierre Garidel, Albrecht von Haller, Philibert Commerson, and above all by Pierre Berard, a library of 4,000 volumes. With the example of Jean-François Séguier in mind, whose library had been preserved by the city of Nîmes, he had hoped that his library would be taken by the public library in Grenoble; but no such arrangement had as yet been concluded. To all of those in the region where he had collected plants, who had sheltered and encouraged him, he offered his thanks and best wishes. "While penning these lines," he concluded, "my eyes have filled with tears; and my heart has weakened at the very moment my reason asks all its strength to separate me from the Alps."[874]

Even though by all accounts his professional career in Strasbourg would be distinguished for the remaining ten years of his life, he never reconciled himself to the divorce from Dauphiné. Letters to his botanical correspondents leave no doubt that he found Strasbourg, especially its climate, most uncongenial. He would have done better to have lived out his days as a country doctor in Champsaur, as surely Chaix would have advised him.

The fate of Villars' herbarium was little more fortunate than that of Chaix. The city of Grenoble did not acquire Villars' principal herbarium until 1837, more than twenty years after his death, for preservation in the Musée d'histoire naturelle. An examination of it in 1859 revealed only about 2,000 species and in such condition that it would have been difficult to prove that they had been collected by Villars. Yet, his herbarium had been reputed to number at least 4,000 plants; and his flora had been based upon 2,744 species, at least 2,600 of which he had collected personally. It was obvious that his herbarium had already been partially dismantled by 1859.[875] Today, after the destructive effects of successive wars, very little of original material has survived, meaning the absence of types for many of his species.[876]

It is known that a medical student named Honnorat acquired a herbarium of 1300 species from Villars. Whether these were Villars' duplicates, or, more probably, whether this was one of those collections prepared by Chaix at Villars' request, is unsure. Dr.

～

[871] Edouard Timbal-Lagrave, "Villars et Lapeyrouse, extrait de leur correspondance." **Bull. Soc. Bot. France** 7 (1860): 689-690.

[872] Pierre-Joseph Amoreux, "Notice historique sur Antoine Gouan." **Mém. Soc. Linn. Paris** 1 (1822): 702-703.

[873] Manteyer, **Les Origines de Dominique Villars**, p. 198.

[874] Ibid., p. 200.

[875] Ernest Faivre, "Jardins-des-plantes et herbiers de Grenoble." **Bull. Soc. Bot. France** 6 (1859): 822-823.

[876] C. R. Fraser-Jenkins, "Nomenclatural notes on *Dryopteris* 4." **Taxon** 29 (November 1980): 609.

Honnorat died in Digne in 1853, and the collection was bought by Reinaud de Fonvert of Aix-en-Provence. He lent it to J.-B. Verlot when the latter was preparing his catalogue of the vascular plants of Dauphiné.[877] Verlot's notes provide comments on both Villars' and Chaix's plants.

Villars had exchanged plants with many other botanists. It is known that the herbarium of Etienne-Pierre Ventenat contained numerous plants from Villars. The Ventenat herbarium was purchased by Benjamin Delessert in 1809.[878] As the great Delessert collection went ultimately to Geneva, all was not lost despite the virtual disappearance of the collection in Grenoble.

---

[877] J.-B. Verlot, Cat. Pl. Dauph., p. viii; Alphonse de Candolle, La Phytographie, ou l'art de décrire les végétaux considérés sous différents points de vue (Paris: G. Masson, 1880), p. 457.

[878] Antoine Lasègne, Musée botanique de Monsieur Benjamin Delessert (Paris: Fortin, Masson, 1845), p. 70.

The Chaix Eponymy

1. *Androsace chaixii* Gren. & Godron.
Collected by Chaix in the Loubet Woods above Les Baux and below Mont Aurouze.
Lumped by Villars, **Hist. Pl. Dauph.** 2: 481 (1787), with *Androsace septentrionalis* L.
Segregated by Grenier and Godron in 1853.

2. *Centaurea* x *chaixiana* Rouy.
[*Centaurea calcitrape* L. x *Centaurea aspera* L. ssp. *aspera*.] Collected by J.-P.-F. Deleuze
near Ventavon on flat ground. Published by Chaix as *Centaurea hybrida* Chaix, non All.,
Pl. Vap. 62 (1785); Chaix in Vill., **Hist. Pl. Dauph.** 1: 365 (1786). Republished by Rouy
in 1905.

3. *Centaurea chaixii* Briq.
[= *Centaurea scabiosa* L.] Collected by Chaix at Manteyer near Combe noire. Published
by him as *Centaurea menteyerica* Chaix, **Pl. Vap.** 61 (1785); Chaix in Vill., **Hist. Pl.
Dauph.** 1: 365 (1786). Republished by Briquet in 1902.

4. *Euphorbia chaixiana* Timb.-Lagr.
[= *Euphorbia characias* L. ssp. *characias*.] Collected for Chaix by Danthoine near Le Poët.
Published as *Euphorbia rubens* Chaix & Vill., Chaix, **Pl. Vap.** 44 (1785); **Hist. Pl. Dauph.**
1: 348 (1786); but recognized later as *Euphorbia characias* L. in **Hist. Pl. Dauph.** 3: 831
(1788). Timbal-Lagrave based his new species on fragments examined in the Chaix
herbarium (1856), finding them different from *Euphorbia amygdaloides* L.

5. *Hieracium chaixii* Serres.
Verlot (1872) believed this was *Hieracium villosum* Jacq. var. B in Vill., **Hist. Pl. Dauph.**
3: 106 (1788): *Hieracium villosum alpinum magne flore alterum*. Collected by Chaix at La
Grangette in the Clos de Tiniets, and later in the same location by Colonel Serres (1857).

6. *Hieracium chaixianum* Arvet-Touvet & Gaut. (*Hieracium pseudocerinthe* x *H.
leiopogon*.) From S.W. Alps. This may be the same species Grenier sent to Verlot. See
Hieracium leipogon Gren. ex Verlot, **Cat. Pl. Dauph.** 396 (1872), from the Alps near Gap:
Rabou and Mont Céüse.

7. *Lactuca chaixii* Vill.
[= *Lactuca quercina* L. ssp. *chaixii* (Vill.) Celak.] Collected in the forest near Les Baux,
Rabou, and Chaudun.

8. *Poa chaixii* Vill.
[See *Poa sylvatica* Chaix in Vill., non Poll.] Collected by Chaix in woods around Chaudun.

9. *Verbascum chaixii* Vill.
Collected by Chaix around Les Baux and Rabou *in rupestribus*. He placed it between
Verbascum lychnitis L. and *V. nigrum* L., where it still belongs today as *Verbascum chaixii*
Vill. ssp. *chaixii*.

10. *Verbascum* x *nigro-chaixii* Timb-Lagr. Based upon material in the Chaix
herbarium that Chaix had labeled *Verbascum meyeri* Chaix, but had not published. Meyer
was curé de Lachau in the Baronnies and an amateur botanist.

11. *Veronica chaixii* Lap.
[= *Veronica teucrinum* L. ssp. *dubia* (Chaix) Nyman] [= *Veronica kindlii* Adamovi ssp.
vahlii (Gaudin) D. A. Webb]. Based upon an examination of *Veronica dubia* Chaix, an
unpublished species in the Chaix herbarium.

Catalogue of Botany Books
with their purchase prices[879]

In-*Fol.*

Jacques Barrelier, **Plantae per Galliam, Hispaniam et Italiam**, 1714 .... 24#
Carlo Allioni, **Flora pedemontana**, 3 vols., 1785 .... 90#
John Ray, **Historia plantarum**, 2 vols., 1686-1688 .... 24#
Jacques Daléchamps, **Histoire générale des plantes**, 2 vols., 1615 .... 20#
Dominique Chabrey, **Stirpium icones**, 1677 .... 9#
[These are the figures for J. Bauhin and J. H. Cherler, **Historia plantarum universalis,***
 3 vols., 1650-1651]*

total 167#

In-*4*

Carlo Allioni, **Specimen pedemontana**, 1755 .... 12#
Marci Mappi, **Historia plantarum alsaticarum**, 1742 .... 9#
Caspar Bauhin, **Pinax theatri botanici**, 1623 .... 6#

total 194#

In-*8*

Linnaeus, **Species plantarum**, 3rd ed., 2 vols., 1764 .... 18#
Genera plantarum, 6th ed., 1764 .... 9#
Materia medica, 1749 .... 5#
Philosophia botanica, 1751 .... 6#
Mantissa plantarum, Vol. I, 1767 .... 6#
Systema plantarum, 13th ed., Vol. I = John A.
 Murray, **Systema vegetabilium**, 1774 .... 11#
Louis Gérard, **Flora gallo-provincialis**, 1761 .... 8#
Antoine Gouan, **Hortus regius monspeliensis**, 1762 .... 6#
F. W. Weiss, **Plantae cryptogamicae flora gottengensis**, 1769;
 and G. H. Weber, **Spicilegium florae gottengensis**, 1778 .... 10#
J. A. Pollich, **Historia plantarum in Palatinatu**, 3 vols., 1776-1777 .... 18#
J. F. Séguier, **Plantae veronenses**, 3 vols., 1745-1754 .... 18#
Carlo Allioni, **Stirpium praecipuarum littoris et agri Nicaeensis**, 1757 .... 4#
Albrecht von Haller, **Nomenclator ex Historia plantarum**
 indigenarum Helvetiae, 1769 .... 2#
 Flora jenensis, 1745 .... 6#

≈

[879] A.H.-A. Série L. 1007. Chaix had supplied the departmental authorities with a list of his books in 1794 in his own handwriting. The list below, also found in the departmental archives, was more complete but not in Chaix's handwriting, very possibly a copy made by Dr. Serre who sent the original to Villars. Most of the authors and titles were given in abbreviated forms.

Lindern, **Hortus Alsaticus**, n.d. .... 5#
Pierre Magnol, **Hortus regius monspeliensis**, 1697 .... 2#
    Botanicum monspeliensis, 1676 .... 2#
                                                                  total 330#
Dominique Villars, **Histoire des plantes du Dauphiné,**
    3 vols., 1786-1789 .... 40#
Nicolas Lémery, **Pharmacopée universelle**, 1697; and
    Dictionnaire universel des drogues simples, 1698 .... 24#
J. F. Rozier, **Dictionnaire d'agriculture**, 2 vols., 1781 .... 5#
Manuel des Dames de la Charité, n.d. .... 2#
Madame Fouguet, [title unknown] .... 2#
Laurent Blanc, **Essai de botanique pratique**, 1784............................... 2#
J. J. Rossignol, **Botanique élémentaire**, 1781 .... 1#
                                                                   total 406#

Index Of Plant Genera

Index Of Names

ARCHIVES INTERNATIONALES D'HISTOIRE DES IDÉES
*
INTERNATIONAL ARCHIVES OF THE HISTORY OF IDEAS

43. P. Dibon: *Inventaire de la correspondance (1595-1650) d'André Rivet (1572-1651)*. 1971   ISBN 90-247-5112-8

44. K.A. Kottman: *Law and Apocalypse*. The Moral Thought of Luis de Leon (1527?-1591). 1972   ISBN 90-247-1183-5

45. F.G. Nauen: *Revolution, Idealism and Human Freedom*. Schelling, Hölderlin and Hegel, and the Crisis of Early German Idealism. 1971   ISBN 90-247-5117-9

46. H. Jensen: *Motivation and the Moral Sense in Francis Hutcheson's* [1694-1746] *Ethical Theory*. 1971   ISBN 90-247-1187-8

47. A. Rosenberg: *[Simon] Tyssot de Patot and His Work (1655–1738)*. 1972   ISBN 90-247-1199-1

48. C. Walton: *De la recherche du bien*. A study of [Nicolas de] Malebranche's [1638-1715] Science of Ethics. 1972   ISBN 90-247-1205-X

49. P.J.S. Whitmore (ed.): *A 17th-Century Exposure of Superstition*. Select Text of Claude Pithoys (1587-1676). 1972   ISBN 90-247-1298-X

50. A. Sauvy: *Livres saisis à Paris entre 1678 et 1701*. D'après une étude préliminaire de Motoko Ninomiya. 1972   ISBN 90-247-1347-1

51. W.R. Redmond: *Bibliography of the Philosophy in the Iberian Colonies of America*. 1972   ISBN 90-247-1190-8

52. C.B. Schmitt: *Cicero Scepticus*. A Study of the Influence of the *Academica* in the Renaissance. 1972   ISBN 90-247-1299-8

53. J. Hoyles: *The Edges of Augustanism*. The Aesthetics of Spirituality in Thomas Ken, John Byrom and William Law. 1972   ISBN 90-247-1317-X

54. J. Bruggeman and A.J. van de Ven (éds.): *Inventaire* des pièces d'Archives françaises se rapportant à l'Abbaye de Port-Royal des Champs et son cercle et à la Résistance contre la Bulle *Unigenitus* et à l'Appel. 1972   ISBN 90-247-5122-5

55. J.W. Montgomery: *Cross and Crucible*. Johann Valentin Andreae (1586–1654), Phoenix of the Theologians. Volume I: Andreae's Life, World-View, and Relations with Rosicrucianism and Alchemy; Volume II: The *Chymische Hochzeit* with Notes and Commentary. 1973   Set ISBN 90-247-5054-7

56. O. Lutaud: *Des révolutions d'Angleterre à la Révolution française*. Le tyrannicide & *Killing No Murder* (Cromwell, *Athalie*, Bonaparte). 1973   ISBN 90-247-1509-1

57. F. Duchesneau: *L'Empirisme de Locke*. 1973   ISBN 90-247-1349-8

58. R. Simon (éd.): *Henry de Boulainviller - Œuvres Philosophiques*, Tome I. 1973   ISBN 90-247-1332-3

For Œvres Philosophiques, Tome II *see below under Volume 70.*

59. E.E. Harris: *Salvation from Despair*. A Reappraisal of Spinoza's Philosophy. 1973   ISBN 90-247-5158-6

60. J.-F. Battail: *L'Avocat philosophe Géraud de Cordemoy (1626-1684)*. 1973   ISBN 90-247-1542-3

61. T. Liu: *Discord in Zion*. The Puritan Divines and the Puritan Revolution (1640-1660). 1973   ISBN 90-247-5156-X

62. A. Strugnell: *Diderot's Politics*. A Study of the Evolution of Diderot's Political Thought after the *Encyclopédie*. 1973   ISBN 90-247-1540-7

ARCHIVES INTERNATIONALES D'HISTOIRE DES IDÉES
*
INTERNATIONAL ARCHIVES OF THE HISTORY OF IDEAS

ARCHIVES INTERNATIONALES D'HISTOIRE DES IDÉES
*
INTERNATIONAL ARCHIVES OF THE HISTORY OF IDEAS

ARCHIVES INTERNATIONALES D'HISTOIRE DES IDÉES
*
INTERNATIONAL ARCHIVES OF THE HISTORY OF IDEAS

ARCHIVES INTERNATIONALES D'HISTOIRE DES IDÉES
*
INTERNATIONAL ARCHIVES OF THE HISTORY OF IDEAS

125. R.M. Golden (ed.): *The Huguenot Connection*. The Edict of Nantes, Its Revocation, and Early French Migration to South Carolina. 1988   ISBN 90-247-3645-5
126. S. Lindroth: *Les chemins du savoir en Suède*. De la fondation de l'Université d'Upsal à Jacob Berzelius. Études et Portraits. Traduit du suédois, présenté et annoté par J.-F. Battail. Avec une introduction sur Sten Lindroth par G. Eriksson. 1988
ISBN 90-247-3579-3
127. S. Hutton (ed.): *Henry More (1614-1687). Tercentenary Studies*. With a Biography and Bibliography by R. Crocker. 1989   ISBN 0-7923-0095-5
128. Y. Yovel (ed.): *Kant's Practical Philosophy Reconsidered*. Papers Presented at the 7th Jerusalem Philosophical Encounter (December 1986). 1989   ISBN 0-7923-0405-5
129. J.E. Force and R.H. Popkin: *Essays on the Context, Nature, and Influence of Isaac Newton's Theology*. 1990   ISBN 0-7923-0583-3
130. N. Capaldi and D.W. Livingston (eds.): *Liberty in Hume's 'History of England'*. 1990
ISBN 0-7923-0650-3
131. W. Brand: *Hume's Theory of Moral Judgment*. A Study in the Unity of *A Treatise of Human Nature*. 1992   ISBN 0-7923-1415-8
132. C.E. Harline (ed.): *The Rhyme and Reason of Politics in Early Modern Europe*. Collected Essays of Herbert H. Rowen. 1992   ISBN 0-7923-1527-8
133. N. Malebranche: *Treatise on Ethics* (1684). Translated and edited by C. Walton. 1993
ISBN 0-7923-1763-7
134. B.C. Southgate: *'Covetous of Truth'*. The Life and Work of Thomas White (1593–1676). 1993   ISBN 0-7923-1926-5
135. G. Santinello, C.W.T. Blackwell and Ph. Weller (eds.): *Models of the History of Philosophy*. Vol. 1: From its Origins in the Renaissance to the 'Historia Philosphica'. 1993   ISBN 0-7923-2200-2
136. M.J. Petry (ed.): *Hegel and Newtonianism*. 1993   ISBN 0-7923-2202-9
137. Otto von Guericke: *The New (so-called Magdeburg) Experiments* [Experimenta Nova, Amsterdam 1672]. Translated and edited by M.G.Foley Ames. 1994
ISBN 0-7923-2399-8
138. R.H. Popkin and G.M. Weiner (eds.): *Jewish Christians and Cristian Jews*. From the Renaissance to the Enlightenment. 1994   ISBN 0-7923-2452-8
139. J.E. Force and R.H. Popkin (eds.): *The Books of Nature and Scripture*. Recent Essays on Natural Philosophy, Theology, and Biblical Criticism in the Netherlands of Spinoza's Time and the British Isles of Newton's Time. 1994   ISBN 0-7923-2467-6
140. P. Rattansi and A. Clericuzio (eds.): *Alchemy and Chemistry in the 16th and 17th Centuries*. 1994   ISBN 0-7923-2573-7
141. S. Jayne: *Plato in Renaissance England*. 1995   ISBN 0-7923-3060-9
142. A.P. Coudert: *Leibniz and the Kabbalah*. 1995   ISBN 0-7923-3114-1
143. M.H. Hoffheimer: *Eduard Gans and the Hegelian Philosophy of Law*. 1995
ISBN 0-7923-3114-1
144. J.R.M. Neto: *The Christianization of Pyrrhonism*. Scepticism and Faith in Pascal, Kierkegaard, and Shestov. 1995   ISBN 0-7923-3381-0
145. R.H. Popkin (ed.): *Scepticism in the History of Philosophy*. A Pan-American Dialogue. 1996   ISBN 0-7923-3769-7

ARCHIVES INTERNATIONALES D'HISTOIRE DES IDÉES

*

INTERNATIONAL ARCHIVES OF THE HISTORY OF IDEAS

146. M. de Baar, M. Löwensteyn, M. Monteiro and A.A. Sneller (eds.): *Choosing the Better Part*. Anna Maria van Schurman (1607–1678). 1995   ISBN 0-7923-3799-9

147. M. Degenaar: *Molyneux's Problem*. Three Centuries of Discussion on the Perception of Forms. 1996   ISBN 0-7923-3934-7

148. S. Berti, F. Charles-Daubert and R.H. Popkin (eds.): *Heterodoxy, Spinozism, and Free Thought in Early-Eighteenth-Century Europe*. Studies on the *Traité des trois imposteurs*. 1996   ISBN 0-7923-4192-9

149. G.K. Browning (ed.): *Hegel's* Phenomenology of Spirit: *A Reappraisal*. 1997
     ISBN 0-7923-4480-4

150. G.A.J. Rogers, J.M. Vienne and Y.C. Zarka (eds.): *The Cambridge Platonists in Philosophical Context*. Politics, Metaphysics and Religion. 1997 ISBN 0-7923-4530-4

151. R.L. Williams: *The Letters of Dominique Chaix, Botanist-Curé*. 1997
     ISBN 0-7923-4615-7

152. R.H. Popkin (ed.): *Scepticism in the Enlightenment*. 1997   ISBN 0-7923-4643-2

KLUWER ACADEMIC PUBLISHERS – DORDRECHT / BOSTON / LONDON